Entomology and the Law

Used particularly when there has been a suspicious death, insect-related evidence is one of the most powerful, but least understood examples of modern forensic science. *Entomology and the Law* provides a detailed roadmap that can be followed from crime scene to courtroom by entomologists, law enforcement personnel and lawyers preparing for trial. Part I focuses on carrion flies as forensic indicators, exploring relevant biology clearly, and concisely illustrated by real-life cases. Flies are usually first on the scene of a death, and knowledge of their habits and lifestyles can help to reveal time of death, weeks or even years later. Part II provides a thorough examination of the law of scientific evidence worldwide, complete with caselaw and applicable code provisions, and legal issues relevant to the admissibility and use of forensic entomology in litigation. It will prepare both scientists and lawyers for real-world forays into the world of forensic entomology.

BERNARD GREENBERG is Professor Emeritus in the Department of Biological Sciences at the University of Illinois at Chicago. He is an internationally recognized fly biologist. One of the founding fathers of the modern field of forensic entomology, he has been an expert witness and consultant for many murder trials.

JOHN CHARLES KUNICH is a Professor in the Roger Williams University School of Law in Rhode Island. Trained originally as an entomologist, he changed track and gained his Juris Doctor degree cum laude at Harvard Law School, followed by a Master of Laws degree summa cum laude from George Washington University School of Law. He now teaches and publishes in the areas of scientific evidence, trial advocacy, law and science, environmental law, and natural resources law.

Entomology and the Law

Flies as Forensic Indicators

Bernard Greenberg
University of Illinois, Chicago

AND

John Charles Kunich
Roger Williams University School of Law, Rhode Island

PUBLISHED BY THE PRESS SYNDICATE OF THE UNIVERSITY OF CAMBRIDGE
The Pitt Building, Trumpington Street, Cambridge, United Kingdom

CAMBRIDGE UNIVERSITY PRESS
The Edinburgh Building, Cambridge CB2 2RU, UK
40 West 20th Street, New York, NY 10011-4211, USA
477 Williamstown Road, Port Melbourne, VIC 3207, Australia
Ruiz de Alarcón 13, 28014 Madrid, Spain
Dock House, The Waterfront, Cape Town 8001, South Africa

http://www.cambridge.org

First published 2002

Printed in the United Kingdom at the University Press, Cambridge

Typeface Utopia 9/13 pt *System* QuarkXPress™ [SE]

A catalogue record for this book is available from the British Library

ISBN 0 521 80915 0 hardback

Dedication

To all my grandchildren and those to be.
BERNARD GREENBERG

and

To my wife and daughters Christina Laurel and Julie-Kate, from whom all that is good in my life flows
JOHN CHARLES KUNICH

Contents

Glossary

ADH – accumulated degree hours: the number of heat units × time required for a specific development to occur.

alpine – a biotic zone occurring roughly at 3500–4800 m.

anautogenous – requiring a protein meal to develop the ovaries and eggs.

apodeme – an invagination of the body wall in arthropods that serves for muscle attachment or for strengthening the body wall.

apolysis – separation of the new skin from the old one, as in the fly when the pupa is no longer attached to the inside of the puparium.

asynanthropic – flies and other organisms that live entirely independently of the human environment.

autogenous – the ovaries and eggs mature without a prior protein meal.

bionomics – the habits, breeding and adaptations of organisms.

blowflies – see Calliphoridae.

Calliphoridae – a family of large, generally metallic blue or green carrion-feeding flies commonly called blowflies, bluebottles, and greenbottles; some species are parasitic.

carbohydrase – an enzyme that splits carbohydrates.

carrion flies – sarcosaprophagus flies (feeding on dead animals) belonging to various families including Calliphoridae, Sarcophagidae, Muscidae, Phoridae, and Piophilidae.

cephalic – toward or referring to the head.

chorion – the outer shell of the insect egg.

chrysalis – the pupa in moths and butterflies.

chymotrypsin – an enzyme from the mammalian pancreas that digests proteins in alkaline solution.

cicatrization – formation of a scar at the site of a healing wound.

circadian rhythm – an organism's behavior/physiology closely tied to a 24-hour cycle.

collagenase – an enzyme in blowflies that splits collagen, the main constituent of fibrous tissue, in alkaline solution.

corroboration – additional evidence that supports some other item of evidence.

crop – an inflated part of the foregut of insects, functioning as a temporary storage organ for excess food in fly larvae; also present in many adult flies.

cross examination – the process by which the opposing party's attorney asks questions of the other side's witness in court. Leading questions are generally permitted during cross examination.

cryptocephalic pupa – the stage following larval-pupal apolysis within the fly puparium; adult structures are not yet visible.

demonstrative evidence – evidence other than actual testimony of witnesses, which relates to the senses. Examples include photographs, charts, diagrams, maps, models, computer-aided animation, etc.

deponent – a person who gives testimony in a deposition.

deposition – the testimony of a witness taken not in open court, but in advance of the in-court phase of a trial; a discovery device in which the attorney for one party asks questions of the other party or the other party's witnesses. The questions and answers can be written or oral. If oral, a verbatim transcript is made of the deposition.

Dermestidae – a family of beetles the larvae of which attack a wide variety of dead animal substances, usually exploiting middle- or late-stage carcasses.

diapause – arrested development in the egg, larvae, or pupa, or dormant period in the adult.

direct examination – the process by which the attorney who called a witness to testify asks questions of that witness in court. Generally attorneys may not ask leading questions during direct examination.

dura mater – the outermost and toughest of three membranes which envelope the brain and spinal cord; closely applied to the bone in the cranium.

ecdysone – hormone from the prothoracic gland that initiates new cuticle formation. In maggots, it stimulates the growth of the imaginal discs which produce adult structures, e.g. legs, wings, and antennae.

eclosion – the process of adult emergence; in flies, from the puparium.

endemic – restricted to, or native to, a particular region or area.

endophily – characterized by a readiness to enter homes, etc.

eschar – dry crust or scab at site of wound.

eurythermal – adapted to a wide range of temperatures. (*See* stenothermal.)

eusynanthropic – closely tied to and dependant on the human environment.

exophily – preferring the outdoors and characterized by a reluctance to enter homes, etc.

expert witness – a witness qualified by virtue of specialized knowledge, education, training, or experience to offer an opinion in court or in a deposition on some matter in controversy.

exuvium – the cast skin.

facultative imaginal diapause – the adult may or may not exhibit arrested development or prolonged cessation of activity.

fecundity – potential reproductive capacity.

fleshflies – *see* Sarcophagidae.

forensic – used in or belonging to the courts of justice.

fuscous – dark brown, approaching black.

gravid – full of eggs.

hearsay – an out-of-court statement offered in court to prove the truth of the matter asserted in the statement. Some forms of hearsay are admissible as evidence, usually as specific exceptions to the general rule against admissibility of hearsay.

hemisynanthropic – overlapping the human environment but capable of functioning independently of it.

hemolymph – the blood of arthropods, which circulates in a system where it is only partly confined to blood vessels.

Holarctic – the northern regions of the Old World (Palaearctic) and the New World (Nearctic).

holometabolous – complete metamorphosis, i.e., egg, larva, pupa, adult.

holoptic – the compound eyes touch, or nearly so, dorsally.

hypertrophy – abnormal enlargement or excessive development.

imaginal discs – saucer-shaped embryonic clusters of cells in maggots destined to give rise to adult structures, e.g., legs, wings, antennae.

instar – the stage between molts in the larva, e.g. the first instar is the stage after hatching from the egg and before the first molt. Flies typically have three larval instars.

interrogatories – a set or series of written questions used during discovery, pre-trial, to gather information about the case, as in a deposition. The answers are generally given under oath and subject to the penalties for perjury.

juxta – or apical plate, shovel-like and sclerotized at the apicoposterior part of the phallus.

larva – the stages subsequent to the egg and before the pupa , which are usually mobile and actively feeding.

larviparous – eggs hatch within the female.

larviposit – the female deposits larvae instead of eggs, as in sarcophagids.

leading question – a question which suggests or implies a particular answer. Leading questions often begin with such phrases as "Isn't it true," "Is it a fact that," and other words that attempt to lead the witness to agree with the questioner. Generally leading questions are only permitted to be used during cross examination.

lipase – an enzyme that splits fats.

livor mortis – blood circulation ceases after death, blood collects in the lowest regions where a red-purple discoloration appears.

maculae – spots.

material – a quality of evidence that tends to influence the trier of fact because of its logical connection with the issues in the case.

medicolegal – in forensic entomology, pertaining to a homicide, suicide, or accidental death under questionable circumstances.

metabolic heat – heat produced by an organism(s) in excess of the ambient temperature.

metamorphosis – the changes that occur during development to the adult stage. There are three categories: (1) holometabolous – complete metamorphosis (egg, larva, pupa, adult) e.g. flies, bees, beetles; (2) paurometabolous – gradual metamorphosis (egg, nymphal stages and adult, usually winged), e.g., grasshoppers; and (3) ametabolous – as in (2) but with wingless adult, e.g., silverfish.

micropyle – the minute opening at the anterior end of the egg through which sperm enter.

montane – a biotic zone occurring roughly at 2700–3500 m.

Neotropical region – includes South and Central America, the Caribbean, and the coasts and southernmost parts of Mexico. The rest of Mexico is in the Nearctic region.

neuroendocrine system – its input is sensory signals, its output is effecter signals and hormones.

nival – a biotic zone occurring roughly at over 4800 m.

non-responsive – an objection to a witness's answer to a question in court on the basis that the answer failed to address the matter at issue in the question. Rambling, discursive, or argumentative answers are often vulnerable to this objection.

objection – a formal legal protest against something in a trial or deposition, used to call the court's attention to improper or illegal evidence or procedure. Objections often must be made at trial to preserve a legal issue for subsequent appeal.

obligate heliophile – showing a strong preference for light and avoiding shade or darkness.

ommatidium – a photosensitive unit of the arthropod eye with a set of lenses, retinal cells and optic nerve fibers

oocyte – female gamete before maturation.

osteomyelitis – inflammatory disease of bone that may involve the marrow, cortex, or periosteum, caused by an infectious agent.

oviduct – one of the paired tubes through which the egg passes from the ovarian tubules into the vagina.

oviposition – the act of laying eggs.

pepsin – an enzyme that digests proteins in an acid solution (pH 2.5–3.5).

peritrophic membrane – secreted at the anterior of the midgut and envelopes the incoming food.

phanerocephalic pupa – the stage within the fly puparium where head, legs, antennae, and wing buds are everted.

pharate adult – the adult fly prior to emergence from the puparium.

phenotype – the detectable expression of the interaction of an organism's genotype and environment.

Phoridae – a family of small flies of widely different habits, including carrion feeders e.g., the coffin fly.
phragma – an invagination of the exoskeleton to which a muscle is attached.
pilose – hairy.
Piophilidae – a family of small dark flies, some of which feed on late-stage carrion, cured hams, and cheese, hence the name 'cheese skipper' for the mature larvae of some species which exhibit a remarkable method of propulsion.
Pleistocene – the period on earth from 2 million to 100 000 years ago.
pollen – a dusty, grayish surface on flies.
polymorphism – existence of populations of a species with different forms and/or behaviors.
postfeeding larva – the nonfeeding stage of the fly larva prior to pupariation, during which the gut contents are digested and a single layer of fat cells is deposited on the inner surface of the cuticle.
prepupa – the brief maggot stage within the puparium before larval–pupal apolysis.
probabilistic evidence – evidence that relies on statistics and/or probability estimates. Many courts disfavor such evidence, at least in the absence of corroboration.
protease – an enzyme that splits proteins.
prothoracic gland – endocrine gland generally located in the thorax that secretes ecdysone. It is part of the ring gland in the maggot.
psychrophilic – cold loving.
ptilinum – an ephemeral bladder-like structure between the compound eyes of the emerging adult fly which alternately inflates and deflates.
pupa – in insects with complete metamorphosis the stage between larva and adult, wherein extensive destruction of larval tissues occurs to make the adult structures.
pupariation – the process that shrinks the postfeeding maggot and makes its skin dark, shiny, and brittle.
pupation – formation of the pupa within the puparium of flies (see chrysalis).
respiratory horns – breathing tubes on the fourth segment, only in phanerocephalic pupae, that are accessory to the posterior spiracles.
relevance – a tendency of any evidence to make any fact in issue in a case more or less probable than it would be without that evidence. All evidence, at a minimum, must be relevant to be admissible at trial.
rigor mortis – rigidity of muscles after death.
ring gland – in higher Diptera the corpus allatum, corpus cardiacum, and prothoracic gland form a composite structure which rings the aorta, and is closely associated with the brain toward the rear part of the foregut (*see* Fig. 2.2).

Sarcophagidae – a family of black and grey checkered flies, mostly carrion feeders, thus the common name 'fleshfly'. Some species are parasitic.
sarcosaprophagous – consuming dead meat.
sclerotization – the process that leads to hardening and usually darkening of the arthropod cuticle.
SEM – scanning electron microscope: an electron beam is focused on the gold-coated surface of the object; secondary electrons are emitted from the surface and picked up by a secondary electron detector. The effect is a view of the surface, as in the light dissecting microscope.
seta – hollow bristle or hair developed from specialized cells in the epidermis of the arthropod cuticle.
spiracle – an opening in the insect's body through which air passes into the tracheae or air tubes.
standard error – the standard deviation of the sample mean.
stenothermal – adapted to a narrow range of temperatures (see eurythermal).
succession – in forensic entomology, the sequence of arthropods visiting and exploiting a carcass.
testaceous – brownish yellow; bearing a hard covering.
tracheae – a system of ramifying tubes that bring oxygen from the outside to the cells. The maggot has two parallel tubes that run the length of the body and give off branches in each segment.
trier of fact – the person or persons who determine which facts will be considered to be established as true for purposes of a given case. It is generally the function of the jury to serve as the triers of fact, while the judge determines the law to be applied to those facts.
trypsin – an enzyme that digests protein in an alkaline solution.
voir dire – the process of questioning potential jurors prior to their empanelment as the official jury for a given case. Voir dire can also take place with regard to a potential expert witness, out of the presence of the jury, to determine whether he or she can qualify as an expert witness on the relevant subject matter at trial.

PART 1

"Who saw him die? I, said the fly,
with my little eye, I saw him die."

Anonymous

Preface to Part 1

Plautus, one of the most popular dramatists in ancient Rome, put it best: "This man is a fly, my father, nothing can be concealed from him, whether secret or public, he is presently there and knows all the matter". Even today, we still wish to be "the fly on the wall" at some secret meeting or momentous event. Not even Plautus could have dreamed that one day the intrusive fly would tell its story in court and help tip the scales of justice. For a good part of the 50 years that I studied flies as vectors of disease I had never associated flies with homicides. Few entomologists had during those years, and the name of Mégnin, the founder of medicolegal entomology, was buried in the 100-year-old literature. It all changed in 1976 with my first case – a double murder – in which there weren't even specimens, just photographs taken two years before. Since then, forensic entomology has become a 'growth industry' in countries around the world. Although still viewed by the public as something of an oddity, insect evidence is recognized by the courts and increasingly introduced in cases involving accidents, homicides, and suspicious deaths, especially where time of death is a key issue.

We focus on carrion flies as forensic indicators. Blowflies particularly are usually the first insects at a body, sometimes before the police arrive. As the initial colonizers they may arrive in minutes and lay eggs within a few hours. Their first generation provides a biological clock that more precisely measures the time of death for two or more weeks, than the medical examiner's estimate which is limited to about a day or two. If the discovery of a body is delayed beyond the first generation of flies, the succeeding colonizers – various species of beetles and flies with more variable schedules of arrivals – will still provide a useful but less precise time of death.

The stakes are usually high when a forensic entomologist is hired. Sound science must be applied to the available evidence no matter the charge. In the courts, a healthy skepticism is replacing the acceptance of science on blind faith. To cite an example: that venerable icon of identification, fingerprinting, was never based on good scientific proof and has just fallen from its pedestal in one of the Federal courts. The entomology chapters provide a wealth of useful information on fly biology in a context relevant to entomologists everywhere, and hopefully, clear enough for lawyers to follow. It is a sad fact that science, on its way from laboratory to court, is sometimes transmuted under adversarial heat into pseudoscience. Given the limitations of our knowledge and experience, we have tried to expose the shoals and quicksands of the weaker side of forensic entomology as guides through the litigation process and for future research.

The fly is a magic carpet that has taken me from home base at the University of Illinois at Chicago to Mexico, Peru, Brazil, and Italy to unravel a few of its mysteries. I am deeply indebted to many colleagues and collaborators along the way. I am especially indebted to Hiromu Kurahashi of Japan, František Gregor of the Czech Republic, Juan Carlos Mariluis of Argentina, Baharudin Omar of Malaysia, and James F. Wallman of Australia for providing keys to the flies of forensic importance in their region. Others, after a stint in my laboratory, have returned to their home countries to establish forensic entomology. John Kunich became a Professor of Law. He and my family pried me from semi-retirement to write the entomology portion of this book. Now I am grateful that they did this and it is my sincere hope that the reader will be too.

I am grateful to Lon Kaufman, my Department Head, for his encouragement and essential logistical support, and to Matt Dean for his computer skills.

Bernard Greenberg
January, 2002

(Illustration by Janice Rajecki.)

1

A History of Flies

Travelling back in time we are reminded of the intimate, if disagreeable, association between man and flies, traceable almost to the dawn of recorded history. A 5000-year-old cylinder seal from Mesopotamia depicts a fly above two ibexes and a reclining gazelle (Fig. 1.1). They are carefully and skillfully carved in stone, understandably so in the case of the graceful mammals. But why a fly?

The mounting piles of garbage as human settlements grew, and the carnage of war, generated exploding legions of flies. They could not be ignored by the living and in death, whether king or slave, all carrion was equal. Some time during the 10-year siege of Troy, Achilles worries over the dead body of his dear friend, Patroklis, "that flies might get into his wounds beaten by bronze in his body and breed worms in them, and there make foul the body . . ." His mother, the goddess Thetis answers: " I shall endeavour to drive from him the swarming and fierce things, those flies, which feed upon the bodies of men who have perished . . ." (Homer, *The Iliad*, Book 19).

The first clear reference to blowflies was "published" more than 3600 years ago in the Ḫar-ra-Ḫubulla, a collection of cuneiform writings on clay. Tablet XIV is a systematic inventory of wild terrestrial animals that dates from the time of Hammurabi and is based on even more ancient Sumerian lists. It is the oldest known book in zoology. In it are 396 names of animals inscribed in Akkadian cuneiform on clay tablets, of which about 10 are flies. Here is the first mention of the "green" fly (probably *Phaenicia sericata* or *Chrysomya albiceps*) and the "blue" fly (possibly a *Calliphora*). These lists may be the source for the passage in Genesis where God "formed every beast of the field and every fowl of the air and brought them unto the man . . . And the man gave names to all cattle, and to the fowl of the air, and to every beast of the field" (Genesis 3:19,20).

Given the human propensity to name things, it is reasonable to inquire about the origin of the word 'fly'. There are probably almost as many names as there are cultures but we will only deal with two. The English word 'fly' originated in northwestern Europe where the insect is epitomized by the restless flight of the common housefly *Musca domestica*. In this part of the world, the insect is not so much attracted to us as it is to objects around us, especially food. The word has evolved relatively little in a millennium. In Old English it is *fleoge*; in Middle English, *flie*; in Swedish, *fluga*; Norwegian, *flue* or *fluge*; Danish, *flue*; and in German, *Fliege*.

More widespread is the root that traces back to the Old World tropics. In Sanskrit, the word for fly is maksika, pronounced mukshika. If one slurs the

Fig. 1.1. Mesopotamian cylinder seal with fly, *c.* 3000 BCE. (© The British Museum.)

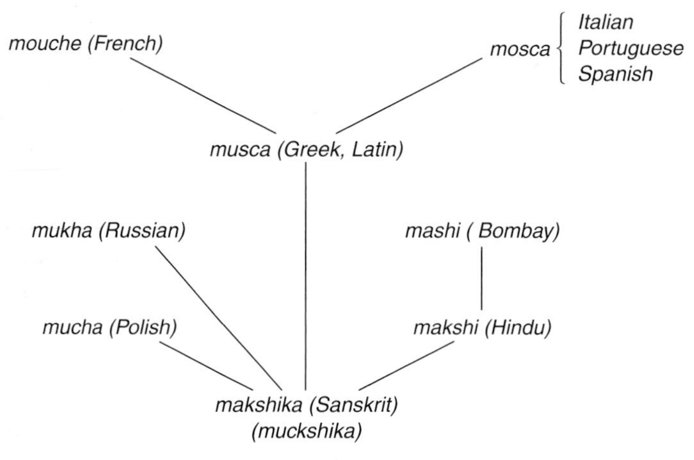

Fig. 1.2. A suggested 'genealogy' for the genus name *Musca*.

pronunciation a bit how easily this becomes mushka and then musca. The root '*muk*' means face or mouth in Hindi and this strongly suggests a link to the eye fly, *Musca sorbens*. Unlike the northern housefly, this fly is relentless in its appetite for the secretions of the eyes and mouth, best described as "in your face." This behavior suggests the nomenclature and the source of the scientific name as set out in Fig. 1.2.

In our time familiarity may breed contempt but among the ancients the fly appears to have bred a kind of reverence. A cylinder seal, dating from the Old Babylonian or early Kassite period (c. 1700 to 1400 BCE) depicts an oversized fly with the god Nergal holding a scimitar (Fig. 1.3). Other small creatures such as grasshoppers and frogs were depicted in these early Kassite seals (Porada, 1948; J. Brinkman, pers. comm.), but flies were deified. The Assyriologist, Elizabeth van Buren (1936–7, 1939), writes: "comparisons with flies carry no

Fig. 1.3. Cylinder seal with fly and the god Nergal holding scimitar. Old Babylonian or early Kassite, *c.* 1700–1400 BCE. (Courtesy the Pierpont Morgan Library, New York. Seal impression #571.)

stigma, and the gods themselves are compared to these pestilential insects. In the Epic of Gilgamesh (Tab. IX, 162) we are told that the gods gathered like flies around Utnapistim when he offered sacrifice." Another passage relates: "The gods of strong-walled Uruk are changed into flies and buzz around the streets". It is interesting to note that in the Epic of Gilgamesh (Tablet XI), after the flood has receded, Belitili/Aruru, the mother goddess, lifts up a necklace with carved lapis lazuli fly heads representing her dead offspring and vows: "You gods, as surely as I shall not forget this lapis lazuli around my neck, may I be mindful of these days [days of the flood] and never forget them" (Kovacs, 1985). Kilmer (1987) believes that the necklace suggests the iridescence of blowfly wings that are like the rainbow that brings peace and an end to the flood. She is supported by a passage in Genesis where the flood has receded and God makes a covenant with Noah: "I have set my bow in the cloud and it shall be for a token of a covenant between Me and the earth. And it shall come to pass, when I bring clouds over the earth, and the bow is seen in the cloud, that I will remember my covenant which is between Me and you and every living creature of all flesh; and the waters shall no longer become a flood to destroy all flesh" (Genesis, 8–17). But Homer's view of rainbows is less rosy: "As when in the sky Zeus strings for mortals the shimmering rainbow, to be a portent and sign of war or of wintry storm . . ." (*The Iliad*, Book 17).

From the royal tombs of Queen Puobi at Ur, dating from 2600 to 2500 BCE, comes an exquisite string of beads with flies of gold and lapis lazuli (Plate 1).

The work might have been done in different workshops because the technology of fashioning gold and carving stone is quite different and so is the rendering of the flies. The ones in lapis lazuli are literal, with typical head, thorax, abdomen and wings, as in Figs. 1.1 and 1.3. Those in gold are highly stylized and abstract. Never before had the lowly fly attained such esthetic glory.

In the Chaldean, Philistine, and Phoenician pantheon, flies were deified as Baalzebub, Lord of Flies. Appearing as a fly, he could ward off plagues of the insect or send his winged legions as punishment. He would reappear thousands of years later as Beelzebub, First Lieutenant of Satan, in the witchcraft atmosphere of Colonial New England.

The Greek god, Apollo, had many duties – as pasture god and protector of herds, the function of Myiagros, or fly-chaser, belonged to him, because the hordes of flies could lead to plagues (Noury, 1932). Sometimes Zeus, himself, insisted on chasing away the flies, and he saved Hercules when the latter was almost vanquished by these insects. Such was the mythic power of flies.

The Egyptians, too, were obsessed with flies. They wore carved amulets of flies to ward off evil, and burial beads to avert the destruction of the body (Plate 2). Fly whisks were a more practical defense (Fig. 1.4) and even today they are used in widely separate cultures in Africa and the Pacific islands as symbols of rank (Riegel, 1979). In ancient Egypt the fly also symbolized impudence, persistence and courage, and a necklace of flies of purest gold was awarded to soldiers who distinguished themselves in battle – a kind of Congressional Medal of Honor or Victoria Cross (Plate 3). No fly in that region epitomized persistence more than *M. sorbens,* the eye fly, whose avidity for eyes and mouth still torments people on that continent.

When death occurred, could flies be far behind? In Egypt, carrion flies may have even assisted in the embalming process. The late Bruce Ralston (pers. comm.), a neurosurgeon, considered excerebration by mechanical and chemical means insufficient for the task of removing the brain although this method has been accepted as dogma since Herodotus, some 2500 years ago. Herodotus wrote: "First they draw the brain through the nostrils with an iron hook, taking part of it out this way, the rest by pouring in drugs". In fact the ancients, including Aristotle, considered the heart, not the brain, the seat of intelligence, so the brain was not worth saving. According to Ralston, telltale scratch marks made by an instrument inserted into the skull to remove the brain and the more resistant dura mater are absent and "no chemical has been found to this day which will remove all tissues from the bone without damaging it" (Babin *et al.*, 1990). A recent simulation of the embalming process faithfully follows the ancient practice but overlooks this issue in the following description of the entry into the cranium: "We inserted a long bronze instrument, shaped like a miniature harpoon, inside the nasal passage and hammered it through the cribiform plate into the cranium with a wooden block" (Brier, 2001). Leek (1969) successfully removed a sheep brain by this method and in an appendix

Fig. 1.4. Defeating flies might be as difficult as vanquishing lions in this drawing by Francesco del Tuppo, Aesop, *Vita et Fabulae*, Naples, 1485. (Courtesy The Pierpont Morgan Library, PML,15404.)

to the same paper, Sudhoff and Patterson reported success on human bodies. There is no mention in these papers that the procedure removed the dura mater.

Professor Arthur Aufderheide, University of Minnesota Medical School, (pers. comm.), has examined 49 Roman period mummies excavated from the Kellis-1 cemetary at Ismant el Kharab in the Dakhelh Oasis, Egypt. Of these, 35 were sufficiently preserved to permit reconstruction of their mummification process. "In only five of the twelve anthropogenically mummified bodies did resin gain access to the cranial cavity and, in two of these, multiple insects (some of which were identifiable as mature dermestids) were found mired in the hardened resin. Dura mater was not identifiable in the cranial cavities of any of either anthropogenically or spontaneously [desiccated] mummified bodies" (A. Aufderheide, pers. comm.).

Maspero (1904) has calculated the number of bodies undergoing a 60- to 80-day embalming procedure in ancient Thebes as a minimum of about 1200 to 1500 at one time. Hamilton-Paterson and Andrews (1979) mention black swarms of flies in the charnel houses. If such conditions existed, and it is hard to believe otherwise, carrion flies would have had a picnic cleaning out the brain, exiting the skull as postfeeding larvae, and leaving the remnants of drier tissue to dermestid and other beetles and piophilid flies (Pettigrew 1834; Curry 1979; Babin et al. 1990). In homicide cases the brain is often the first organ attacked and destroyed by flies, given their predilection for eyes, mouth, and nose as sites for oviposition. Furthermore, according to Tapp (1984), "the ancient Egyptians tended to leave the brain to liquefy for a few days before attempting to remove it . . ." (see also Isherwood *et al.*, 1979). Under these conditions it would have been extremely difficult to exclude flies. The intervention of flies would have reduced the process of decerebration to just a few days, given the warmth of the charnel house and the metabolic heat produced by the actively feeding maggots. David and Tapp (1984, p. 73) describe an endoscopic examination of the Asru mummy skull, which revealed several intact larval skins lying free within the skull that were removed and identified as *Chrysomy[i]a*. The species is probably *albiceps*, a common fly in the region (see also Cockburn *et al.*, 1975).

Whether fly eggs or larvae were ever deliberately introduced into the head by the embalmers is not known. With or without encouragement, flies could have done the job.

Since the brain was considered an unimportant organ, the object of embalming was to protect all organs except the brain from the ravages of insects and time. This concern was expressed in the Book of the Dead, Chapter 154: "That my body will not become prey to maggots" (Huchet, 1995). At death, the guardian spirit, or Ka, remained with, or periodically returned to, the body. A horde of maggots leaving the body and possibly taking the Ka with them would have defeated the purpose of the sacred ritual. A slip of paper found in the mouth of a mummy contains the inscription: "The maggots will not turn into flies within you" (Papyrus Gizeh no. 18026:4:14) – a recognition of the metamorphosis of flies, and what might have been a widespread worry that the body's destruction by maggots might thwart the Ka's return. A similar exhortation is in the Book of the Dead, Chapter 154. Kritsky (1985) mentions a belief that still persists in parts of rural Egypt: "shiny, metallic calliphorid flies found inside houses represent the spirit of a deceased individual who once lived there. That association is enough to keep some people from killing flies for fear they might harm an ancestor." It suggests the persistence of the ancient, ambivalence toward flies as both sacred and a scourge.

The Greek demon of decay Eurynomos, appears both as a vulture and as a fly. In the Persian book *Vendidad*, it is written that as soon as a person dies the demon of death throws himself upon the corpse in the form of a fly.

The ubiquity of flies remained unabated down to early modern times. Here is the personal account of the famed military surgeon, Amboise Paré, following "The Battel of S. Quintin, 1557," in a contemporary translation: "We saw more than halfe a league about us the earth covered with dead bodyes; neither could we abide long there, for the cadaverous scents which did arise from the dead bodyes, as well of men, as of dead horses. And I think we were the cause, that so great a number of flyes, arose from the dead bodees, which were procreated by their humidity and the heate of the Sunne, having their tayles greene and blew; that being up in the ayre made a shadow in the Sunne. We heard them buzze, or humme, which was much mervaile to us. And I think it was enough to cause the Plague, where they alighted".

During the Renaissance, death and the devil were personified by the fly, depicted in two paintings by Carlo Crivelli, a fifteenth-century Venetian painter. In Plate 4, the insect, probably a bluebottle, rests on a human skull next to a vase of flowers – a metaphor of life and death. In Plate 5, the Christ child clutches a bird, a symbol of resurrection, while recoiling from a fly, a symbol of Satan and death. The same attitudes were prevalent in Northern Europe as shown in Plates 6, 7, and 8. Dr. Marcel Dicke of Wageningen University, The Netherlands, has supplied the following observation of Jan de Kok of the Rijksmuseum. In the sixteenth century the fly on the white cap of the woman was often added to the painting after she had died (Plates 6 and 8; Dicke, 2000). For superb reproductions of Renaissance paintings depicting flies and other insects the reader is referred to Chastel (1984).

Given the ancient and enduring intimacy with flies it is surprising that in the West, murder was never associated with the triad – flies, death and decay – until the mid-nineteenth century. Not so in China. We are indebted to Professor Fang Jianming of the Shanghai Institute of Entomology, Academia Sinica, for the following case and translation. The crime is set sometime between 907 and 960 and is described by Cheng (1890).

An officer of the court suddenly heard a woman endlessly weep and wail. The officer asked her what had happened. The woman said that her husband was killed by fire, but the officer discovered many flies clustered on the head of the corpse. At autopsy, there was a snag in the head of the corpse. The woman confessed that she and another man had put a snag in the head of her husband.

In 1247, Sung Tz'u, a high-ranking judicial administrator, published *The Washing Away of Wrongs*, a training manual for death scene investigators. In it he relates a murder investigation in a rural, agricultural village. A man was found dead in the road with numerous slash wounds in his head that appeared to have been made by a sickle. An investigator assembled the men of the village and lined them up with their sickles on the ground in front of them. It was summer and flies were numerous. The investigator went up and down the line and finally stopped in front of one man, turned to face him and accused him of the murder, which the man denied. When the investigator confronted him with the fact that his was the only sickle with flies clustered on it, the man confessed. He had not washed the sickle thoroughly enough to fool the flies (McKnight, 1981).

The third case occurred in the eighteenth century and, like the first, is supplied by Professor Fang Zianming in translation: "A merchant was killed and his silk was robbed in the road. A retired policeman was in charge of investigation. After two days the investigator saw the boat on which there were plenty of flies clustered on the washed silk. Then the police arrested men of the boat. Those men had to own up. Because the silk had traces of blood" (Cheng, 189?).

In all three cases, the flies performed like bloodhounds, with the investigators relying on their acute sense of smell to zero in on the blood evidence. Given that sense, which is actually more acute than that of bloodhounds, it is not surprising that carrion flies can flourish in a landscape where corpses are few and far between.

In mid-nineteenth century Europe, faunal inventories of decaying corpses laid the scientific groundwork but did not link flies with murders. Orfila (1848), a pathologist, listed 30 insects and other arthropods that visited a corpse to feed and oviposit. He included *Calliphora vomitoria* (L.), *Lucilia caesar* (L.), *Musca domestica* L., and *Sarcophaga carnaria* (L.), which were succeeded by a number of species of beetles. Like Motter (1898) and Hough (1897), he focused on the fauna of the grave, and may have been the first to systematize the knowledge of arthropod succession in a human corpse. Bergeret (1855) credits Orfila with noting the presence of insects on dead bodies, but points out that Orfila failed to apply their development to medicolegal matters, as he did in the following case.

In March, 1850, the mummified body of an infant was discovered in the bricked-up space behind a fireplace. During the prior three years, four families had successively rented the apartment. Which one was guilty? Bergeret investigated and found that the baby had been born to term, but there was no evidence of the cause of death. He found large numbers of empty puparia in several cavities of the body which he determined as *Musca* (= *Sarcophaga*) *carnaria* (L.) , the fleshfly. Based on what he knew of its habits, he concluded that larvae were deposited on the body (the fly is larviparous) soon after death in 1848, and adults emerged the following year. The insects had entered through a tiny fissure in the bricks, small enough for them but insufficient for air to circulate. This resulted in the gradual mummification of the body and the subsequent infestation with clothes moth larvae. Bergeret's reconstruction of the insects' natural history led to the exoneration of three families. His work was an important contribution to the infant science of forensic entomology. It was the systematic use of life cycle events of the two species on the body that enabled him to estimate the time of death. At the time, it was a novel application of biology to the practice of criminal justice and it remains our *modus operandi.*

There were other glimmerings of the beneficial use of carrion flies, none more unusual than as healers. This is the account of D. V. Larrey, Napoleon's military surgeon, during the Syrian–Egyptian campaigns of 1799."There remains an unusual matter which we do not believe ought to be passed without mention; it concerns that which we have had occasion to observe in Syria, during the Egyptian expedition, among the majority of our wounded. While their sores were suppurating, these wounds were bothered by worms or larvae of the blue fly, common in this climate. These insects formed in a few hours, developing with such rapidity that, from one day to the next, they were the size of a small quill which frightened our soldiers very much, despite all we could do to reassure them in this regard: . . . that, far from being harmful to their wounds, these insects, by accelerating cicatrization, shortened the work of nature, and also by lessening the cellular eschars which they devoured. These larvae, in effect, have an avidity only for putrefying matter, always sparing the living parts; also I have never seen in these circumstances, evidence of hemorrhage, the insects are carried only to that depth which is the extent of the wound." A very sound scientific observation, unappreciated by contemporary surgeons who viewed maggots (as most still do) as disgusting creatures with no redeeming qualities. Even in the severest winter weather in a Pennsylvania hospital, Coates (1842) had to battle to keep his patients' wounds clean.

Credit for the first deliberate, beneficial use of flies probably goes to J. F. Zacharias, a surgeon in the Confederate Army during the US Civil War when he was serving in the hospital at Danville, Virginia. This is what he wrote:

> "I first used maggots to remove the decayed tissue in hospital gangrene and with eminent satisfaction. In a single day, they would clean a wound much better than any agents we had at our command. I used them afterwards at various places. I am sure I saved many lives by their use, escaped septicemia, and had rapid recoveries."

Why did it take almost a century to re-discover the therapeutic benefit of maggots? And longer still, to recognize their forensic value when found on a dead person? Needed were life cycle studies of common carrion flies initiated in the 1930s in the United States during the "golden age" of maggot therapy. At that time, maggots were raised and shipped everywhere to treat osteomyelitis. The surgical use of maggots was soon eclipsed by the advent of sulfonamides and antibiotics. In the following decades, renewed interest in flies as vectors of human and animal pathogens led to studies of their distribution, habits, and life cycles which strengthened the scientific foundations of forensic entomology (Greenberg, 1971,1973).

Mégnin (1894) established the science of forensic entomology but the seed lay dormant until after the mid-twentieth century. More than 15 years of experience at the Paris morgue gave him the expertise to link the stages in the decomposition of the human body with specific insects. He described eight stages and their time of onset: (1) fresh – month 1; (2) decomposition begins – months 1 to 3; (3) fatty acids – months 3 to 6; (4)caseous products – months 3 to 6; (5) ammoniacal fermentation, black liquefaction – months 4 to 8; (6)dessication – months 6 to 12; (7) extreme dessication – years 1 to 3; (8) debris – over 3 years.

Mégnin went on to list the groups or "squads" of arthropods (mainly insects and mites) specific for each stage, and unknowingly established the concept of the corpse as a dynamic, though finite, ecosystem with a faunal succession. Flies were the first to attack the fresh corpse and he named the housefly *M. domestica*, the bluebottles *Calliphora vomitoria* and *C. vicina*, and the false stablefly *Muscina stabulans*, in the first squad. In the second period, when decomposition had commenced, the greenbottle *L. caesar*, and several species of sarcophagid fleshflies arrived. Then the beetles moved in and dominated for a period of about five months, followed by mites as the remains dried. During the final two periods, clothes moths, dermestids and other beetles completed the recycling of the remaining soft tissues, leaving bones and dust. In a brief note, Hough (1897) confirmed and extended the observations of Mégnin. He listed the succession of arthropods in "Europe" and in "America" that came to a variety of corpses including horses, dogs, cat, rats, and men in the course of decomposition.

Mégnin's contemporaries expressed amazement at his often accurate predictions. Johnston and Villeneuve (1897) wrote: "There is almost something uncanny in the way which M. Mégnin could state for instance after examining a few bones and some dust that a murder had been committed during the latter part of February of the year before last – and then be absolutely justified by the dying confession of the suspected party." But, they caution: "The chief danger to be feared by Mégnin's imitators is that they might tend to apply rules to countries and climates where they were inapplicable." Prophetic words.

Niezabitowski (1902) was the first to study insects on cadavers in the Russian Empire. His observations differed from those of Mégnin and cast doubt on the suitability of forensic entomology. This created a pall over the field in Russia that was not lifted until the 1930s.

Mégnin established a biological framework that subsequent experience has proven to be too rigid. The decomposition stages can be fewer and the insect traffic can vary. In biology, the strait-jacketing of complex biological interactions is bound to be fallible. Marchenko (1988) reviewed about 50 papers dealing with the rate of decomposition of the human corpse and concluded that it is a highly variable process subject to many factors, e.g. age, sex, fat or lean, clothed (light or dark) or naked, mode of death including the nature of wounds, weather (wind, rain, temperature), body in sun or shade, indoors or outdoors, condition of substrate, time of year, etc. And the list does not even include insects.

Over-reliance on rigid schedules of insect arrivals and departures can lead to false estimates of the time of death. For example, Mégnin noted that blowflies arrived first at the body and disappeared as conditions became more favorable for beetles. However, blowflies may return at a much later stage when rains have made the soft tissues once again attractive for oviposition and larval development. The converse is also possible. Dermestid beetles are typical latecomers to a corpse but they can arrive early, completely off-schedule, to attack the mummified hand of a fresh corpse. In Bergeret's case, the infant had been dead for two years, yet only two species of insects had infested the body. Normally there would be dozens. Of course, the circumstances in this case were special, but so is every case when the entomologist confronts an insect-infested corpse.

The methodology of forensic entomology applies to homicides, suicides, and accidental and untimely deaths, as well as poaching. It is used productively in archeology (Hall, 2001) and paleontology. To the fly a corpse is simply carrion, no matter the cause of death. The key attribute of flies is their uncanny sense of smell that would make a bloodhound blush, if it could. Place meat out at the appropriate time and flies materialize, seemingly from nowhere. This may come as a surprise to those accustomed to a relatively fly-free life.

The most frequent question put to the entomologist is "When did death occur?" The forensic pathologist depends on rigor mortis, livor mortis, and core temperature of the corpse to derive acceptable estimates within 24 to 36 hours of death. Insects offer a much longer time scale measured in hours, days, weeks, or season, and sometimes years. This can lead to the identity of a missing person whose decomposed body is beyond recognition, ironically because of the insects.

All valid uses of forensic entomology are based on accurate identification of the species. The movement of a body from one site to another may result in an improbable fly fauna and open a new avenue of investigation. There are increased reports of maggot infestations of the elderly or the infirm. This results from abusive conditions or neglect by relatives, or in nursing homes and hospitals. Here again the species and its growth rate can be used to pinpoint culpability. This takes taxonomic skill and knowledge of fly habits. Chapter 3 contains keys, accompanied by illustrations, that enable the entomologist to classify eggs, larvae, and adults of forensically important flies in different parts of the world. The reader then proceeds to the tables and graphs of development for these species. Because these data are not engineering or physical constants, Chapters 2, 3 and 4 provide a detailed discussion of fly biology, with a perspective of the problems and pitfalls in its forensic application. Chapter 5 moves into the adversarial arena of the courtroom with actual cases. The relevant details of each case are described, opposing entomological arguments are presented, and readers are asked to judge.

Thus prepared, the forensic entomologist and the litigator will be better able to maximize the strength of their entomological arguments and minimize those of the opposition. What a strange fate for the lowly fly, alternately revered and despised throughout history, to become the one we now turn to in matters of life and death.

2

Forensic Biology of Flies

Introduction

Forensic experience bears out Mégnin on a key point. In the decomposition of the human body, which can take weeks or years and usually involves scores of diverse arthropods, blowflies are almost invariably the first arrivals – if the body is exposed above ground or shallowly buried. Like it or not, insects in the family Calliphoridae are the law's best friend. The family name means 'beauty bearer' in Greek. To a sympathetic eye, many of the species are quite beautiful with blue, green or copper iridescence.

Are calliphorids first to the feast regardless of the species and geography? Can forensic entomologists and the courts rely on them in Europe, Australia, Africa and Asia, as in North America? And in different seasons? Furthermore, do these "early birds" leave their calling card as eggs? If not, they have little forensic value. The significance of blowflies, and sometimes fleshflies, as the primary and most precise indicators of time of death, hinges on the answers to these questions. The evidence is persuasive, but first let us sketch some of the broader aspects of calliphorid biology.

Most members of the family are remarkable for their ecologically vital role in waste management. Put succinctly, a human cadaver can lose 60% of its mass to maggots in a week. Vivian B. is a prime example. On the afternoon of September 10 she walked away, or was kidnapped, from a nursing home outside of Chicago. On September 18, her body was found in a wooded area just three-tenths of a mile (*c.* 483 m) away. The remains were white with maggots and almost completely skeletonized. An estimate of the age of these maggots suggested that the body had been deposited outdoors just five days before! A medical examiner or pathologist who overlooks or minimizes this awesome power of maggots runs the risk of extending an estimate of time of death well beyond what it actually is. By the time other insects have finished what the maggots started, there is little left of a corpse, and most of the insects themselves have been recycled. In the larger scheme of things it is fortuitous that some of our most common calliphorids are waiting to serve as the foremost indicators of the time of death. The family is comprised roughly of 130 genera and about 1000 species worldwide. Although relatively few species are now used forensically, the number will grow as flies are globally enlisted in the pursuit of criminal justice.

To a certain extent, carrion flies are "fingerprinted" by the way they partition space, time, and season. In terms of space, eusynanthropic flies, e.g., *Phaenicia sericata,* the greenbottle (New World dipterists prefer *Phaenicia sericata* whereas Old World dipterists use *Lucilia sericata*; For purposes of this publication *Lucilia* (= *Phaenicia*) will be used when used by an author), and *Calliphora vicina,* the bluebottle, are closely tied to the human environment, and the latter's specific name aptly reflects this habit. Hemisynanthropic flies like *Lucilia illustris* and *Calliphora vomitoria,* another greenbottle and bluebottle, respectively, are less habituated to the human environment, and asynanthropic flies the least. Povolný (1971) provides an excellent analysis of the synanthropy of mid-European flies, and Nuorteva (1963), in Finland, is one of the early proponents of the concept. For obvious reasons, the preponderance of medicolegal cases have usually involved eusynanthropic and hemisynanthropic flies. As the discipline reaches into the fields of wildlife conservation and biology, and into such illegal activities as poaching, the roster of asynanthropic indicator species will probably increase. But the major obstacles in these cases will not be the science but inadequate legal authority and enforcement.

Stenothermal blowflies live in an environment with a comparatively narrow temperature range. They are selected for the phenotype that offers the greatest efficiency in that environment because they do not have to maintain variable thermal responses. The greenbottle *Phaenicia pallescens* is confined mainly to southeastern United States and fits this niche. *Phaenicia sericata,* a sibling species, is eurythermal, cosmopolitan and more versatile. It is adapted to sharply different temperatures in its life cycle with several functional polymorphisms – cold-hardiness, variability in developmental rates, and a diapause, or arrested development, under adverse conditions (Ash and Greenberg, 1975).

We studied niche partitioning in 26 species of South American blowflies and found that the ecological niches included altitude, synanthropy, and endophily/exophily. Flies were collected from the Peruvian coast, across the Andes, down to the Amazon (Baumgartner and Greenberg, 1985). The transect crossed diverse climatic zones, temperature regimes, vegetation types, and habitats. As altitude increases, conditions for life become more rigorous – vegetation diminishes, food is more scarce, humidity and temperature fall, daily temperature fluctuations are larger, and wind velocity increases. With respect to altitude, our results showed a clear partitioning of species, with no overlap between 16 tropical and subtropical species and the four species that occupy the more rigorous montane, alpine and nival zones (2700 to >4800 m). Although a few species straddled two or three biotic zones, only *Calliphora peruviana* (= *nigribasis*) occurred in all five zones from the subtropics to the nival. We found that the six species confined to the lower rain forest are all asynanthropic. Among the six species that occupy both lower and upper rain forests, by far the most abundant are the eusynanthropes: *Cochliomyia macellaria, Chrysomya putoria,* and *Phaenicia eximia.* It is perhaps not surprising that these flies have the widest food preferences, including animal and human feces, fruits, and carrion. They are also endophilic, readily entering houses.

The studies of Kano, Kurahashi, Shinonaga and Nishida (see "References") provide valuable information on the carrion flies of Japan and the Pacific Rim. The monographs edited by Fan (1992), and Xue and Chao (1996), offer monumental coverage of Chinese Diptera. With the above works as guides for those regions, the forensic entomologist must, nevertheless, be familiar with the habits of the local flies. The presence of a species on a corpse in an unlikely habitat is reason to suspect that the body was moved.

Flies, like flowers, have their season. The greenbottle flies mentioned above are summer flies, while Holarctic *Calliphora* are psychrophyllic, or cold adapted. In temperate regions, species of *Calliphora* are among the first carrion flies in spring and the last to depart in fall. *Calliphora vicina* is generally reported to be a cool-weather fly throughout its distribution (Greenberg, 1971,1973). Thus, in the sub-tropics it is active in winter, while in the subpolar zone it is present in summer. Sychevskaîa (1965) confirmed the data of Smirnov (1940), that showed that the ovaries of this fly become inactive during hot summer months in Central Asia, and this inhibition ceases toward the end of August. In fact, populations of this fly crash or go into hiding in regions with hot summers. Therefore, its presence on a body in such a region suggests that the body became infested where it was cool and then was moved. Recent evidence, however, complicates the picture.

The summer of 1988 was one of the hottest and driest in the Midwest, with more than 40 days above 32 °C in the Chicago area. Yet in studies we conducted that summer, *C. vicina* was not only present in July but was active at night and laid fertile eggs (Greenberg, 1988).

Blowflies are generally indigenous to the regions where they are found, surviving freezing winters as postfeeding larvae and pupae in the ground, or as adults in protected places. A notable exception is *Cochliomyia macellaria,* the secondary screwworm, endemic to the tropics and subtropics of the New World. In the spring it disperses northward from overwintering sites in Mexico and the southernmost regions of the United States. When we surveyed flies coming to rabbit carcasses in urban and rural habitats in the Chicago area, we found no *C. macellaria* from April 9 until June 5. Some time in June in the Midwest, the fly reaches the level of southern Lake Michigan and Lake Erie. In parts of Colorado the pattern appears to be similar (De Jong, 1994). The fly continues northward, proliferating throughout summer by virtue of a relatively brief life cycle, and by late fall it is gone. It behaves the same in the cooler regions of South America where it is common enough in the warm season, but could not be found in Curitiba, Brazil, in winter (Ferreira, 1978).

On May 28, 1985, a man was collecting wild mushrooms on an abandoned homestead, 8 miles (*c.* 13 km) west of Aberdeen, South Dakota. He had been coming to the farm for years but this time he noticed the door to the ice-house was missing. He looked inside and found the door on the ground with a body beneath it. The deceased was identified at the morgue and was last seen alive on

May 8. All the larvae belonged to just one species, *C. macellaria,* and the oldest ones, in paperboard next to the body, were about one day into the postfeeding stage. There were no pupae. Putting aside the contending entomological arguments regarding time of death, the matter that concerns us here is the problematic presence of the fly "out of season". In the entomology collections at the University of South Dakota, among the 50 or so specimens of this fly, none had been taken in the state before August. What was this eusynthropic fly doing out of place and out of season? There are several possibilities. Perhaps warmer weather that year permitted earlier arrival of the flies. Or there could be flyways we know nothing about. It is known that fly dispersal in open country may be canalized by features of terrain and microclimate, in addition to olfactory trails. In Texas, for example, *Cochliomyia hominivorax,* the primary screwworm fly, dispersed along dry gullies, fence lines, and natural openings in brush (Hightower and Alley, 1963). It seems reasonable that a patchy distribution may be characteristic of these emigrating populations, easily missed by the casual collector. Or the body may have been infested farther south before it ended up in the ice-house.

Flies partition the seasons to minimize competition, and they also partition time. *Calliphora* species take longer than the greenbottles (and *Phormia regina*) to complete metamorphosis. The latter flies may be into the second generation on a body while the *Calliphora* are still first-generation pupae. Such a case involved the final victim of a serial murderer in Waukegan, Illinois, and is discussed in Chapter 4 under the subject of accumulated degree hours.

The Case for Blowflies: A Global Sampling

When Payne (1965) placed thawing carcasses of baby pigs outdoors in summer in rural South Carolina, sarcophagids were on the carcasses within five minutes and calliphorids, chiefly *Cochliomyia macellaria,* within 10 minutes. He does not mention when eggs were laid. In a study from June to November 1968 in the same region, he buried baby pigs in wooden boxes at a depth of 90 cm. No blowflies or fleshflies, in any stage, were present on the carcasses when they were exhumed.

Six unembalmed human cadavers were buried in soil trenches at depths of 0.3 m, 0.6 m, and 1.2 m, in an open wooded area, and at different times of the year, in east Tennessee. Care was taken to avoid stray oviposition before and during the burial process. The bodies from the shallow graves were exhumed after about one to three months, and from the deeper graves, after up to one year. Larvae, pupae, and adults of blowflies and fleshflies were found on cadavers only from the shallow graves. Following heavy rains, blowflies were seen ovipositing in cracks of the lightly packed soil, and it is assumed that the newly hatched larvae were able to burrow through the soil to reach the corpses (Rodriguez and Bass, 1983, 1985).

In Mississippi, a summer and winter study of insect succession on rabbits, opossums, and fish demonstrated that calliphorids were the first arrivals. In summer, *Phaenicia caerulaeviridis* was the dominant blowfly, while in winter, (average temperature, 8 °C), the bluebottle, *Cynomyopsis cadaverina*, dominated (Goddard and Lago, 1985). When large animal carcasses (sheep, goats, cattle, horses) were exposed in southwestern Texas and in southern Arizona during January and February (average ambient temperature above 6 °C), two blowfly species were the first and dominant exploiters – *Phormia regina*, the bluegreen blowfly, and *Cochliomyia macellaria* (Deonier, 1940). In a forested area in northeastern Illinois, carcasses of small mammals were first attacked by the bluebottle, *Calliphora livida*, from March to May. With warmer weather, *Phaenicia sericata* and *Phormia regina* took over (Johnson, 1975). On islands in the Gulf of Maine, colonization of the fresh carcasses of harbor seals was initiated by *P. regina* and *L. illustris* from May to October (Lord and Burger, 1984).

In Cracow, Poland, Niezabitowski (1902) recorded the arrivals of insects on human bodies from May to September. Besides the housefly and another species of *Musca*, the following blowflies were in the first wave: *Lucilia caesar, Calliphora vicina*, and *C. vomitoria*.

Mouse corpses were set out in woodland and grassland areas in Berkshire, England, during early July. Adult *C. vicina* and *Lucilia richardsi* predominated on mice in grassland, while the former and *L. caesar* were most common on woodland mice. Calliphorid eggs were laid within 24 hours in and around all the orifices of the mice. The larvae were identified only as *Calliphora* and *Lucilia* (Lane, 1975).

Alexandria, Egypt, is in the Mediterranean coastal belt and has a higher annual rainfall and a richer flora than the rest of the country. Outdoor studies on rabbit corpses there revealed a number of blowflies and fleshflies as initiators of decomposition. The species mix changed with the seasons. In summer, the sarcophagid, *Wohlfartia nuba*, was the only primary fly on the carcasses. In fall, it was joined by *Sarcophaga aegyptica, Sarcophaga argyrostoma*, and *Phaenicia sericata*. In winter, the sarcophagids were strangely absent, leaving the carcasses to *Calliphora vicina* and *P. sericata*, and only to the latter in spring. *Chrysomya albiceps* was present as a secondary fly in all seasons except winter (Tantawi *et al.*, 1996). Its later arrival is linked to the predatory habit of its larvae that feast on both meat and maggots, including their own.

Braack and Retief (1986) investigated the corpses of impala and kudu in summer and winter in Kruger National Park in the Republic of South Africa. In summer the first to the bodies was *Chrysomya marginalis*, arriving within minutes and laying eggs within 16 hours. *Chrysomya albiceps* arrived next and laid its eggs by the second day. Winter oviposition was delayed a few days and up to a week, respectively. Under similar semi-arid conditions in Tsavo National Park, Kenya, these two species arrived in the same sequence on the first day on an elephant corpse, and numerous unspecified maggots were present by day 4 (Coe, 1978).

Research on blowflies has been ongoing for much of the twentieth century in Australia, spurred by the needs of the livestock industry. An excellent early study was that of Fuller (1934). Her field investigations took place in the Canberra region in spring, summer, and winter. In spring, several fresh sheep heads, exposed for 6 hours, yielded almost 2000 larvae of *L.* (=*P.*) *sericata, Calliphora stygia,* and *Calliphora augur.* The same procedure in summer produced *Sarcophaga* spp., *C. augur,* and some *C. stygia,* while in winter, only the latter two blowflies were the primary species. O'Flynn (1983) found a somewhat different mix of blowflies in her experiments near Brisbane, southern Queensland. Working with large carcasses (sheep, pigs, dogs), the primary flies were *Lucilia cuprina, Chrysomya rufifacies,* and *Chrysomya saffranea* in summer, and *L. cuprina* and *Calliphora stygia* in winter (also O'Flynn and Moorhouse, 1979). Monzu (1977) worked with wallaby carcasses in the southwestern region of Western Australia. There, species of blowflies partitioned the seasons and virgin (asynanthropic?) versus human habitats differently. During winter–spring, in virgin country, *Calliphora varifrons* began laying eggs as soon as the animal was dead; *Calliphora albifrontalis* began ovipositing during the second day. In summer, the species changed, with *Calliphora nociva* the first to larviposit at death, followed by *Chrysomya rufifacies* and *Chrysomya varipes* during the first and second days. The situation was entirely different in the urban environment. Here, in all seasons, *P. sericata* was the only significant carrion fly, and it, too, oviposited at time of death.

In Oahu, Hawaii, insect succession was followed in pigs in a tropical rainforest, and in cats in a xeric and in a semi-moist region. In all three habitats, the same two blowflies, *Chrysomya megacephala* and *C. rufifacies,* were the first flies to initiate decomposition. In the pig study, eggs were found by the second day (Tullis and Goff, 1987); and in the cat study, eggs were not reported for the first two days (Early and Goff, 1986).

On plantations near Kuala Lumpur, Malaysia, fresh carcasses of cats and cynomolgus monkeys attracted the blowfly *Lucilia sinensis* within 15 minutes, but the time of oviposition was not noted (Omar *et al.*, 1994a). In another study, *C. megacephala* was at a monkey carcass within 30 minutes; although no eggs were found on day 1, there were larvae of this fly by day 2 (Omar *et al.*, 1994b).

Blowflies also led the succession in Japan (Utsumi, 1959). Calliphorids were consistently the first on domestic dogs, rabbits, and albino rats. The species included *Lucilia illustris, P. sericata, Calliphora lata,* and *Aldrichina grahami.*

Are blowfly arrival times on small mammals valid for human corpses? Although carcass size does not seem to effect arrival times, it can influence actual numbers of species that breed in a carcass. For example, the carcass of a large mammal, e.g., a pig, may decompose slowly enough to support a sequence of primary, secondary and even tertiary blowflies. On the other hand, a mouse or rat, can be quickly consumed by the maggots of the first flies (Kneidel, 1984). That is why studies of small mammals are not transposable to a human corpse if one is looking for key indicator species that are active later in the insect succession.

In any general discussion of the forensic role of blowflies it would be amiss not to mention the groundbreaking book by K. G. V. Smith entitled *A manual of forensic entomology* published in 1986. It is well worth studying as a general introduction to the field.

HOW SOON DO FLIES LAY EGGS?

While some of the above studies fail to note the time of oviposition, there is little doubt that the carcasses were soon heavily infested with the same fly species first noted. Murders however, are not field experiments, and flies often lay eggs that hatch before experts arrive at the crime scene. Thus the time of oviposition is only an approximation and this can become one of the pivotal points in a trial. The following controlled field studies focus on this question.

Fifty-pound (*c.* 23 kg) pigs were shot twice and immediately exposed outdoors in summer in a rural area 50 miles (*c.* 80 km) east of Vancouver, Canada. *Lucilia illustris* adults were at the carcasses within minutes and eggs were laid within an hour (Anderson and VanLoerhoven, 1996). In another study near Olympia, Washington in late June, two 22-kg pigs were shot in the back of the head and cut on the neck. One pig was placed in shade, the other in sun. Both attracted blowflies within 20 minutes and oviposition occurred around each head in two to three hours. The larvae were subsequently identified as *Phormia regina, Calliphora vomitoria,* and *L. illustris* (Shean *et al.*, 1993). In east Tennessee, four nude, unembalmed human bodies were placed in an open, wooded area at different times of the year. Adult blowflies were on the corpses within two to three hours and eggs were found in facial cavities "shortly afterward". The flies were not identified (Rodriguez and Bass, 1983).

At the Police Mortuary near Kampala, Uganda, Lothe (1964) examined the unembalmed bodies of 33 non-selected medicolegal autopsies for fly eggs and larvae. The windows were open during the day and there were numerous flies around the bodies. Twenty two bodies were infested, 11 were not. Of the latter, nine had been dead for more than 24 hours, five for 48 hours, and one for 72 hours (oddly, the figures do not add up). As for the other bodies, eggs were seen as early as 12 hours after death and larvae 18 hours after death. The flies were *Chrysomya albiceps* and *Chrysomya chloropyga* (form *putoria*).

Following are two cases in which the complete absence of flies became the issue.

The no-fly case At 19:15, on a very hot day in Rock Island, Illinois, police found a woman strangled on her bed, her face and bed linens bloodied. Windows and screens are open in the living room and bedroom. The victim and her male companion had been out until 23:00 on Friday, the night before. The man returns next evening to find her dead and calls the police. The pathologist estimates that she died 12 to 24 hours before. According to police engineers, given the heat and open windows, the apartment would have been uncomfortable in two to three hours or less. At the trial, the police testify that they found the apartment comfortable. There are no flies in the apartment, and no eggs, larvae, or any other insects on the body. Within 75 yards (*c.* 69 m) of the building is a ravine with vegetation, and beyond the ravine, a snack bar with a dumpster that is emptied once a week. The habitat is suitable to maintain a resident population of *Phaenicia sericata* and other blowflies. Three witnesses – the building manager, the newspaper boy, and the mail carrier – testify that the windows were not open when they made their rounds 03:00 and 13:00 on Saturday. The boyfriend is charged with first degree murder. The prosecution's theory is that he killed her the night before, returned to the apartment next evening, and opened the windows and sliding door to suggest the work of an intruder. Then he called the police. If she had been killed the night before or the next morning, and the windows were open all that time, there should have been blowflies in the apartment and eggs or larvae on the body. There were none. But what if she had been murdered only two or three hours before, contrary to the pathologist's opinion, possibly by an intruder? Would there have been flies or eggs? We don't know.

Another no-fly case. A vehicle flipped over at a turnoff on an isolated road at night, and the driver was flung into tall grass. He was dead when he was found by a road crew one and a half days later. There are numerous ant bites but no fly evidence on the body or in the photographs. Might the victim have survived if the crew had searched the area more diligently right after the accident? The entomological argument is that the two flies, *Chrysomya megacephala* and *Chrysomya rufifacies*, are "aggressive" and will lay eggs soon after death, but not on a living person. It would follow that the absence of blowfly eggs or larvae on the body proves that the victim was still alive for some time after the accident, perhaps almost to the time the body was found. At first glance, the argument appears reasonable. If the victim died soon after the accident why was there no blowfly evidence on the body? Even if one rejects night-time oviposition, there is still the entire next day for flies to lay eggs. But the published evidence of the flies' oviposition behavior tells a different story. It is true that the flies are aggressive and may come to a body within minutes to feed, but they don't lay eggs then. Egg laying may not occur until the second day, if not later, according to published reports. The absence of specific insect evidence – eggs, larvae or pupae – is seldom case determinative. Explaining their absence is more difficult than dealing with their presence. Under these circumstances, the entomologist does well not to stray beyond the science or yield to the siren song of litigation.

FACTORS EFFECTING WHEN EGGS ARE LAID

Based on the foregoing studies there is little doubt that blowflies rarely relinquish their priority of arrival on a body. But Lothe's observations signal the need for caution in estimating the time of oviposition. The variability he observed is no artifact and has been reported by others. In Ives' (1991) field experiments, the carcasses of laboratory rats and mice, selected for uniformity, were differentially attractive to flies. Macleod (1947) sought to avoid the problem of variability among individuals by using the same live sheep throughout his oviposition experiments. He states, "It has been found, however, that susceptibility varies, not only from sheep to sheep, but in the same sheep at different times . . ." Even fly traps that are uniformly constructed and baited may have individual idiosyncracies manifested as disparities in the number of flies caught. Or, consider the following: "we placed the baited trap in a field . . . for two hours on a sunny day, and failed to catch any carrion flies. The trap was immediately moved 100 m in the same field and in two hours caught more than 460 adults of *Lucilia caesar* as well as other carrion species" (Blackith and Blackith, 1990).

A female fly visits a carcass to feed, to copulate with a male waiting nearby, or to feed and lay eggs. Normally a virgin female requires a protein meal in order to develop her ovaries (anautogenous), and it may take a week after ingesting a suitable meal before she produces progeny. Studies in Australia show that the vast majority of wild-type females of *Lucilia cuprina* are anautogenous. A few, however, are capable of maturing their oocytes without a prior protein meal because they carry over sufficient protein and fat from the larva (Williams *et al.*, 1977). Putting the forensic implications aside because they are probably remote in this regard, it is interesting to wonder why autogenous populations are not more prevalent. One observation is that anautogenous females survive somewhat longer than autogenous females in the absence of sucrose and water, and this may give them an advantage in the arid areas occupied by the species (Gerwin *et al.*, 1987).

A gravid female's readiness to oviposit requires appropriate chemical cues and environmental conditions to consummate a chain of behaviors that start with attraction at a distance, followed by oriented flight, landing, searching, and finally, egg laying. Weather is an important determinant of fly activity. Although flies can be on the wing in light rain or drizzle, strong winds, heavy rains, or freezing temperatures will ground them, except, of course, if they happen to be indoors. Chemical cues include ammonia-rich compounds, e.g., ammonia, ammonium carbonate, and indole, and sulfur-rich volatiles. They are all products of animal decomposition. We need to know more about the specific chemistry of the fresh corpse that attracts flies, and also the nature of the volatiles that sometimes drive flies to oviposit on the living (Hall, 1995). Persons in a vulnerable and neglected condition, e.g., intubated patients in hospitals, and those with suppurating sores, whether indoors or outdoors, are potential targets. Following are three examples.

A homeless man came into the emergency room of a Chicago hospital complaining of a rumbling in his head. The rumbling stopped when a score of maggots was washed from his ears.

A nurse in a public hospital in Chicago lifted the bandage of a tracheotomy patient and discovered a cluster of third stage larvae of *Phaenicia sericata* feeding in the wound. It was summer, the unscreened windows were open and a dumpster in the courtyard below was the depository for miscellaneous trash, and a likely source (Greenberg, 1984).

In a Colorado hospital, as a daughter kept vigil beside the bed of her terminally ill mother, several mature maggots, 1.9 cm long, issued from the nose of the patient.

Although such cases of myiasis are unusual, the forensic entomologist has to consider the possibility that the victim was infested with maggots before death occurred. This could make a difference of several days or more in an estimate of the time of death. Dr. Henry Disney, University of Cambridge, has sent this personal observation. "a sample of *M*[*egaselia*]. *scalaris* from a corpse . . . was sent to me for identification. I responded by pointing out that in view of the dates of collection, etc., this must be a case of pre-mortem myiasis and not post-mortem oviposition! Re-examination of the detailed record of the post-mortem examination confirmed this." (Dewaele *et al.*, 2000).

Hall and Doisy (1993) asked the question, "will a corpse dead 3 d[ays] but protected from insect activity by wrapping in plastic or placement in a closed automobile trunk attract blowflies in the same manner as a fresh corpse will?" Their experiments were conducted on chickens, 0–96 hours dead, during two summers in the midwestern United States. They concluded that *Phormia regina* and *Phaenicia sericata* were reliable indicators on carrion of that age-range, although the latter was not well represented the first day.

Flies are most active around midday, and much less so in the morning and evening, even under favorable conditions (Greenberg and Szyska, 1984; Baumgartner and Greenberg, 1985; Mariluis and Schnack, 1985/86). This has led to the notion that flies are not active at dusk and during the dark hours of the night and do not lay eggs then. The following has been written concerning *P. sericata*, the exemplar of this behavior, "Therefore, if eggs of this species are found on a human corpse lying on a place that is in shadow during the entire day, the finding may be interpreted as indicating that the corpse has been removed from an area that was sunnier earlier" (Nuorteva, 1977). It is even suggested that this fly is an obligate heliophile, usually ovipositing between 11:00 and 14:00 at maximum sunshine (Hall,1948). And it requires "a high surface temperature (85 °F) in order to receive the proper egg releasing stimuli" (Lord, 1986). None of these statements corresponds with our experience nor with an analysis of 42 human death investigations in British Columbia in which *P. sericata* was collected indoors in 86% of cases (Anderson, 1995). The following cases from our file throw further light on this aspect of blowfly behavior.

Two bodies in a car trunk. Two bodies were discovered when a car trunk was pried open near Chicago, in May, 1981. On the bodies and elsewhere in the trunk were postfeeding larvae of *Phaenicia sericata,* various larval instars of *Phormia regina,* and pupae of *Muscina assimilis.* All doors and windows were shut and the presumed entry of the flies into the dark trunk was through the drainage hole in the spare tire well. The weather was cool and ambient temperature reached 26.5 °C only once (for one hour in May) during the entire spring that the bodies were in the trunk.

The Mafia case. Two Mafia hit men were found in the closed trunk of a Volvo in a Chicago suburb on 15 July, 1983. All windows and doors were shut. Second and third instars of *Phaenicia sericata, Phormia regina,* and *Cochliomyia macellaria* were recovered from the bodies. Larvae on one body were about one day older than those on the other body, raising the possibility that one had been killed a day earlier than the other. The weather was hot, with air temperature reaching the mid-30s (°C). A few days later when the weather was similar, the temperature in the trunk registered 46 °C.

Girl on the bathroom floor. The partially decomposed body of a nine-year-old girl was found on a tiled bathroom floor on the ground level of a boarded-up, abandoned building in Waukegan, Illinois, on 19 June, 1984. The boarded-up bathroom window faced a dimly lit light shaft and with the door closed (as it was when the victim was found), the room was completely dark. Oviposition by *Phaenicia sericata, Phormia regina,* and *Calliphora vicina* had occurred. There were numerous dead and recently emerged adults of the first two species all over the apartment, and various instars of their larvae were on the body. On the bathroom floor, not yet emerged, was a large number of *C. vicina* pupae. Average outdoor temperatures during the relevant period ranged from 19 °C to 26 °C, but the temperature of the apartment and particularly the bathroom fluctuated little and remained near the low end.

A double murder. The bodies of a man and a woman, both infested with third instar larvae of *Phaenicia sericata,* were discovered in a house in a Chicago suburb on 20 July, 1986. Doors and windows were shut, an air conditioner was running, and the temperature was 24 °C. The woman's body was on a couch in the living room. The man's body, more heavily infested, was on the bedroom floor. The bedroom had two windows facing east and north, and with the curtains drawn and the shades pulled down, the light was subdued.

In each of the above cases, sunlight was excluded and the temperature was well below the presumed threshold necessary for oviposition (Greenberg, 1990).

Because *P. sericata* is a cosmopolitan fly, found on all continents except Antarctica, it is desirable to sample its oviposition behavior in other regions. For example, Smeeton and his co-workers (1984) in Auckland, New Zealand, report that this fly laid eggs on 18 of 20 human corpses discovered indoors. Under these conditions, exposure to the sun was limited or nil. In English slaughterhouses, where the same fly was among the dominant blowflies in the environment, small numbers of gravid female blowflies penetrated the hanging and chilling rooms for the purpose of laying eggs (Green, 1951). Elsewhere, however, Green says that these flies seldom, if ever, oviposited on the meat at night.

Other blowflies may be even more willing to enter dark places to oviposit, as in the following cases

Skeleton in a closet. In July, 1986, the naked, partially skeletonized body of a female was found in a closed closet of an abandoned house in Chicago. The room faced north, the windows were shut and the light was reduced. Adults of the bluebottle *Calliphora vicina*, were flying about the room and their empty puparia were scattered on the floor. Of tangential interest was the presence of two species of small phorid flies – *Megaselia scalaris*, usually later in the insect succession, and *Megaselia abdita*, a species not previously recorded from North America.

Body in a bag. On 5 April, 1990, the body of a woman was discovered in the vacant basement of an occupied two-storey apartment building in Chicago. The body was wrapped in a quilt, covered with plastic garbage bags and propped, upside down, in the recess of a brick wall. The basement was dark, although some light came through a few small, dirty windows at ground level. Pupae were found on the floor near the body, and larvae and pupae were collected from the body at the morgue. They all belonged to *Calliphora vicina*. The fly not only oviposited in a dark place, but did so in early February, in the middle of a typical Chicago winter.

The missing child. The body of a six-year-old girl was found in an abandoned house in Covington, Kentucky on May l, 1989. She had been reported missing on April 21, 1989. The room in which she was found had boarded-up windows and it was dark inside. Closer examination revealed large masses of maggots in the head area and in the groin close to where dogs had chewed the thigh down to the bone. The maggots belonged to *Calliphora vicina* and *Phormia regina*.

Dr. Zakariah Erzinçlioglu, in England, has published extensively on blowfly biology. He has seen *C. vicina* and *Calliphora vomitoria* infesting human bodies in dark places, such as World War II pillboxes (bunkers) and beneath the floorboards of houses (Z. Erzinçlioglu, pers. comm.).

Adult blowflies are not early risers. Although air temperatures may be optimum, they are slow to get started during the single hours of the morning, often resting on vegetation with no apparent interest in the nearby carcass that hummed with their activity the day before. By 10:00, the flies are active, reaching maximum numbers by midday to early afternoon. Activity declines in late afternoon, probably in response to diminishing light intensity (in the tropics temperatures often continue to be optimum) and flies settle in for their nocturnal rest by evening (Baumgartner and Greenberg, 1985; Mariluis and Schnack, 1989).

The above examples indicate that blowflies will enter dark places to satisfy the egg-laying urge, but they do not tell us when oviposition actually occurred. The fly may be willing to enter a dark place during the day but not at night. In the laboratory, blowflies will oviposit any time of the day or night, with lights on or off. They will even mound their eggs in cracks and crevices of the fly cage, in the absence of a suitable substrate. But these are caged flies with no choice and a disrupted circadian rhythm.

Are flies on the job when murders occur in the dark of night? It has been generally accepted that blowflies are not active at night and do not lay eggs then. If true, this eliminates a block of about 10 to 12 hours from calculations of the time of death. But what if egg laying can occur? Inclusion or exclusion of 10 or 12 hours can make or break an alibi.

In English slaughterhouses, *Calliphora* (*vicina* and/or *vomitoria*) "quite commonly flew and oviposited during the night, spreading their eggs in ones, twos, and threes over livers and the cut surfaces of carcasses" (Green, 1951). Presumably, the slaughterhouse was illuminated, although it isn't stated. The question is worth pursuing.

During the summer of 1988, we placed skinned rat carcasses under bushes in a Chicago suburb from 01:00 to 04:00. Each carcass was then put into a zip-lock bag with wood shavings in a fly-tight cage. The bags were opened about a month later and any dead flies were tallied. Four of eleven trials were positive, yielding *Phaenicia sericata, Phormia regina,* and *Calliphora vicina.* Dr. Devinder Singh, a visiting entomologist from Punjabi University, repeated the experiment in a limited trial in the latter part of May, 1993. Skinned rat carcasses were put out from 21:00 to 04:00 at three places on the University of Illinois campus in the heart of Chicago. There was artificial illumination from street and campus lights. Under these conditions, the same species laid eggs that produced adults, although, as before, fewer eggs were laid at night. The reduced number suggests that single females were involved. Tessmer *et al.* (1995), in Louisiana, could not confirm these results. They found no nocturnal oviposition on chicken carcasses in urban habitats with lighting, or in rural habitats without lighting. Also in Alexandria, Egypt, nocturnal oviposition did not occur on rabbit carcasses when *Chrysomya albiceps, P. sericata,* and *Wohlfahrtia nuba* were active in the area during the day (T. Tantawi, pers. comm.).

Our experiments raised an important question. Since the meat was placed on the ground, flies that happened to be resting nearby could have walked to the bait. Singh and Bharti, (2001) avoided this problem in the following experiment in Patiala, India. They placed the meat on top of 2-m poles and wrapped the base of each pole with a 23-cm long band of sticky material to prevent the flies from walking up the pole. The meat was exposed from 22:00 to 03:00 daily for a week during March and September. Altogether, there were 14 trials and oviposition occurred in five trials; the number of eggs (judged by the number of reared adults) was consistently less than the normal number. The active flies were *C. vicina, Chrysomya megacephala,* and *Chrysoma rufifacies.* In England, under caged conditions, *Calliphora vicina* laid large quantities of eggs in complete darkness at a temperature of 9.1 °C. The flies did not have access to the bait but could smell it, and eggs were laid through perforations in the container (Barry Fitz-Gerald, pers. comm.).

WHERE FLIES LAY EGGS

As we have seen, the roster of species visiting a body varies regionally and seasonally. But once on a body most flies intent on laying eggs have one goal – to find moist sites, preferably with access to the interior. This can be summed up with the acronym EENT – eyes, ears, nose, and throat – the usual portals of entry. Other sites include the anus, knife and bullet wounds, breaches in the body wall, and the space between the body and the ground. In summer, when flies are numerous, the scalp may have clusters of eggs that resemble grated cheese. In field studies near Kuala Lumpur, Malaysia, Omar *et al.* (1994a) found that four *Chrysomya* species – *megacephala, rufifacies, pinguis,* and *chani* – preferred to oviposit in the fur all over the bodies of cats and monkeys. But carrion flies do not stop there. They will go underground in their quest to oviposit. This is particularly true of the smaller species in the family Phoridae, e.g., *Dohrinophora, Metopina,* and *Conicera.* It is assumed that these adults can burrow as much as 50 to 100 cm or more (Motter, 1898; Payne and King, 1968) to reach a buried body; *Conicera tibialis,* the coffin fly, is the most famous example (Oldroyd, 1964). It is assumed in these cases that the overlay of soil was relatively loose and not compacted. Larger flies are generally confined to the surface (Lundt, 1964), or take advantage of cracks in the soil to oviposit near to or directly on the carrion. For example, VanLoerhoven and Anderson (1999) recorded the muscids, *Hydrotaea, Morellia, Fannia,* and *Ophyra,* and the calliphorids, *Calliphora* and *Eucalliphora,* on pigs buried 30 cm deep in British Columbia. The larger flies will lay eggs on the surface of the soil, on blood-soaked materials, or in response to decompositional gases emanating from the remains. Laboratory experiments by Barry Fitz-Gerald (pers. comm.) suggest that *C. vicina* can oviposit on the soil surface in response to carrion buried at a depth of at least 11 cm. Furthermore, if the first and second instar larvae feed on bloody materials at the surface, they can then penetrate through uncompacted soil to a depth of 15 or 16 cm. Presumably, third instars are even more capable of burrowing.

The Egg

Blowflies typically lay up to 300 eggs per batch, three or four times in the life of a fly. Mounds, sometimes consisting of thousands of eggs are the communal output of many females, often of different species. The massing of eggs appears to result from the tendency of females to oviposit near other females in the process (Cragg, 1955; Browne, 1958). What drives females to "put all their eggs in one basket"? Flies are not as stupid as they look, and we shall see later that there is a method in their frenzy. *Calliphora* species have another strategy – they scatter their eggs singly or in small numbers.

The blowfly egg is white, shaped like a grain of rice, and about 1.1 to 1.4 mm in length (Figs. 2.1a,b). The outer shell, or chorion, is tough enough to withstand immersion in 3 to 5% sodium hypochlorite (bleach) solution for 10 minutes without affecting the embryo's viability; but the embryo has limited ability to withstand dessication. Given a choice, the female typically selects moist sites, as already stated. Sperm enter through the micropyle at the cephalic end of the egg as it travels down the oviduct. Dorsally, a median area typically runs the length of the egg. Seen under the scanning electron microscope (SEM) the median area is a sponge-like network of interconnected struts and pillars – the aeropyle. The borders of the median area are structurally weak and split during hatching, enabling the maggot to squirm free from the chorion. Usually, by the time a corpse is discovered, the first batch of eggs has hatched. To establish time of death within narrow limits based solely on eggs requires knowledge of egg development. Consider the following case in Burlington, North Dakota.

The body of a young adult male was discovered at 13:00 in August lying near farm machinery at the edge of town. There were mounds of eggs among more than 100 stab wounds on the body. The victim was seen alive at 03:00 and sighted again at 11:00 by someone else. Was either sighting credible? Toward late afternoon, when eggs were collected, placed in 70% alcohol and refrigerated, none had yet hatched. Other batches were placed on meat in the laboratory and yielded *Phaenicia sericata* (the most numerous), *Lucilia illustris,* and *Phormia regina. Phaenicia sericata* is urban, *L. illustris* frequents open woodland and meadow, and *Phormia regina* straddles these habitats. The presence of all three flies reflected the proximity of urban and rural environments where the body lay.

Microscopic examination revealed that the oldest eggs belonged to *Phaenicia sericata*; they contained incipient larvae with brown pigmented posterior spiracles and tracheal trunks. To estimate their age, we took freshly laid eggs (≈15 min) from a colony of this fly, placed them at 22 °C (the average prevailing temperature at the crime scene) and examined hourly samples. One batch of eggs began hatching at hour 19, another batch, 20 minutes later. Under the microscope, diagnostic markers became visible at hours 15 to 16. At hour 17, there were distinct mouthparts, and spinose bands encircled the segments, but none of the victim's eggs had reached that stage. It was possible to reject the later eyewitness account and to conclude that the murder occurred earlier that morning. This was subsequently corroborated (Liu and Greenberg, 1989; Greenberg, 1991).

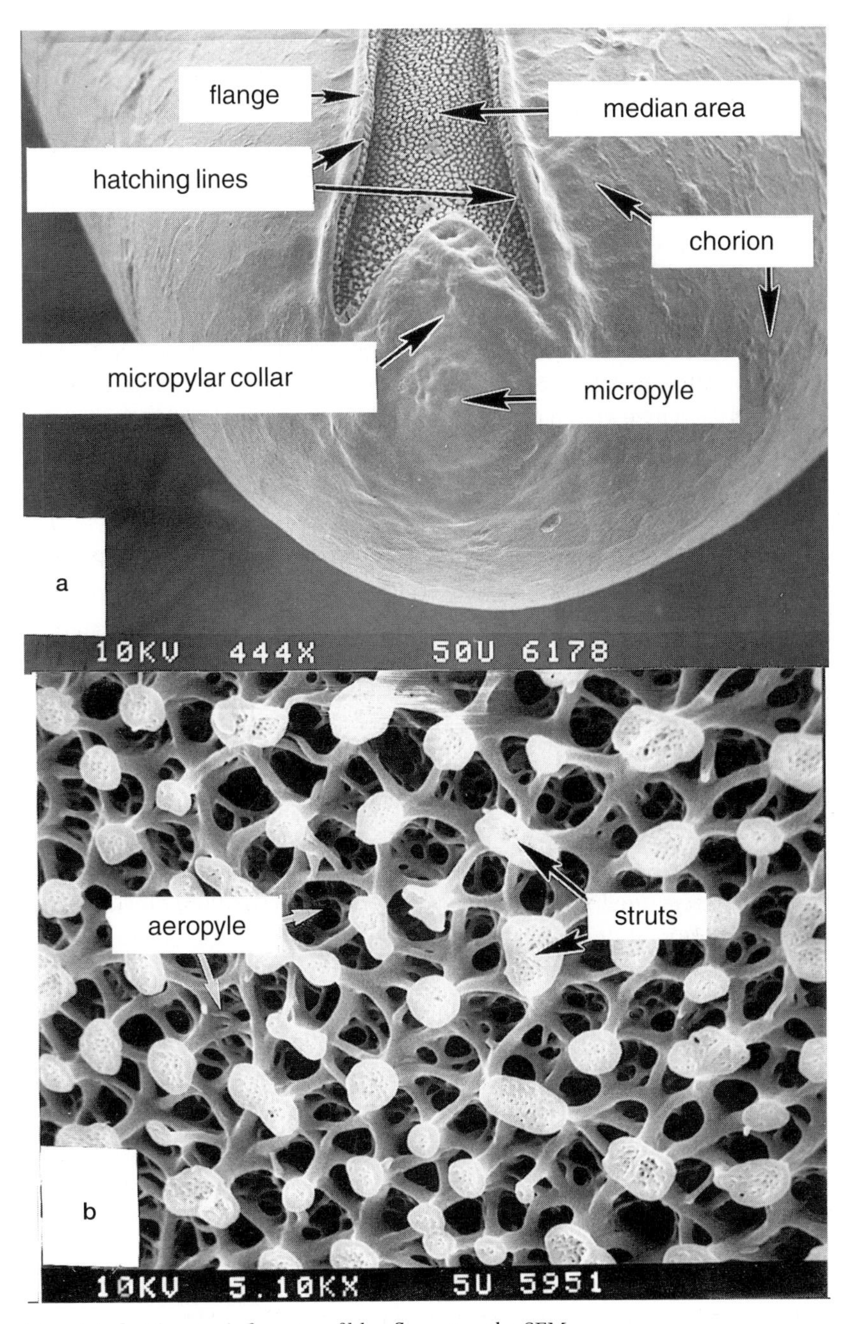

Figs. 2.1a,b. Diagnostic features of blowfly eggs under SEM.

If preserved eggs are all that is available, reliance on SEM analysis of the chorion's surface becomes necessary for identification. As Zakariah Erzinçlioglu points out (pers. comm.), once these features are understood, they can be readily interpreted through light microscopy. In any case, the following features of the median area are diagnostic (Figs. 2.1a,b). (1) are its borders raised (*Muscina* and *Megaselia scalaris*), or flat (calliphorids)?; (2) are the apices of the plastron flat and somewhat coalesced (Calliphorinae), or spiked and generally separate (Chrysomyinae)?; (3) does the median area bifurcate at the micropylar collar to form an inverted Y (*Cochliomyia, Phormia, Protophormia,* and *Phaenicia/ Lucilia*), or does it end at the top of the micropylar collar (*Calliphora*)?

These structures have been illustrated and their diagnostic value has been discussed (Greenberg and Szyska, 1984; Erzinçlioglu, 1989a; Liu and Greenberg, 1989; Greenberg and Singh, 1995). Figures 3.1 to 3.5, provide SEM views of the eggs of 19 forensically important flies to accompany the key to the eggs. Because a species determination based on chorionic topography may not be possible, it always helps to know the habits and seasonal distributions of the local flies. For example, in the local *Calliphora* that we have studied, the eggs of *vomitoria* and *livida* are indistinguishable from those of *vicina* except for a slight widening of the median area at the micropylar collar in the former. But adult *vicina* are eusynanthropic and endophilic and that helps to narrow the classification of their eggs. We are unable to distinguish the eggs of three common *Lucilia* species – *sericata, cuprina,* and *illustris* – but differences in their geography and synanthropy will be useful here, as well (Greenberg, 1971).

Eggs can be inundated by rain or by fluids issuing from the body. Yet the embryo does not drown. While submerged, it continues to breathe air that is trapped in the aeropyle of the median area. The trapped air film functions as a passive lung, operating on the same principle as the fine pile of hairs on aquatic beetles or the "diving bell" of aquatic spiders. As the embryo consumes oxygen from the aeropyle, more oxygen flows into the aeropyle from the atmosphere via the water. As long as the aeropyle retains an air film (of nitrogen) it will continue to serve as a conduit. The aeropyle can withstand several atmospheres of pressure before collapsing, and it also resists wetting. This takes care of most situations eggs are likely to encounter (Hinton, 1960).

What if eggs are laid on a body and the body is then submerged? How long can submerged blowfly eggs remain viable? In our preliminary trials, the eggs of *P. sericata* and *Phormia regina* gave identical results. Separated eggs were submerged in test tubes with tap water that had reached equilibrium with the atmosphere and room temperature. Samples of submerged eggs were pipetted at intervals onto moist filter paper and scored for hatching. Eggs that were six to twelve hours old, survived 66 hours under water although the survival rate was lower than that of controls placed on moist filter paper without submergence. In another trial, eggs began hatching under water after about 62 hours. Ironically, the maggots that hatched under water, drowned.

If blowfly eggs are found on a submerged body it indicates that at least that part of the body was previously in air. Several lines of inquiry follow. (1) Are the eggs still viable? If so, how soon will they hatch at uniform temperature? If not, can their stage of development be determined by microscopic examination? (2) What is (are) the species? (3) Was the body in sun and/or shade? (4) What were the estimated air and water temperatures? We have seen no published work on these or other aspects of submerged eggs as forensic indicators. Even when larvae are found on a body that has been submerged, it is still necessary to know the capabilities of submerged eggs.

Eggs are normally fertilized at the time of oviposition. Erzinçlioglu (1990a) describes an anomalous situation in *Calliphora vicina* in which a fertilized egg was retained in the vagina until the female found a suitable substrate. When this egg was finally deposited with a batch of eggs, it had a head start and hatched sooner than the rest. This observation has been extended to include *Calliphora terraenovae, Calliphora vomitoria* and *Lucilia* (=*Phaenicia*) *sericata* (Wells and King, 2001). In the latter two species it occurred infrequently but in *C. terraenovae*, 49 out of 55 gravid females had single eggs in their vagina and in 34 the larvae had developed where the spines were visible through the chorion. In conducting an analysis of a sample that includes an outlier or "precocious" specimen, it should be recognized and reported but not used to skew the result. Generally, this is not a problem because it is rarely difficult to obtain an adequate sample at the death scene or at the morgue. Reliance on a single specimen is shaky science.

The Larva

The account that follows describes structures and functions typical of the larvae of common forensic blowflies. Calliphorids vary widely in diet from saprophagous feeders on rotting seaweed, dung, and carrion, to living earthworms and snails, to intermittent bloodsuckers of nestling birds and mammals, and, finally, to obligate parasites of mammals (Ferrar, 1987). Some species are larviparous, depositing larvae, or even macrolarviparous, depositing advanced stage larvae. It would be foolhardy to generalize given the catholic menus and habits of the group. Our focus is on the sarcosaprophagous species.

THE MAGGOT SHAPE

The shape is almost universal among higher Diptera, adopted by thousands of species, and copied in arrowheads and bullets. The pointed anterior and blunt posterior ends are ergonomically the most efficient for burrowing, aided by a skin bristling with spines and heavy-duty locomotor pads for traction. Maggots in a single egg batch will often hatch within minutes of each other and they are driven to penetrate the carcass as quickly as possible. In the interior, they find adequate moisture, a more even temperature, and protection from downpours, parasites and predators. Here, too, are tissues more amenable to attack than the tough outer wall of the corpse with its multiple layers of skin and muscle. When confronted by the latter, there isn't much that a solitary, newly hatched larva can do with its delicate mouthparts. In the absence of an opening or breach it helps to have the collective effort and digestive enzymes of many maggots to facilitate an entry.

GROWTH AND FEEDING

The newly hatched maggot is already longer than the eggshell from which it has just emerged. There will be two molts and three instars, or stages, as it grows from about 2.5 mm to 4 mm (first instar), sheds its skin and grows to about 8 mm (second instar), sheds once again and grows to about 15 to 22 mm (third instar). Size, especially in the second and third instars, depends on the species and is influenced by nutrition and sometimes by temperature. Maggots, like all insects, lack thermostats. Their growth rate depends on ambient temperature, and is linear within physiological limits. These limits will be discussed later when we deal with the developmental timetables of individual species. When the third instar larva has reached a programmed size, it stops feeding and wanders for several days as a postfeeding larva. Externally, there are no structural changes except for a possible shrinkage in size as it turns a creamy-white. The maggot is still a third instar larva, but internally, there are forensically important changes in its digestive tract. When it stops wandering, the maggot becomes immobile and contracts its anterior and posterior ends into a football-shaped puparium in preparation for metamorphosis. This is dealt with in the next section on the pupa.

The maggot feeds voraciously in all three stages, driven by the need to consume before it and the carrion are consumed by the competition. Competitors include maggots of the same or other species, and carrion-feeding vertebrates, e.g., vultures, and a host of carnivorous mammals. In Africa, about 90% of carcasses are consumed by vertebrate scavengers. The latter can quickly finish off a carcass, maggots and all. This is also true in North America (Ives, 1991). Maggots have evolved ways to get around this. One counter strategy is the ability of undernourished larvae to develop successfully into adults. The minimum effective larval weight is known for a few flies. For example, *Calliphora nociva, Calliphora augur, Chrysomya megacephala,* and *Chrysomya rufifacies* produce small but fertile adults from larvae that are only about 45%, *Phaenicia cuprina,* 60% and *Calliphora stygia* about 73% of full size (Ullyett, 1950a,b; Levot et al., 1979). It is obviously better to produce diminutive adults, albeit with reduced fecundity and possibly reduced longevity and dispersal ability, than none at all. Zdarek (1985) has described a variation of this strategy. *Calliphora vomitoria* larvae that are starved from the moment of the last larval ecdysis (young third instars) survive for many days without any sign of further development. When these larvae are allowed to feed, they resume development leading to pupariation. It is noteworthy that these undernourished larvae have a considerably longer postfeeding period.

Urbanization has diminished competition from vertebrate scavengers and replaced it with efficient garbage disposal, but the fly's ingrained patterns persist. Both the egg and the feeding larva have been selected in the course of evolution for rapid development. For example, the egg stage takes only about 6% of the total period from oviposition to adult emergence, and the feeding larva takes about 20%. The rest of the time is spent in the postfeeding maggot and pupal stages "rehabbing" larval structures to make the adult.

The maggot has evolved yet another way to beat the competition – fast food. Toothless and with a small mouth, it must first liquefy its food externally which it then gulps down. The flow rate through the gut was calculated by feeding maggots a food covered with carbon black particles. Specimens were dissected at 5- and 10-minute intervals and average rates were determined by measuring the farthest distance moved by the carbon particles. At 23 °C, gut motility is 1 to 2 mm per minute, with a transit time from mouth to anus of 65 minutes. By comparison, the rate in the human small intestine is approximately 33 mm per minute. Ambient temperature has little effect on gut motility in warm-blooded animals, but it is important in cold-blooded animals. When maggots were kept at 31 °C instead of 23 °C, the motility rate was three times faster, and the particles reached the anus only 20 minutes after ingestion!

It should come as no surprise that the maggot's digestive tract is as highly differentiated as our own (Fig. 2.2). It secretes a battery of digestive enzymes that include proteases, lipases, and collagenase. A weak carbohydrase may be present or absent. Pepsin-like enzymes were reported to be absent in insects until they were found in houseflies and blowflies in the 1950s (Greenberg and Paretsky, 1955; Lambremont *et al.*, 1959; Fraser *et al.*, 1961).Trypsin-like, chymotrypsin-like, and pepsin-like enzymes have been identified in *Calliphora vicina* (Pendola and Greenberg, 1975) and are undoubtedly widespread in blowflies.

As mentioned above, digestion is both *in situ* and *in vivo*. Enzymes are excreted onto the carrion and the collective output of large numbers of maggots quickly creates carrion soup that turns to froth with their churning. In heavy infestations, maggots are crammed side-by-side and almost completely immersed, with only their posterior spiracles exposed to the air. These spiracles may lead the layman astray, as communicated to me by Dr. R. Henry Disney (1998) , an entomologist at Cambridge University, "Twice Zak [Dr. Zakariah Erzinçlioglu] had a maggot farmer, who supplies angling shops with maggots that are used to bait fish hooks, put up against him [in court trials]. The measure of the man's ignorance is that his logo on his business notepaper is a *Calliphora* larva with a crown on one end. However, the crown is on the tail end as the poor fool thinks the posterior spiracles are eyes!" This anecdote would be humorous were it not that testimony by pseudo-experts may sway naive jurors and thwart justice.

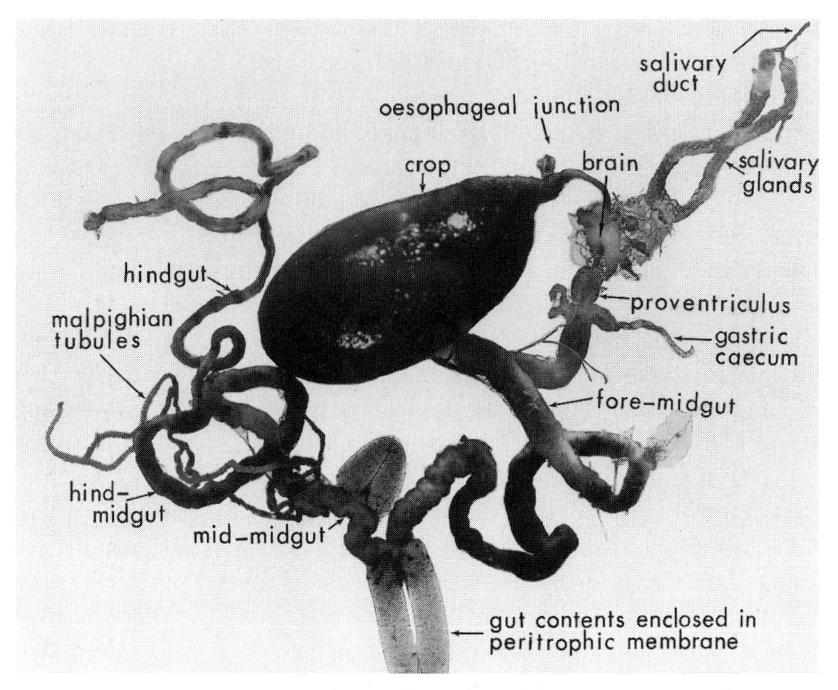

Fig. 2.2. Digestive tract of third instar *Calliphora vicina* larva at peak of feeding. Note engorged crop and midgut with latter's contents extruded but retained within peritrophic membrane.

Phenomenal feeding results in phenomenal growth. A newly hatched bluebottle larva weighs 0.1 mg; five days later it weighs about 0.84 g, an increase of 800 times. There are only two molts during the rapid growth phase and it is the cuticle's plasticity that accommodates the nearly ten fold increase in larval length and the increase in bulk. In *C. vicina,* a maggot 1 cm long has an intestinal tract about seven times longer, more than twice as long as that of the adult. Feeding is continuous and as the maggot grows it ingests faster than it digests, and the excess food is shunted into the crop. During the last day of feeding, as the larva attains maximum length, the crop doubles in length (Fig. 2.3). In *C. vicina,* a fully distended crop is 7 mm long, a little less than one-third the maggot's length. At the peak of feeding, it is shaped like a blimp, dominating the anterior half of the maggot, and is visible through the translucent skin as a dark red oval (Plate 9). The postfeeding larva wanders for several days before it pupariates. Its length diminishes during this time, but the extent of shrinkage varies within and between species. Therefore, age estimates based solely on length are tenuous. Biochemical estimates also have thus far been inconclusive. For example, although there is a rapid increase in the DOPA decarboxylase titer in postfeeding larvae of *Phaenicia cuprina,* reaching three times the baseline at pupariation, there is considerable variability and overlap, not to

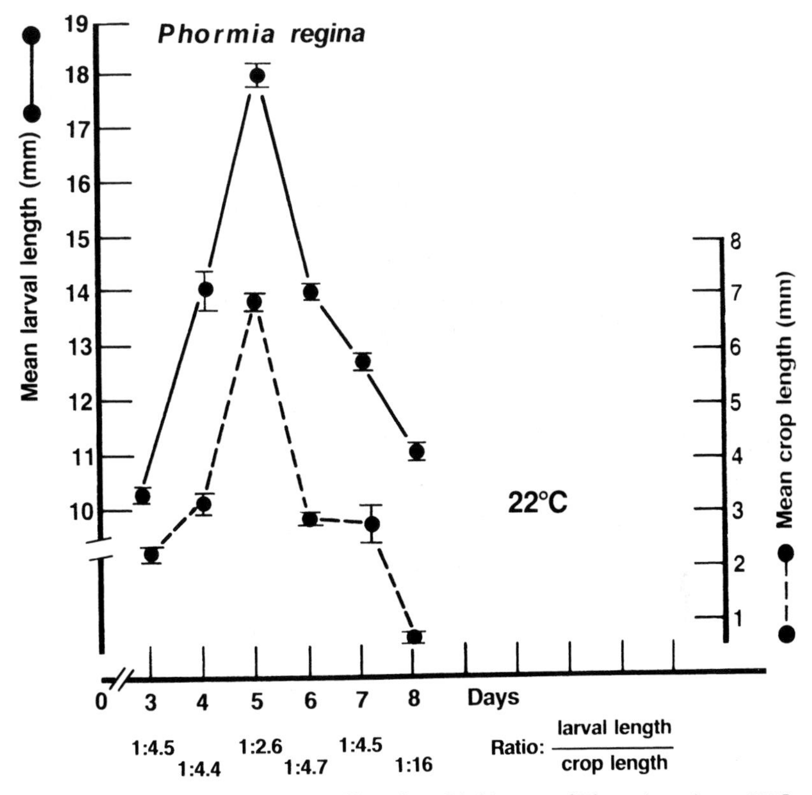

Fig. 2.3. Ratio of crop length to larval length in third instars of *Phormia regina* at 22 °C. Relative crop length declines significantly only in the postfeeding stage (Greenberg, 1991).

mention the assay's technical difficulties (Turnbull and Howells, 1980; Skelly and Howells, 1987).

In some species, crop length may be a more reliable indicator of the age of the postfeeding larva. As the larva attains maximum size during the last day of feeding, the crop doubles in length (Fig. 2.2). When the maggot stops feeding and becomes a wanderer, the crop functions superficially like the rumen of a cow, slowly discharging its contents into the digestive tract to be processed in a leisurely way (Fig. 3.42a). During the first postfeeding day, the crop of *P. sericata* empties rapidly (Fig. 3.34), and significant age estimates can be based on crop length during that interval. During the next two days the crop empties very slowly and the data points overlap. One day before pupariation the crop undergoes final shrinkage. *Phaenicia regina* follows a similar pattern to that of *P. sericata*. In postfeeding *C. vicina*, crop length tracks maggot age at forensically useful intervals, e.g., three hours, nine hours, etc., out to 39 hours (Fig. 3.36). In the first day, the crop shrinks from 7 mm to 3 mm. Thereafter, until pupariation at about 87 hours, the decline in crop length is too gradual to be particularly useful. The crop of *Chrysomya rufifacies* has less to recommend it. Fully engorged, it is smaller than in the other species, ranging from about 2.5 to 3.5 mm, and it empties gradually, without a pronounced initial discharge (Greenberg, 1991).

The crop is part of the foregut and has a cuticular covering that preserves its shape even after the maggot dies. Desiccated maggots that are taken from the remains or from clothing years later and carefully dissected, may still retain a measurable crop that will reveal the maggot's age at the time it was killed.

As the gut empties, the assimilated food is converted into a monolayer of fat cells, which is the primary source of energy during metamorphosis. The dark, distended food-filled coils of intestine that were clearly visible in the feeding maggot, are now white, tight coils obscured by the layer of fat covering the inside of the skin. Where insect evidence has been lost or never collected, good photographs will distinguish the postfeeding larva by its creamy white appearance. This can provide a basis for a rough estimate of the postmortem interval.

DISPERSAL

The dispersal of adult calliphorids has been extensively investigated (summarized in Greenberg, 1973), but it is the dispersal behavior of postfeeding maggots that concerns us here. This is of practical importance in a death investigation where failure to recover the oldest postfeeding maggots or pupae on or away from the body can significantly, and erroneously, shorten an estimate of the time since death. For example, assume a second or third wave of larvae are on the body after the first wave has departed. If sampling is confined to the body, at autopsy or in the field, it is possible to miss the first wave that is now buried in the soil or somewhere in the room, and to derive erroneous estimates based on the younger larvae.

In some situations the vagility of postfeeding maggots raises another problem. Consider the following case. In a drunken fury a son who was denied some money by his mother was accused of murdering her.When her body was found days later there were postfeeding larvae all over the floor of the apartment. The prosecution questioned the age estimate of the larvae and suggested they came from a dead rat. It was a shot in the dark because there was no evidence for this but it raises an interesting, if remote, possibility. There may be situations, indoors or outdoors, where cross-contamination is a possibility. Comparison of the mitochodndrial DNA of the victim with that of the maggots could provide a definitive answer (Wells *et al.*, 2001).

The postfeeding larva that sticks its head out of a carcass faces the perils of predators and parasites. In one of few quantitative field experiments, Putman (1977) demonstrated a 66% predation rate on the postfeeding larvae of *Calliphora vicina.* The larva also faces an uncertain future as regards the physical environment. Zdarek (1985) cites Ohtaki (1966) in pointing out that "Larvae of various fly species, including . . . *Sarcophaga bullata, Sarcophaga crassipalpis,*

Sarcophaga argyrostoma, Sarcophaga peregrina, fail to pupariate [and eventually die] when kept in constant contact with water". Wet weather could result in high mortality in larvae pupariating in exposed situations.

Season is another factor to be reckoned with. Blowflies and fleshflies prepare for the advent of winter in temperate regions with a diapause, or arrested development. In warm regions, dry conditions can be the initiating factor. In blowflies, diapause is most common in the postfeeding larva and/or adult; in fleshflies, it is in the pupal stage. The latter is discussed in the next section on the pupa. Roubaud (1922) was the first to describe larval diapause in *Phaenicia sericata,* and we have often initiated diapause in the lab in this species, sometimes unwittingly, by placing fully fed larvae in dry wood shavings at 30 °C.

Diapause has forensic implications because it can last for months and lead an unsuspecting entomologist down the primrose path. Three factors – maternal influence, photoperiod, and temperature – interact variously on the neuroendocrine system of the adult and larva to produce diapause. The induction of diapause in the larva results from a cessation of brain hormone to the prothoracic gland, which stops producing ecdysone. It is partly under genetic control. In *Calliphora vicina* (Zinovjeva and Vinogradova, 1972), *Lucilia caesar* (Ring, 1967), *Lucilia* (= *Phaenicia*) *sericata* (Cragg and Cole, 1952), and *Lucilia illustris, Lucilia hursutula,* and *Calliphora uralensis* (Zinovjeva, 1978), maternal photoperiod determines whether larval progeny will enter diapause. Short-day treated females are likely to produce diapausing postfeeding larvae. Ring (1967) writes of *L. caesar*: "Early in the breeding season [Glasgow, Scotland] diapause is absent from the offspring but soon after the summer solstice the incidence increases rapidly to 100% even though the larvae are maintained under long photoperiod conditions . . . Conversely, if the larvae from females captured early in the season are reared under short photoperiod conditions there is no increase in diapause incidence relative to the control larvae kept under long photoperiod."

Vinogradova (1986) believes that this mode of larval diapause induction is probably typical for calliphorids. She goes on to say, "As to the factors acting directly on the larvae, the temperature fulfills the main role and the photoperiod a secondary one in diapause induction". The maternal effect, however, can vary in different species, populations, and laboratory generations. In *Calliphora vicina,* which she has extensively investigated, larval diapause "is expressed very distinctly in populations from the north [Soviet Union] and weakens southwards to complete disappearance in the southern strain from Dushanbe [Tadzhikistan]." The "missing link" is the biological clock mechanism that transduces the environmental signal, whether temperature or photoperiod, into a hormonal one. James Wallman (pers. comm.) points out that there is no evidence that any of the native Australian *Calliphora* exhibit diapause, unlike northern hemisphere species, although some are better adapted to certain environmental extremes than others, and this determines their distribution.

We have already mentioned the ability of undernourished larvae to complete development. In *C. vicina,* such larvae may side-step diapause and develop into adults capable of reproducing before winter, while larger larvae from the same cohort enter diapause (Saunders, 1997). Saunders suggests that the larger larvae have enough fat reserves to see them through the winter. The "miniatures" from overcrowded situations would not make it and opt instead for another generation before winter.

Australian workers, investigating sheep myiasis (Fuller, 1934; Waterhouse, 1947; Norris, 1959), have noted that larvae of the subfamily Chrysomyinae (*Chrysomya rufifacies* and *Microcalliphora varipes*) tend to pupariate on or close to their food source and on the surface of the ground (also *Chrysomya albiceps* in South Africa, Ullyet, 1950a, b). On the other hand, larvae of the subfamily Calliphorinae (*Phaenicia cuprina, Calliphora augur,* and *Calliphora stygia*) disperse and dig into the soil before they pupariate (also *Phaenicia eximia* in Brazil, Madeira, 1985). In one experiment, *P. sericata* larvae buried themselves about 11 cm deep in sand (Ullyet,1950a). We discuss the two behaviors – vagility and digging – in that order.

In laboratory experiments, we found three patterns of dispersal among postfeeding larvae of five blowflies and the carrion-feeding muscid *Muscina stabulans* (Greenberg, 1990).

Phaenicia sericata and *Calliphora vicina* were the most dispersive – <1% of the former and only 15.5% of the latter remained at the food source. The rest moved 3 to 8.1 m.

Postfeeding larvae of *Phormia regina, Chrysomya rufifacies,* and *M. stabulans* dispersed the least. Ninety eight percent of *P. regina,* 90% of *M. stabulans,* and 84% of *C. rufifacies* remained at the food site. Among the mobile *P. regina,* none moved beyond 3.3 m.

Cochliomyia macellaria larvae were intermediate, with 40% remaining at the food and about 50% moving no farther than 2.4 m.

How do these laboratory results compare with field observations? Tessmer and Meek (1996) studied dispersal from pig carcasses in pasture and woodland sites in southern Louisiana. In early summer, *Cochliomyia macellaria,* the predominant fly, mainly moved a distance within 0.9 m at both sites, with a considerable number moving up to 2.7 m. In spring, the majority moved up to 2.7 m, with considerable numbers moving up to 4.6 m. In UK, Cragg (1955) reported that the majority of postfeeding *Phaenicia* (=*Lucilia*) larvae (including *P. sericata*) moved 3.7 to 4.6 m, and some moved 6.4 m. *Calliphora vicina* was equally mobile. Dispersal distance often depends on the nature of the substrate. Under conditions of hard ground, larvae of *Phaenicia* and *Calliphora* in England traveled up to 34.5 m before pupariating (Green, 1951). In Malaysia, *Chrysomya megacephala* pupariated a few meters away from carcasses on soil, but up to 15 m away from carcasses on a cement floor (Omar *et al.*, 1994b). True to form, *Chrysomya rufifacies* pupariated on or close to the carcass in Guam (Bohart and Gressitt, 1951) , in Malaysia (Omar *et al.*,1994a,b), and in our laboratory. This is also true of *Protophormia terraenovae* (Cragg, 1955) and of *Phormia regina;* these pupae we have observed in huge numbers on human corpses (Plate 10). In a Finnish woodland with an average temperature of 16.8° C, the larvae of *Lucilia illustris* and *Lucilia caesar* dispersed rather slowly, about 1 m per day, with a maximum distance of 3 m (Nuorteva, 1977).

The field observations of Omar and his colleagues (1994a) in Malaysia, are presented *verbatim* as they represent a distinct blowfly fauna of potential forensic importance in that part of the world.

"Late third instar larvae of *C[hrysomya] rufifacies, C[hrysomya] villeneuvi, C[hrysomya] chani, C[hrysomya] nigripes* and *O[phyra] spinigera* remain at or nearby the food site (carcass). Both *C. rufifacies* and *C. villeneuvi* are the least dispersive and were found on the ground in masses. *C. chani* pupariate on the exposed bones and skin but *C. nigripes* tended to migrate vertically and pupariate on twigs, leaves and wire mesh above ground at or nearby the decomposing animal. *O. spinigera* wander deep inside the soil at the food site and pupate there. *C[hrysomya] megacephala, C[hrysomya] pinguis, L[ucilia] sinensis* and *Hem[ipyrellia] ligurriens* are the most dispersiv[e] and pupariate in the soil far from the food site."

The energy expended in dispersing and digging must have a payoff. Evidence suggests that calliphorine puparia have 'thinner skins' and are more vulnerable to attack by parasites. The following calliphorine puparia that we measured have an average thickness of 0.019 mm: *C. vicina,* and four species of *Phaenicia – cuprina, eximia, ibis,* and *sericata.* The puparium of *Chrysomya rufifacies,* is 0.074 mm, four times thicker! *Mormoniella vitripennis,* a widespread hymenopterous parasite of fly pupae, can penetrate the thin-skinned puparium of *Phaenicia sericata* to oviposit, but has difficulty penetrating the thick puparium of *Chrysomya albiceps* (Ullyet, 1950a,b; Madeira, 1985). Despite the thick skin, pupariating in the open is still a risky business and the shorter the pupation period the less the risk. It turns out that the pupation periods of the chrysomyine species we checked are shorter than those of the calliphorines (Table 2.1). Why do the calliphorines have a longer pupation period and therefore a longer generation time? Given the intensity of competition among carrion flies one might expect the opposite. It could be that the chrysomyine species we selected are largely tropical where the evolutionary forces are chiefly biotic with tremendous intensity of competition leading to a shorter pupation period (Dobzhansky, 1950).

The Pupa

Webster's dictionary defines 'pupa' as doll. The pupae of butterflies and moths fit this definition better than the pupae of flies. The formers' appendages are glued to the outside of the chrysalis and resemble Egyptian mummy cases or Russian matriushka dolls. The fly pupa is within an unadorned barrel.

Table 2.1. *Pupation periods of flies of two calliphorid subfamilies reared at approximately room temperatures*

Chrysomyinae		Calliphorinae	
Species	Pupation period (days)	Species	Pupation period (days)
Chrysomya putoria	4, 4	*Calliphora peruviana*	12, 12
Chrysomya megacephala	4.0–4.3	*Calliphora vicina*	11, 12
Chrysomya rufifacies	4, 5, 5.5	*Phaenicia cuprina*	7, 8, 8.5, 10
Cochliomyia macellaria	5, 6, 5	*Phaenicia eximia*	15, 15, 15
Paralucilia fulvinota	5, 5	*Phaenicia ibis*	12.5, 15
Phormia regina	5.9	*Phaenicia sericata*	6

In blowflies, the pupal stage may last longer than the previous stages combined. For example, the pupa of *Calliphora vicina* occupies about 60% of the time compared to all previous stages, and this is consistent at rearing temperatures from 10°C to 25°C; in *Calliphora vomitoria,* it is about 50%. In *Protophormia terraenovae,* it is 33% at 12.5°C, and 44% at 23° to 35°C (Greenberg and Tantawi, 1993). The pupa occupies a third of the time in *Phaenicia sericata* and slightly less than half the time in *Phormia regina.* In Peru, we performed 29 rearings of 13 species of blowflies (Chrysomyinae, Calliphorinae, and Toxotarsinae) collected in the subtropics at an elevation of 1000 m, and found the average pupal duration to be 50% (S.D. 8.2). In eight rearings of four species of blowflies collected at 3550 m or above, the average duration was 45.2% (S.D. 3.1). Consider the following for two populations of *Sarconesia splendida,* one collected above 3500 m, the other at 1000 m. The nival population spends 44.7% of its development in pupation, and the subtropical population, 42.7%. This is a fundamental feature of the development of these flies, regardless of temperature.

Despite its temporal dominance, the pupa has been a "black box", as forensic research has focused more on the larva. This is probably because of the greater frequency of larval evidence in homicides, and the relative ease with which feeding larvae can be measured and their age estimated. The pupa requires dissection to search for landmarks that correlate with age.

To understand the genesis of the pupa we briefly return to the larva. The word "larva" means mask, and in the days before people thought to look, it was assumed that the adult fly was already fully formed inside the larva. The facts are more interesting and complicated. Dissect a mature larva and you will find minute discs, often flask-shaped, that are attached to the brain, tracheae, and the body wall. These embryonic clusters of cells, or imaginal discs, are destined to become the adult's antennae, eyes, legs, wing muscles, gonads, etc. (Perez,

1910; Ranade, 1977). Growth of the discs is suppressed during the larval feeding stages until a new hormonal program is initiated in the wandering larva. The ring gland, a neurosecretory structure that is attached to the maggot's brain, receives signals from the environment, via the central nervous system, to start the pupariation process. The ring gland responds by secreting a low level of the hormone ecdysone. The concentration rises sharply a few hours before the onset of pupariation and is responsible for the following suite of changes.

1. Slowing of locomotion, shortening and broadening of the larva.
2. Irreversible retraction of the anterior three segments.
3. Longitudinal muscle contraction.
4. Longitudinal shrinkage of the cuticle.
5. Stabilization of the contracted cuticle.
6. Rapid water loss from the cuticle toward the end of the puparial contraction.
7. Tanning (hardening and darkening of the cuticle) (Zdarek and Fraenkel, 1972).

Pupariation transforms the pliable, white cuticle of the larva into a brown-to-black hard, shiny, protective shell (Plate 11). Several factors can delay the onset of pupariation. We have already mentioned starvation, although this seems more likely to occur under experimental conditions because a corpse is usually more than adequate to support normal development of the first wave of maggots. Crowding of larvae – also more likely as an experimental artifact – can delay pupariation. Larvae of various flies, including several forensically important *Sarcophaga* species mentioned in the last section, fail to pupariate when kept in constant contact with water. When they are returned to a dry environment pupariation may be delayed from a little less than a day to almost a day and a half (Zdarek, 1985). It follows that if rain-soaked soil delays development and is overlooked, an estimate of time of death will be inaccurate.

Arrested development, or diapause, occurs in *Sarcophaga argyrostoma* and *Sarcophaga bullata* in the early phanerocephalic pupa shortly after the head is everted and before the antennae are visible (Fig. 3.35). This stage corresponds roughly to a stage of development in non-diapausing pupae at 40 hours and 29 °C (Fraenkel and Hsiao, 1968). Many sarcophagid flies in the temperate region enter pupal diapause, induced by short-day period, as a means to avoid cold weather. Preparation for diapause begins in the postfeeding larvae of *S. argyrostoma*. Brain-ring complexes from short-night larvae implanted into long-night larvae prevented diapause in the pupa. The converse experiment did not induce diapause (Giebultowicz and Saunders, 1983). In the sarcophagid, *Boettcherisca peregrina*, the incidence of pupal diapause is directly related to a north–south gradient, with a line drawn between Tokunoshima and Okinawa (Kurahashi and Ohtaki, 1989). This line is fairly close to the demarcation between temperate and subtropical zones. Above the line pupae entered diapause with few exceptions. Below the line "very few individuals from any tested colonies entered diapause under any light-dark cycle. No photoperiodically-induced diapause was observed in the other five species [of *Boettcherisca*] which are indigenous to the tropical or subtropical habitats." The evidence from other sarcophagid species elsewhere in the world is less clear.

In the studies of Denlinger (1978) and Denlinger and Tanaka (1988), seven species from Tropical Africa consistently use pupal diapause "to circumvent inimical portions of the year." On the other hand, seven Central American species, from five genera and different locales, did not diapause. Of forensic interest is the observation that in the non-diapausing generations of five species from various parts of the world, capable of diapausing, the minimum duration of the postfeeding larval stage is quite brief, generally one day. In six Neotropical species, incapable of diapausing, the postfeeding larva lasts one to four days, depending on the species.

The change in the puparium's color from white to dark brown provides a handy calibration of age for the first 10 hours or so (Plate 11). Sivasubramanian and Biagi (1983) measured the progress of tanning in *S. bullata* as a percentage of light reflectance. At 25 °C, it was completed in 8 to 10 hours. Our first homicide case centered on this brief stage.

Two Bodies in a Basement. In the late afternoon of 3 October, 1974, a janitor was walking through a garbage-littered alley between two buildings in Chicago. An odor more repugnant than that of garbage came from the basement of the abandoned building. Soon the police were at the scene processing what turned out to be a double murder. One victim had been stuffed into a trunk and covered with a quilt, the other lay nearby on the concrete floor. Both had been shot. Three years later, prosecutors with the Cook County State's Attorney of Illinois were preparing for trial. Time of death was crucial and they were concerned that their witnesses were not likely to win the hearts and minds of a jury. The insect evidence consisted of photos taken of the bodies on the Medical Examiner's table that showed large numbers of postfeeding larvae, but there were no specimens. Although the photographs focused on the cadavers, the postfeeding larvae were in every shot. The critical evidence, however, was at the bottom of a few color prints that showed amber colored puparia (Plate 11). Pupariation had evidently occurred just a few hours before. In Chicago in early fall, the most common urban blowfly is *Phaenicia sericata*. With weather data for the period in question and developmental schedules for the fly, I narrowed the time of death to within ±2.5 days. According to the prosecutors, this was the first time in American jurisprudence that entomological testimony based solely on photographs was entered in a capitol case. It was indeed a fortuitous window and oddly enough, it was the first and only case involving this stage.

The stage at the onset of pupariation is properly called the prepupa. This term has been widely and incorrectly used to describe the postfeeding larva and it suggests that pupariation and the formation of the pupa occur at the same time; in fact, they are separate events. In *Phormia regina*, for example, the prepupa is still a third instar larva and the puparium is its skin; it lasts about 10 hours at room temperature. After the color of the puparium is set, the action shifts to the interior. During the transition from the prepupa to the cryptocephalic pupa there is a sharp but brief rise in ecdysteroid titre which appears to be connected with the degradation of the larval epidermis and other tissues, and the synthesis of the pupal cuticle. After the larval-pupal apolysis, or shedding of the larval skin, the pupa is now wrapped in its own skin and is no longer attached to the inside of the puparium. This stage is most notable for the extrusion of the larval mouthparts and the linings of the foregut and hindgut. These are plastered on the inside of the puparial wall. The sclerotized larval mouthparts are tough and can be extricated years later from the puparium; along with external features of the puparium, they help to identify the species.

During the first day of metamorphosis, a punctured pupa will ooze creamy fluid that is easily mistaken for death and putrefaction. On the contrary, the pupa is a 'beehive' of activity, as histolysis of larval tissues and histogenesis of the pupa proceed simultaneously. To study this stage, gentle boiling will solidify the pupa, and the puparium can then be stripped away to reveal a pair of respiratory horns mounted at the anterior border. The head is invaginated and not visible and the segmented body still resembles that of the maggot (see cryptocephalic pupa at 17 hours and 21 °C, Fig. 3.35).

About a day after the formation of the white puparium, a spasmodic contraction of the abdomen causes the increased blood pressure to quickly evert the head. As the head is everted, the respiratory horns are pushed backwards so that each horn lies underneath a delicate, unsclerotized bubble membrane placed postero-laterally in the fourth segment of the puparium (Fig.2.4a). The convulsive movement causes the horn to puncture the membrane and make contact with the exterior (Fig. 2.4b). In their position and function the horns are reminiscent of the respiratory horns of mosquito pupae, except they are auxiliary to the still-functioning posterior larval spiracles. The horns are visible with a hand lens and were first described by Réaumur in 1738. They are the only external indicators of the progress of metamorphosis after the tanning of the puparium. Intact bubble membranes can help distinguish species and are best seen, with no special preparation, under the SEM (Liu and Greenberg, 1989).

Twenty three hours after the white puparium stage, none of the pupae of *Phormia regina*, kept at 22 °C, had everted horns; 3.5 hours later, all had everted horns. At 29 °C, the process was completed in 22 hours. In *Calliphora vicina*, the timing is approximately the same (Fraenkel and Bhaskaran, 1973). What are the chances of a respiratory horn 'finding' and piercing a minute target whose

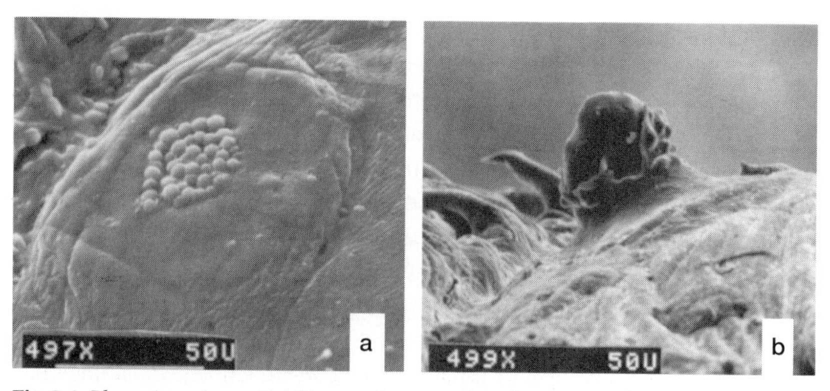

Fig. 2.4. *Phormia regina*. a, Bubble membrane on fourth segment of puparium. b, Same, pierced by respiratory horn.

diameter is only 35 μm? Perhaps it is not surprising that among 400 pupae we examined, 109 failed to evert one or both horns. Pupae with a single everted horn had a 17% mortality rate, which increased to 54% when neither horn was everted. In our experiment, normal pupae had a 1.3% mortality rate. Keister (1953) has shown that pupae with coated horns develop normally. In *Cochliomyia macellaria*, the secondary screwworm fly, the bubble membrane is absent and the respiratory horns do not pierce the puparium. Also in sarcophagids, the horns are weakly sclerotized and do not penetrate to the outside. The horns may be a holdover in higher terrestrial Diptera and it would seem that failure to evert them is symptomatic of a more serious developmental problem (Liu and Greenberg, 1989; Greenberg, 1991).

Accompanying the extrusion of the respiratory horns is the appearance of the familiar head, thorax, and abdomen of the future adult. This is the phanerocephalic pupa, typical of holometabolous insects, with antennae, legs and wings fully exposed (Fig. 3.35, Table 3.9).

The next milestone during metamorphosis is pupal–adult apolysis, or the progressive shedding of the pupal cuticle. This begins on wing buds and legs, and when completed, the pharate adult is enclosed in the transparent pupal skin. The adult is still white, and though all the hairs and setae are present, they are difficult to see (Table 3.9, Chapter 3). Pigmentation first comes to the eyes to protect the delicate retinal cells. In *P. regina*, about 81 hours after pupariation (22 °C), a yellow hue appears in the posterior portion of the compound eye and spreads over the entire eye by 88 hours. The color deepens through various combinations of orange, red, and brown, well before the fly emerges. Transitional eye colors are best recorded from live specimens, as preservatives and dessication can alter them. Table 3.10 summarizes these color changes in *P. regina* at two temperatures, using the color codes of Kueppers (1982). For forensic purposes, the most advanced color for a specific time is given. Tanning of the hairs and setae begins on the head and thorax at 120 hours (22 °C) and is completed over the entire insect by 134 hours. At 29 °C, tanning of the setae begins at 70 hours and is completed by 77 hours (Table 3.9). Additional information can be found in Finell and Jarvilehto (1983) for *Calliphora vicina*, and in Sivasubramanian and Biagi (1983) for *Sarcophaga bullata*.

Dr. Zakariah Erzinçlioglu provides the following information obtained from Dr. Roger Hardie of Cambridge University. The ommatidia (visual units of the compound eye) of *Drosophila melanogaster* elongate during the last 25 to 30 hours of metamorphosis. Before that they are oval in shape and about 10 μm long. Then they undergo a rapid elongation to 70 μm or so after 10 hours, then a slower lengthening to the full length of 100 μm over the last 15 to 20 hours. Metamorphosis takes approximately 100 hours at 25 °C. Dr. Erzinçlioglu observed a lengthening of the ommatidia in preliminary observations on *Calliphora vicina*. This is a promising subject for future research on carrion flies, but it remains to be seen whether it will be forensically useful to calibrate the age of the pharate adult.

In forensic investigations, pupae may be found adhering to or entangled in decaying flesh, hair, or clothes of a corpse that was submerged after the larvae had developed and pupariated (Nuorteva *et al.*, 1974). (The postfeeding larva of some species, e.g., *P. regina*, discharge an adhesive salivary secretion at the time of pupariation.) It is also possible that heavy rains can saturate the soil in which pupae are buried. How does submergence effect pupal viability and adult eclosion? The pupae of five species of blowflies were tested for their ability to withstand submergence – *Protophormia terraenovae, Phormia regina, C. vicina, Phaenicia sericata*, and *Cochliomyia macellaria*. After one day of immersion, survival of the white puparial stage was extremely low in all five species, ranging from 0 to 14%. It was slightly better in pharate adults nearing eclosion, and best in pupae midway during metamorphosis. Survival is inversely correlated with oxygen consumption – it is highest at the beginning and the end of metamorphosis. After four days of immersion of the pharate adult stage, a quarter of the sample of *Protophormia terraenovae, Phormia regina*, and *Phaenicia sericata* produced normal adults. None of the five species survived five days under water. Pupae of *C. vicina* were least able to withstand drowning.

It is noteworthy that in almost all treatments for all species, the pupation period among survivors was extended by approximately the duration of submergence, after they were transferred to suitable conditions (Singh and Greenberg, 1994).

The Adult

The adult pops the cap off the anterior end of the puparium along predetermined lines of weakness. Aiding its escape is the ptilinum, a soft gray bladder between the compound eyes, which inflates and deflates rhythmically. This is one of the most ephemeral structures in the insect world, functioning long enough to enable the fly to bulldoze its way through the soil to the surface. Before the fly's body hardens, the ptilinum is withdrawn into a cavity in the head and sealed with a suture. In anomalous situations, the cuticle hardens before the ptilinum is completely withdrawn. Such flies have short lives.

The empty puparium that is left behind can provide seasonal and archeological insights centuries later. Puparia identifiable to species or genus are not uncommon in Egyptian (Cockburn *et al.*, 1975) and Peruvian mummies. The presence of dung fly puparia, possibly *Spelobia* sp. (Sphaeroceridae), in excavations of Mashantucket Pequot Cemetary in Ledyard, Connecticut, dated at 1670 to 1720, was used to infer cultural burial practices of Native Americans (Dirrigl and Greenberg, 1995). Puparia of *Cochliomyia macellaria,* a tropical species recovered from the La Brea tar pits in California, indicate that the weather was warm at this place 21000 years ago (Gagné and Miller, 1981). Puparia of *Protophormia terraenovae* have been recovered from the skulls and jawbones of wooly mammoths in England that are 12800 years old (Lister, 1993). For other Pleistocene mammal associations with this fly Germonpré and LeClerq (1994) suggest that the deaths occurred in winter or spring. The seasonal associations of this fly apply equally to contemporary medicolegal cases where puparia are the only specimens found (Nuorteva, 1987). Paleotoxicology offers another interesting approach. The drugs, amitryptiline and nortryptiline, have been recovered from phorid fly puparia and dermestid beetle exuviae associated with mummified remains (Miller *et al.*, 1994). Although various techniques exist for dating ancient materials, there is no practical method for dating a puparium that was vacated by a fly just a day or two before. The following case illustrates the need for caution when the entomological evidence rests solely on empty puparia.

On 30 December, 1991, land surveyors in DuPage County, Illinois, happened upon human skeletal remains scattered in a field near a stream. A blanket nearby contained fragments of clothing, hair, and bones, including numerous live and dead insects and a large number of empty fly puparia. Near the skull and other bones was a length of rope with an intact slipknot, duct tape, and a small chain. The evidence suggested murder. The blanket and insect material were brought to our laboratory for further examination. The insects collected from the blanket were typical of the late stages of decomposition, e.g., fly larvae of the cheese skipper, *Piophila casei* and *Ophyra.* But it was the puparia that told the most. The features of the puparia and the larval mouthparts adhering to the inside, were those of *Lucilia illustris,* a greenbottle. The open country, pasture, and semi-wooded sites of the crime scene fit the habitat of this fly. One specimen among the hundreds of empty puparia still contained a pharate adult that matched this species.

Several years before, not far from this site, my students and I had surveyed the flies attracted to rabbit carrion in the spring. No larvae or adults of *L. illustris* were collected between April 9 and May 5. At the same time we collected larvae of three cold-weather blowflies – *Calliphora vomitoria, Calliphora livida,* and *Cynomyopsis cadaverina. L. illustris* appeared during the next 30 days, and, like its urban counterpart, *Phaenicia sericata,* it reached maximum numbers in August and September. Given this seasonal profile, the time of death was bracketed between mid-May and early September. If the murder had occurred earlier in the spring, there should have been some evidence of the cold-weather blowflies. If the murder had occurred later than September, the time would have been too brief for complete skeletonization of the body, even with vertebrate scavenging, before onset of freezing weather. Neither the empty puparia nor the other insect evidence justified a narrower window.

A newly emerged fly is unmistakable with its shriveled wings, tiny abdomen, spidery legs, dull gray color, and pulsating ptilinum. The fly marks its rite of passage by depositing a droplet of creamy meconium. In just a few hours, blood pressure expands its wings and abdomen, the cuticle hardens and the species' color is set. This is the teneral adult, easily mistaken for a reproductively mature fly. Some fly biologists extend the teneral period several days to include internal maturation It is the young first-generation adult flies at the death scene that ought to be preserved as potential evidence.

There are several methods to determine the age of adults, although their forensic applicability has not been tested.

Pteridines are degradation products of purine metabolism and are sequestered in the eyes of a number of flies as a means of storage excretion. Wall *et al.* (1990, 1991, 1992) have shown that pteridine concentration increases linearly with age and temperature. They found no differences between flies fed sucrose only and those fed sucrose plus protein, thus eliminating nutrition as a variable. However, concentrations were significantly higher in males. Furthermore, "Even though the laboratory and field trials used *C[hrysomya] bezziana* from the same stock colony, the fluorescence values on day one of [adult] life were significantly higher for field-caught than laboratory-maintained flies." Even among laboratory-maintained flies during the first few days of adult life, there was considerable overlap in pteridine concentrations.

Another biochemical approach is the quantification of lipofuscin. This is a widely distributed fluorescent lipid-protein complex that accumulates with age in various tissues of a wide variety of animals. Each tissue may have its own form of lipofuscin with characteristic wavelengths (Sohal, 1981). In the head of male *Sarcophaga bullata,* there is a steady increase in lipofuscin over a 24-day period, but it is not clear whether this method will provide significant gradations of differences in the young adult (Ettershank *et al.*, 1983).

In 1963, Neville described the accumulation of daily growth rings in cuticular structures of grasshoppers. When attention was directed to flies, it was found that *Lucilia cuprina* has as many as 10 bands. These are most clearly seen in the mesothoracic post-phragmata – major cuticular attachments for the wing muscles (Tyndale-Biscoe and Kitching, 1974) – and on the hypandrium of

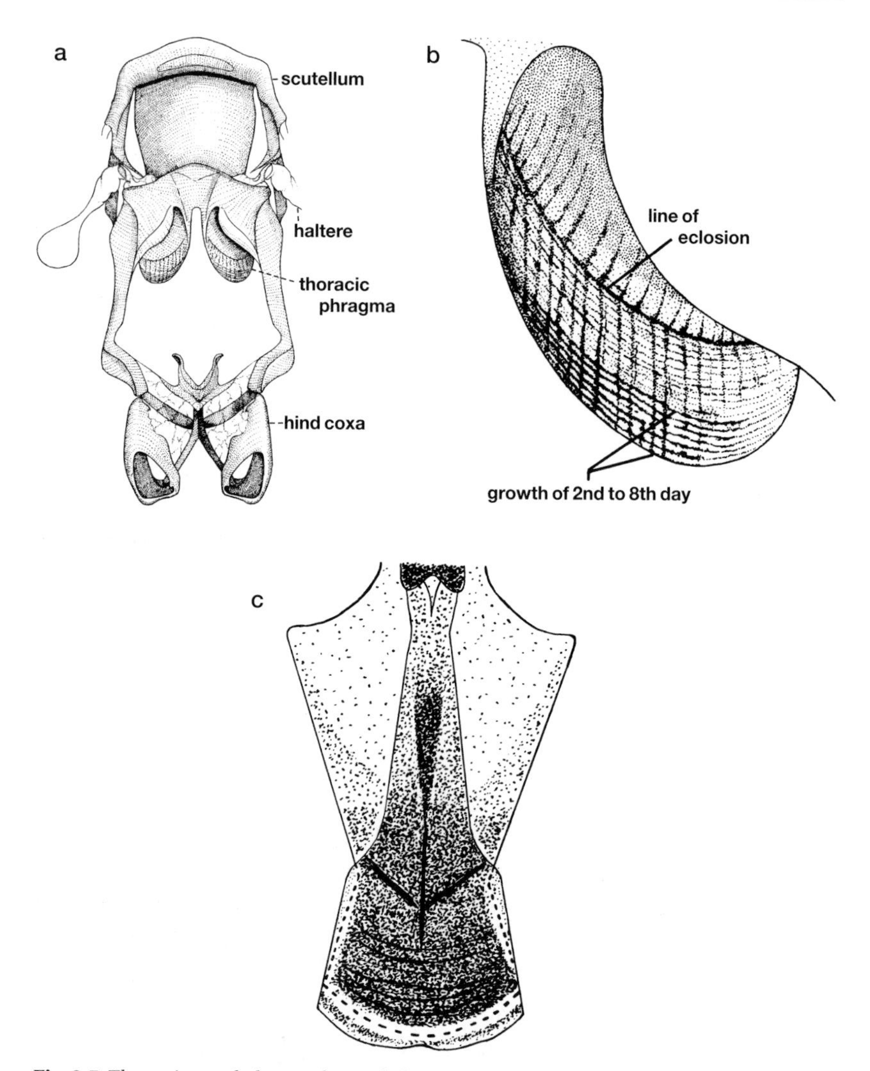

Fig. 2.5. Thoracic exoskeleton of *Anopheles gambiae*: a, Posterior part. b, Lines of growth on thoracic phragma of field-collected specimens (after Y. Schlein, 1979). c, Lines of growth on the hypandrium of the aedaegus in *Cochliomyia hominivorax*. (Baumgartner and Greenberg, 1983.)

the male genitalia of *Cochliomyia hominivorax* (Baumgartner and Greenberg, 1983), and *Calliphora* (Schlein and Gratz (1972) (Fig. 2.5). As Ellison and Hampton (1982) point out, since the flight muscles increase in size following emergence of the adult, it is not surprising that the apodemes to which the muscles attach would also need to grow to provide firm attachment sites. The rate of deposition of these bands is temperature dependent, approaching zero below 15 °C. When day and night temperatures are between 21 °C and 32 °C, growth may be so fast that band delineation is obscured. Trials showed that the

Plate 1
String of beads with flies of gold and lapis lazuli, from the royal tombs of Queen Puabi at Ur (2600–2500 BCE). (Courtesy the University of Pennsylvania Museum of Archeology and Anthropology, #30-12-570.)

Plate 2
In ancient Egypt, fly amulets on mummy beads may have been intended to ward off destruction of the corpse. (Courtesy of the Oriental Institute of the University of Chicago.)

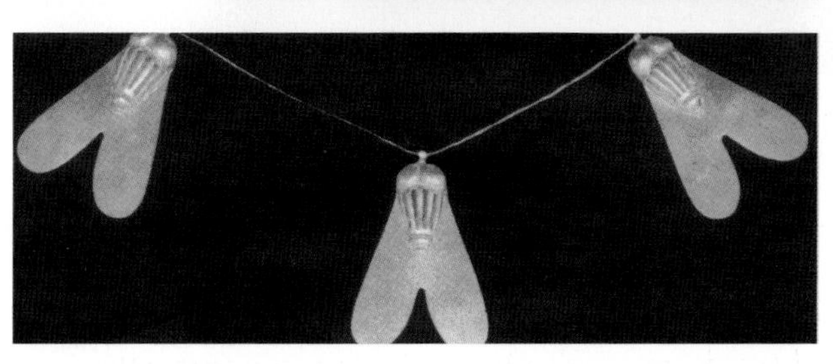

Plate 3
Flies of gold, probably the common housefly, *Musca domestica* or the eye fly, *Musca sorbens*. Awarded by Queen Ahotep to a brave Egyptian soldier.

Plate 4

Vanity. The fly associated with death and decay. North German School, *c*. 1600. (Museum für Kunst und Kulturgeschichte, Dortmund, Germany.)

Plate 5

Madonna with Child. Christ child recoils from the fly as a Satanic symbol, while clutching a bird, symbol of resurrection. Carlo Crivelli, end of fifteenth century. (The Metropolitan Museum of Art, New York.)

Plate 6

Beauty and the bluebottle. Presage of death, or after the fact? *Self-portrait of the artist with his wife.* The Frankfort Master, 1496. Note the fly fecal deposit, a further indignity. (Koninklijk Museum voor Schone Kunsten, Antwerp, Belgium.)

Plate 7

Madonna with Child, Giorgio Schiavone, *c.* 1460. Note fly on cherub, lower left. (Galleria Sabauda, Turin, Italy.)

Plate 8

Portrait of a Woman. Barthel Bruyn the Older, *c.* 1540. (Musées Royaux des Beaux-Arts de Belgique, Brussels.)

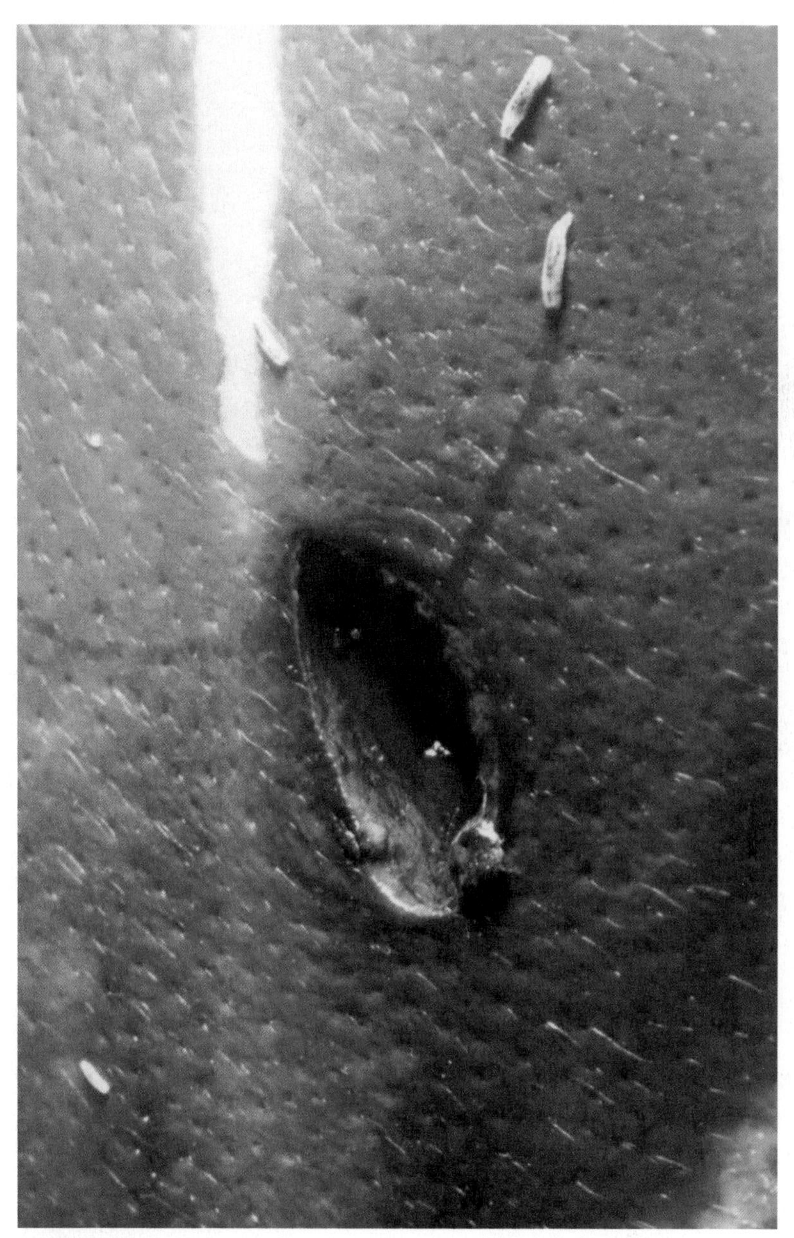

Plate 9

Phormia regina larvae, at the start of the postfeeding stage, exiting a knife wound.

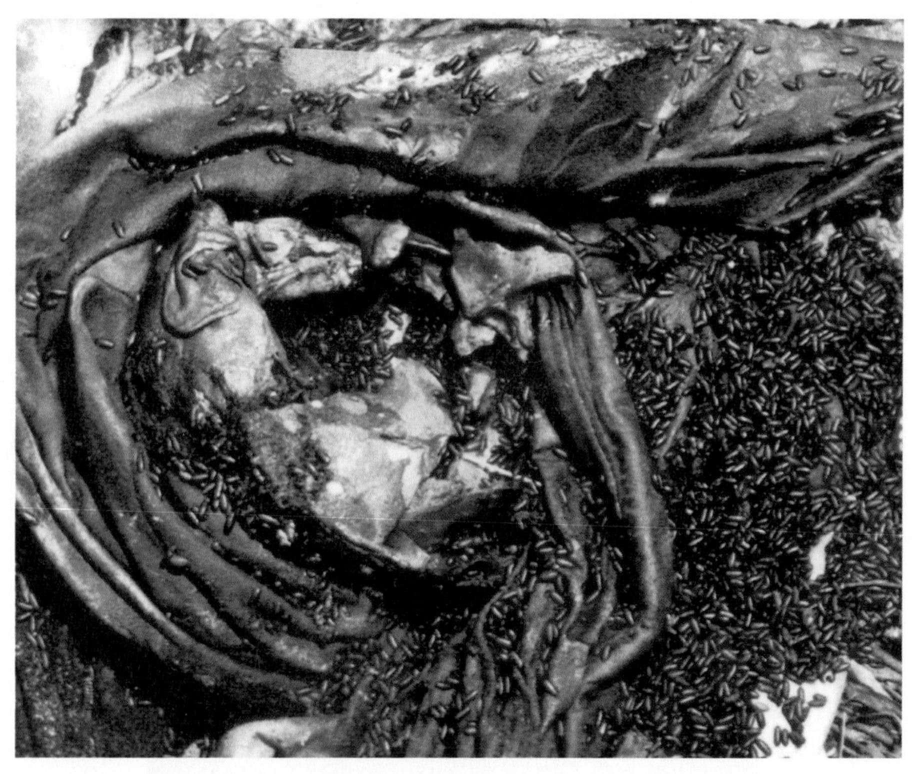

Plate 10
Aggregation of *Phormia regina* pupae on human remains.

Plate 11
Color changes in the puparium of *Phormia regina* during12 hours at 22 °C.

count of growth bands was an acceptable estimate of the actual age in days when temperatures rose 3.5 °C or more above the threshold of 15.5 °C. The greater the diurnal temperature range the greater the accuracy. The authors caution, "For bands to be formed reliably there must be a cessation of growth during the temperature cycle." Furthermore, "there seems little doubt that variation in patterns and sites of deposition of the growth bands occurs both among different sites in the same species and in the same sites among different species." This method has three important advantages: band number equals calendar (not physiological) age; it can be used years later with dried specimens and without sectioning or special illumination (although the bands are best viewed in polarized light); and protein intake has little effect on the deposition of the bands. It appears to be accurate to within plus or minus one day. Although the method seems to be applicable to blowflies, based on the species thus far tested (e.g. *L. cuprina, Calliphora vicina,* and *Cochliomyia hominivorax*), the same cannot be said for *Sarcophaga falculata.* The number of growth layers observed on the apodemes of the aedeagus of this fly did not correspond in days to its age (Schlein and Gratz, 1972).

We investigated the quantification of pupal fat bodies, a large number of which float free in the hemolymph or are loosely attached in the abdomen of a newly emerged fly. Teneral adult *Phormia regina* were maintained on sucrose and water in a 100 cm^3 cage at 22 °C and a 12:12 (L:D) photoperiod. Five specimens were dissected in insect Ringers' solution at 0, 24, 48, 72, and 96 hours after emergence, and the fat bodies were removed from the abdomen and counted. Starting with about 5500 fat bodies at 0 hours, the numbers decreased significantly each day and reached 600, four days after emergence. There was no difference between the sexes of 2-day-old flies. Although the average size of the fat bodies decreased with age there was considerable overlap (Table 3.11). The rate of loss appears to be dependant on nutrition. When *L. cuprina* was fed meat, sugar, and water, fat bodies were lost more rapidly than when meat was excluded (Evans, 1935).

It is evident that each of these aging techniques has difficulties, requiring further study to evaluate their forensic efficacy. If any do find a use, it would more likely be in indoor situations where "contamination" from non-corpse-generated flies would not be a concern. But newly emerged flies could turn out to be prime evidence no matter where the body is found. At present, when the forensic entomologist is confronted with flies of a second or later generation, he must settle for a lower level of precision provided by the insects that follow the first colonizers.

Some flies winter in homes and other buildings. In regions with cold winters, can they function as forensic indicators? Both sexes of the cluster flies *Pollenia pediculata* and *Pollenia vagabunda,* along with the false stable fly *Muscina assimilis,* were collected from October until March in a heated residence in upper New York State. Few *Phormia regina* were also collected. All

Fig. 2.6. Fly specks on glass pane consist of vomit and fecal deposits. Vomit spots sometimes leave a proboscis imprint.

were kept alive at 4 °C until dissected in insect saline solution. The females of all four species had undeveloped ovaries, no sperm in their spermathecae, and remnants of pupal fat bodies. Most of the males had motile sperm in their testes (Greenberg, 1998). The cluster flies are obligate parasites of earthworms and are unable to complete larval development on carrion. We concluded that their presence indoors, often in large numbers, has no forensic importance. One might also dismiss the false stable fly, though it is a carrion feeder, were it not for Kobayashi's (1922) data on *Muscina stabulans*, widely distributed and a sibling species. He collected 1723 flies outdoors in Seoul, South Korea, in February and March. Of 51 dissected females eight had moderate to fully developed ovaries and some of the latter had sperm in their spermathecae. Such flies could be ready for action indoors in winter. As for *P. regina*, both sexes have a facultative diapause in the laboratory that is characterized by hypertrophy of the fat body and by failure of the ovaries to develop (Stoffolano *et al.*, 1974). This is an important forensic fly and a definitive study of its overwintering reproductive readiness would clarify its forensic status.

Fly specks originate from the vomit and feces of a fly (Fig. 2.6). Graham-Smith (1914) counted 1102 vomit spots and nine fecal deposits on an area six inches square (c. 15 cm^2) of a cupboard window. He distinguished three types of spots: "1. faecal deposits, round, opaque, often raised and yellowish, brownish or whitish in colour; 2. 'vomit' stains, round with a small opaque centre and clear peripheral portion, bounded by a darker zone, and 3. proboscis-marks left on half dried material". Readers in developed countries might view Graham-Smith's numbers as history, but readers in developing countries know otherwise. It suggests the possibility of DNA testing, particularly of vomit spots, which have been subjected to less digestive enzyme activity than the fecal deposits. Lambert (1997) points out that DNA can survive for thousands of years under reasonable conditions, especially under cold or dry conditions. By means of the polymerase chain reaction, DNA analysis can be made on a strand of hair. To my knowledge, fly deposits have not been analyzed for DNA or mitochondrial DNA (mDNA). It is admittedly a remote possibility, but fly specks at the crime scene may contain DNA of the victim or perpetrator, critical to a case, after other sources are no longer available.

3

Keys to the Eggs, Larvae, Pupae and Adults of Some Forensically Important Flies

"Why has not man a microscopic eye? For this plain reason: man is not a fly."
Alexander Pope (1688–1744)

The forensic entomologist faces three evidentiary challenges: to know the fly species, its most advanced stage of development on or off the body, and its thermal history. This chapter provides keys and illustrations for the identification of eggs, larvae, pupae and adults of many of the forensically important flies of North and South America, Europe, and the Orient. It also includes graphs and tables of their rates of development at various temperatures. For a good discussion of the collection of evidence and data at the death scene the reader is referred to Smith (1986), Byrd and Castner (2001), and Lord and Burger (1983). Ideally, the entomologist is called to the death scene to process the relevant evidence. This ideal is often short-changed for various reasons. Crime scene investigators frequently regard a maggot infestation as a disgusting nuisance and are repelled by the insects' voracious disregard for the dignity of a human death; or a forensic entomologist may be too far away to process the scene. In a good many cases the entomologist is contacted months and even years later by attorneys who are preparing for trial. At the time of the trial, the forensic value of the available evidence is, in descending order: (1) an adequate number of maggot (or other) specimens have been collected from and/or near the body and properly preserved (Wells *et al.* 2001); (2) desiccated specimens have been collected and preserved; (3) photographs (with a scale) of uncollected specimens are available; (4) photographs of maggots without a scale are available; (5) medical examiner has noted the presence of maggots and has estimated their length; or (6) medical examiner has noted maggot presence without estimating length. Of course, some of these categories can be further subdivided, but the hierarchy of evidentiary value will be obvious to the reader.

With the above as a guide in the collection and evaluation of entomological data we now turn to the keys for the classification of the eggs, larvae, pupae, and adults of some forensically important flies. We recognize that the published literature is not everywhere available to the serious reader, and have tried whenever possible to be geographically inclusive with contributions from regional specialists. Dr. Juan Carlos Mariluis of Argentina has provided a key to the common calliphorid adults of South America. A key with diagnostic illustrations for European flies of forensic importance by Dr. Franticek Gregor (1971) of the Czech Republic, is reproduced here. The key to the forensic adult flies of the Oriental region is the work of Dr. Hiromu Kurahashi of Japan. This is followed by a key to third instar calliphorid larvae of Malaysia by Dr. Baharudin Omar. Finally, Dr. James F. Wallman of Australia has contributed keys to the carrion breeding adults and larvae of forensic importance in south-eastern Australia. Many species are widespread and appear in several keys but there is no redundancy because the regional mix of species is always different. Global coverage is beyond the scope of the book and the inevitable gaps that remain are partly filled with literature citations.

We begin with a key to the eggs of common forensic flies. Figures 3.1 to 3.5 are SEMs of the diagnostic features of the eggs of 19 species. The reader may wish to review the special features of the egg's topography shown in Fig. 2.1a,b. At the present time, the eggs of many common blowflies can be classified only to genus. Hopefully, further research will reveal specific diagnostic differences hitherto overlooked. Nevertheless, a generic designation can often provide useful seasonal and site information. For example, *Calliphora* and *Cynomyopsis* have similar eggs and share the same cool season though not necessarily the same synanthropic site.

Key to Eggs of some Forensic Flies

(Greenberg and Szyska, 1984)

1. – Length 0.6 to 0.75 mm 2
   – Length >1.35 mm 4
2. – Median area flanked by prominent flanges with scale-like projections (Figs. 3.5a,b) .... *Megaselia scalaris*
   – Median area not as above 3
3. – Chorion minutely pebbled (Fig. 3.5c) .... *Megaselia abdita*
   – Chorion reticulated (Figs. 3.5d,e) .... *Piophila casei*
4. – With pronounced flanges and longitudinal ridges (Figs. 3.5f–i) .... *Muscina stabulans* & *assimilis*
   – Not as above 5
5. – Arms of flanges curve around micropylar collar (Fig. 3.3b) 6
   – Arms of flanges straight or slightly diverging (Figs. 3.2h,k) .... *Calliphora vicina, C.livida* & *C. vomitoria*; & *Cynomyopsis cadaverina*
6. – Arms of flanges curve about halfway around micropylar collar (Fig.3.2b) 7
   – Arms of flanges curve less than halfway around micropylar collar (Figs. 3.3b,e,h) .... *Lucilia sericata, cuprina,* & *illustris*
7. – Anterior third of plastron with apices of struts blunt (Fig. 3.2f) 8
   – Anterior third of plastron with peglike struts (Figs.3.1c,f,i,l) .... *Chrysomya albiceps, C.chloropyga* & *C.rufifacies,* & *Cochliomyia macellaria*
8. – Apices of some struts expanded and anastomosed (Figs.3.2c,f) .... *Lucilia ibis, Phormia regina,* & *Protophormia terraenovae*
   – Struts mostly not anastomosed (Figs.3.4h–j) .... *Chrysomya megacephala*

Table 3.1 summarizes data of various authors on the hatching time of the eggs of 13 species of flies found in both the New and the Old World. The Table covers

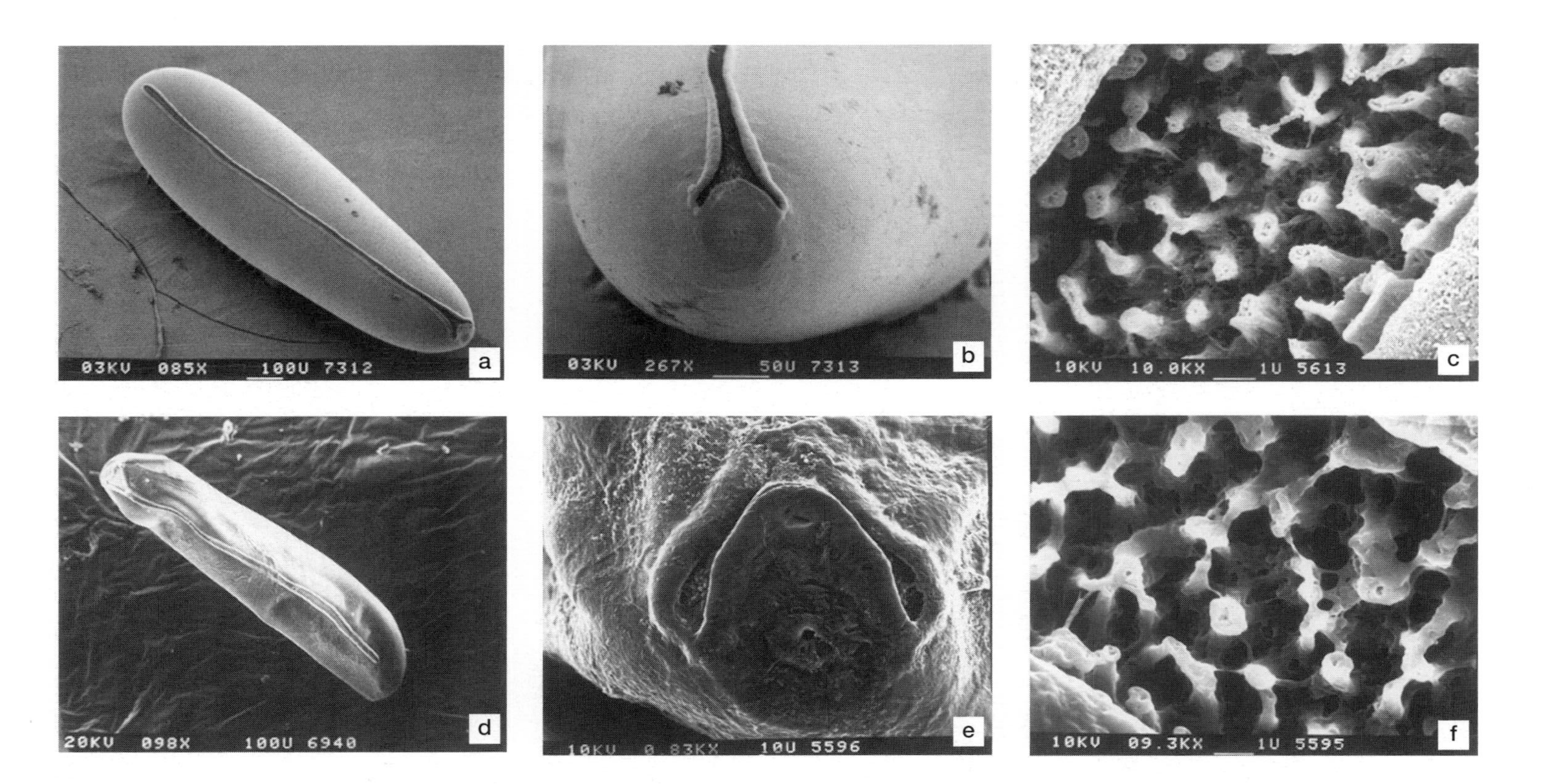
03KV 085X 100U 7312
a
03KV 267X 50U 7313
b
10KV 10.0KX 1U 5613
c
20KV 098X 100U 6940
d
10KV 0.83KX 10U 5596
e
10KV 09.3KX 1U 5595
f

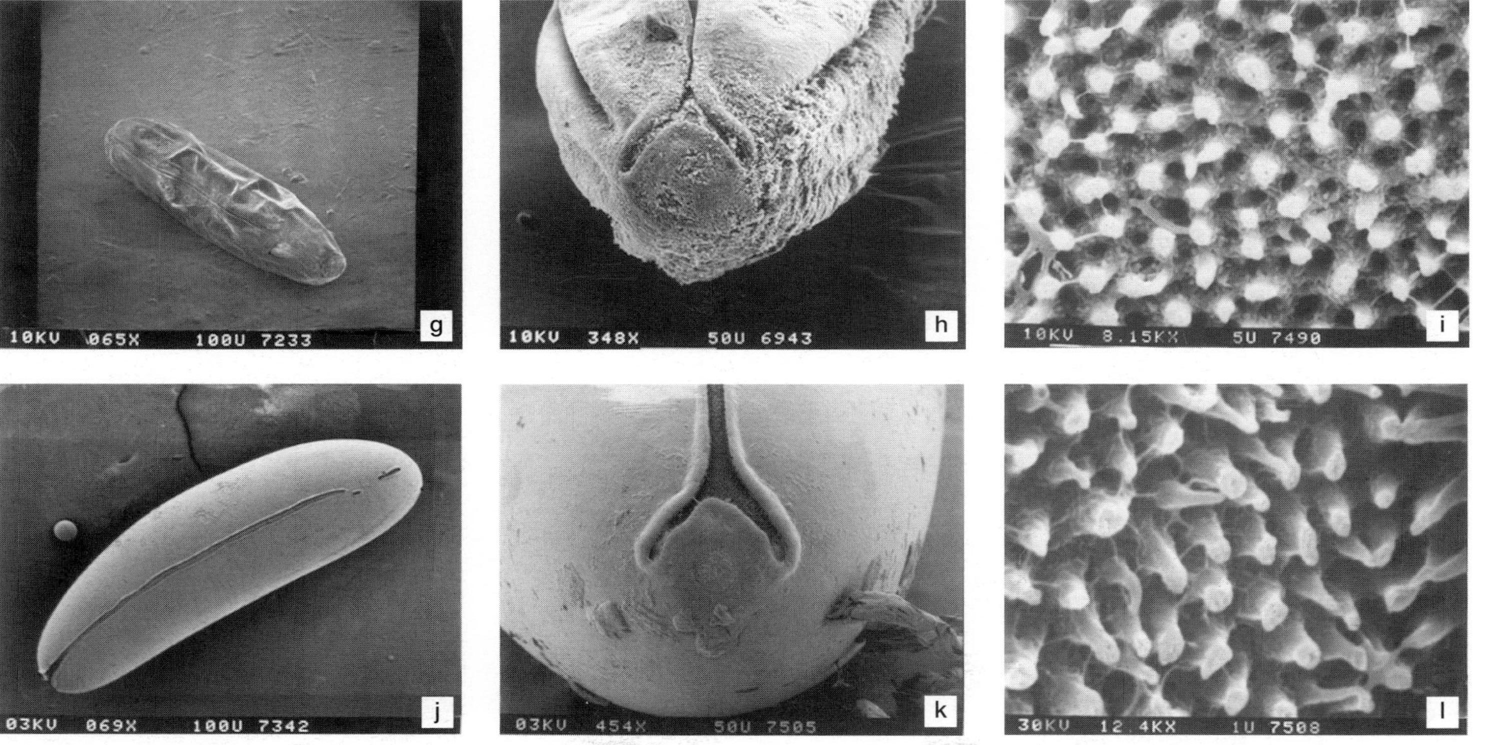

Fig. 3.1. SEM of blowfly eggs. *Chrysomya rufifacies:* a, whole egg; b, micropylar end; and c, plastron. *Chrysomya albiceps:* d, whole egg; e, micropylar end; and f, plastron. *Chrysomya chloropyga:* g, whole egg; h, micropylar end; and i, plastron. *Cochliomyia macellaria:* j, whole egg; k, micropylar end; and l, plastron. (Greenberg and Singh, 1995.)

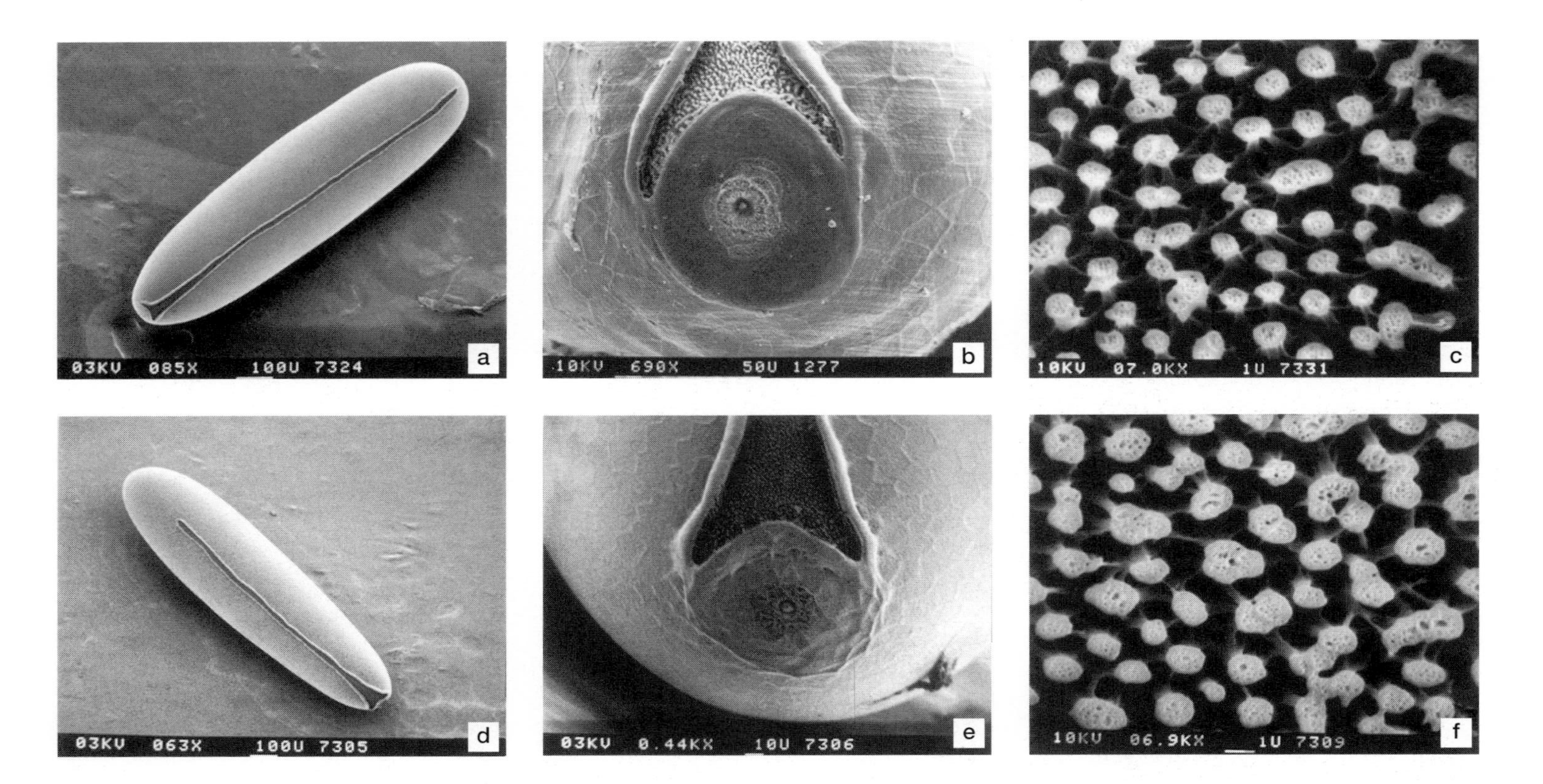
03KV 085X 100U 7324
a
10KV 690X 50U 1277
b
10KV 07.0KX 1U 7331
c
03KV 063X 100U 7305
d
03KV 0.44KX 10U 7306
e
10KV 06.9KX 1U 7309
f

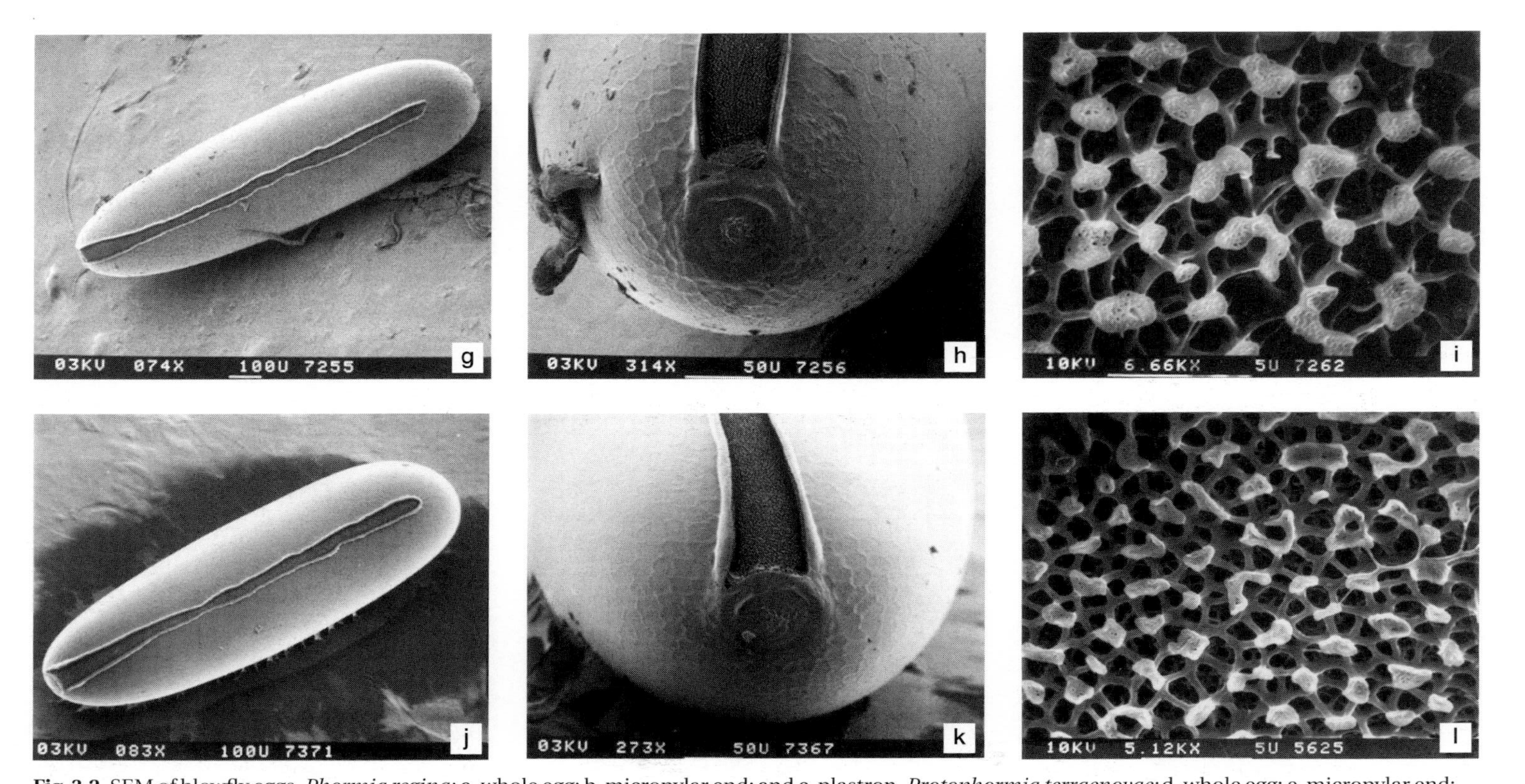

Fig. 3.2. SEM of blowfly eggs. *Phormia regina:* a, whole egg; b, micropylar end; and c, plastron. *Protophormia terraenovae:* d, whole egg; e, micropylar end; and f, plastron. *Calliphora vicina:* g, whole egg; h, micropylar end; and i, plastron. *Calliphora vomitoria:* j, whole egg; k, micropylar end; and l, plastron. (Greenberg and Singh, 1995.)

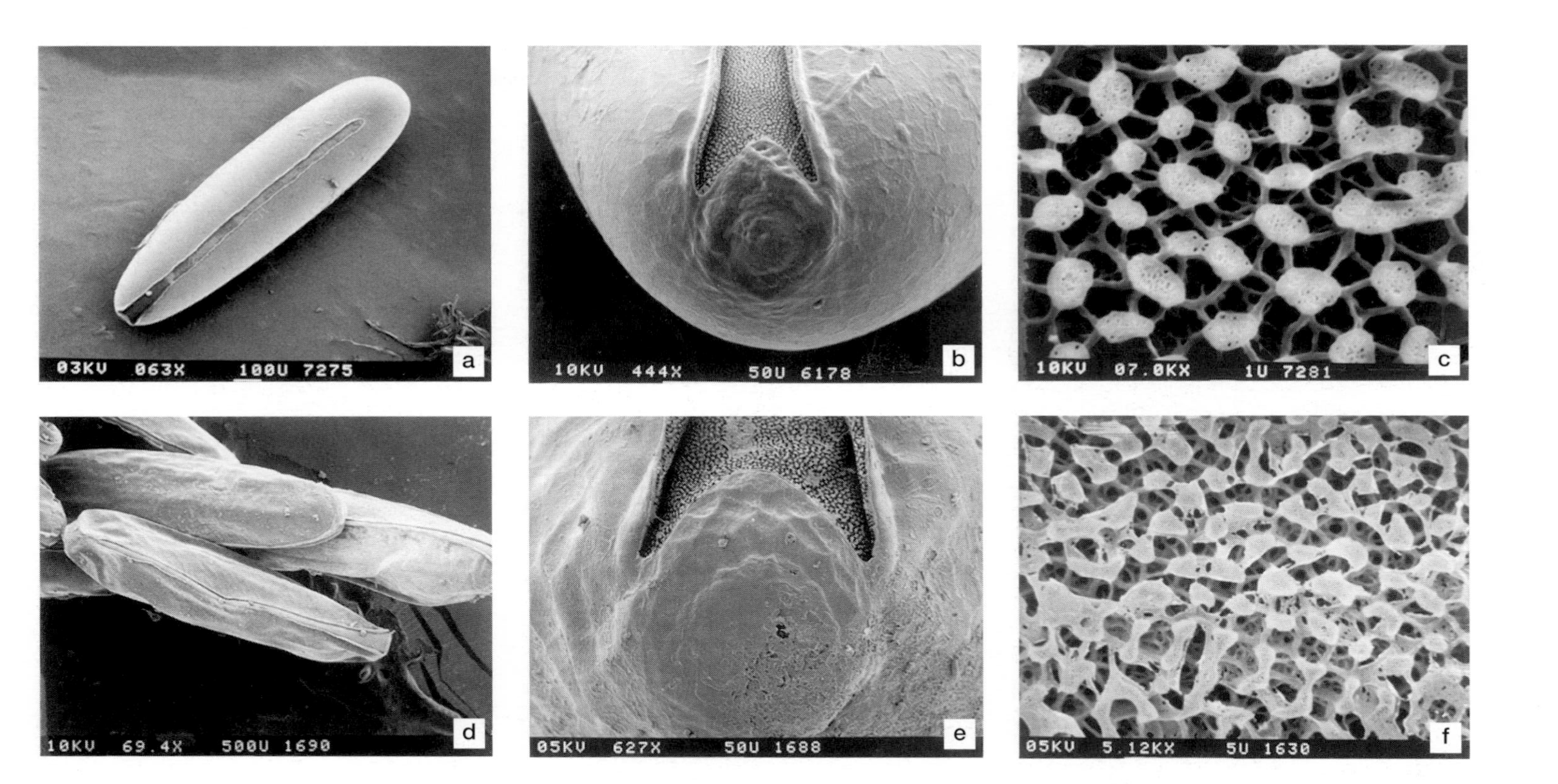

03KV 063X 100U 7275
a
10KV 444X 50U 6178
b
10KV 07.0KX 1U 7281
c
10KV 69.4X 500U 1690
d
05KV 627X 50U 1688
e
05KV 5.12KX 5U 1630
f

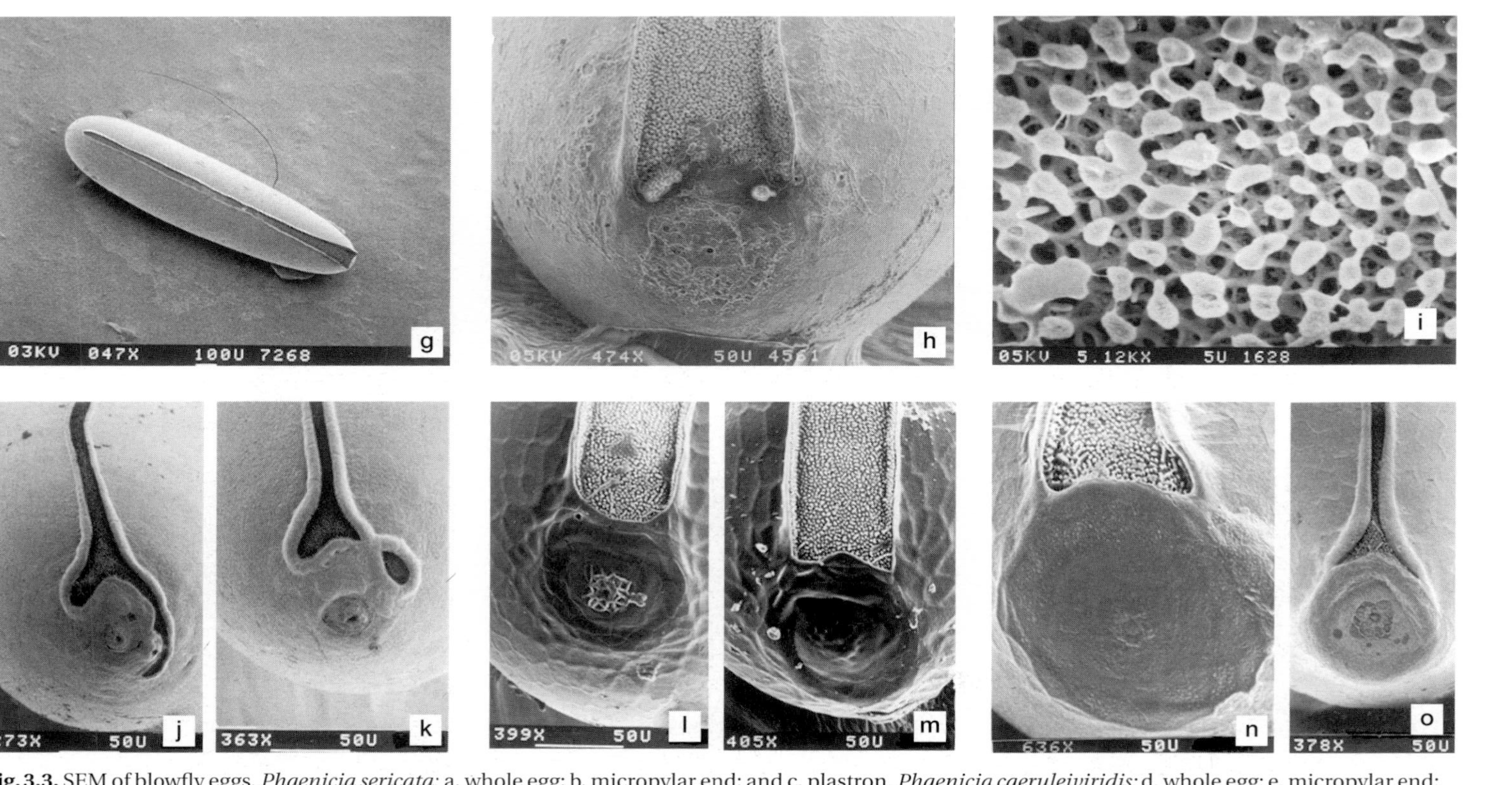

Fig. 3.3. SEM of blowfly eggs. *Phaenicia sericata:* a, whole egg; b, micropylar end; and c, plastron. *Phaenicia caeruleiviridis:* d, whole egg; e, micropylar end; and f, plastron. *Phaenicia illustris:* g, whole egg; h, micropylar end; and i, plastron. *Cochliomyia macellaria:* j and k, micropylar end with anomalies. *Calliphora vicina:* l and m, micropylar end showing variations. *Protophormia terraenovae:* n, micropylar end showing variation. *Phormia regina:* o, micropylar end showing variation. (Greenberg and Singh, 1995.)

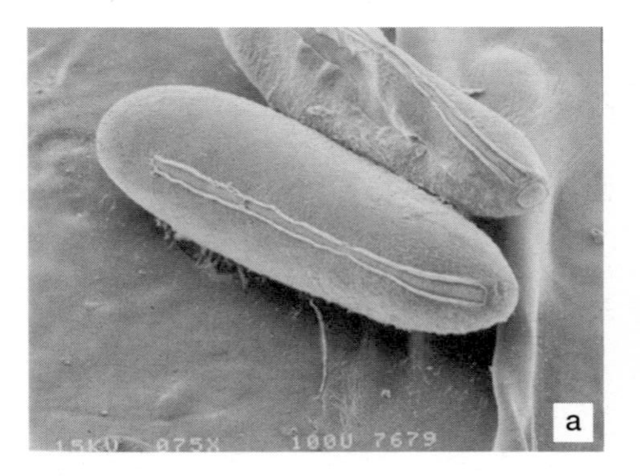
15KV 075X 100U 7679
a

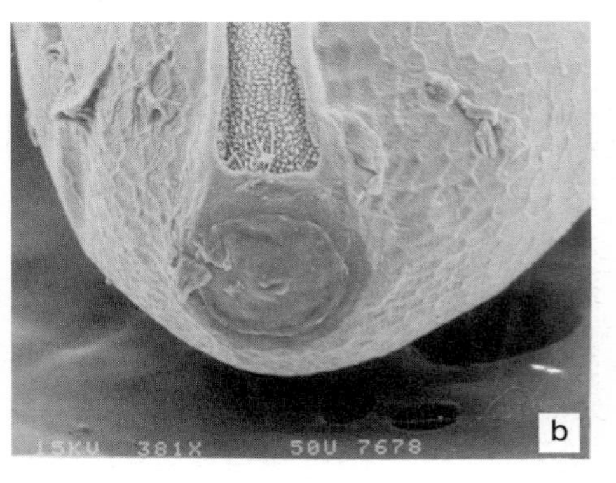
15KV 381X 50U 7678
b

15KV 6.18KX 5U 7677
c

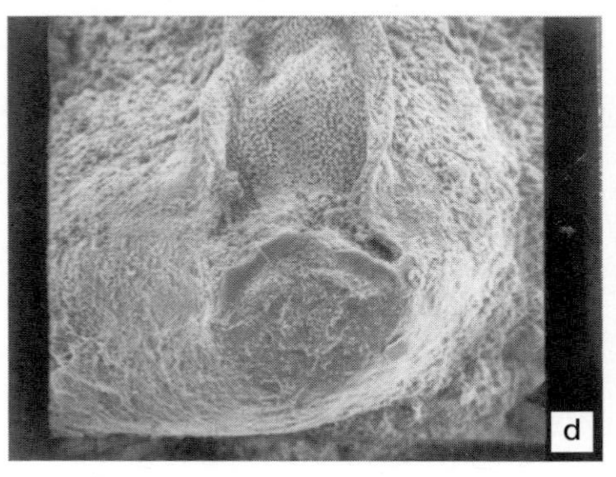
d

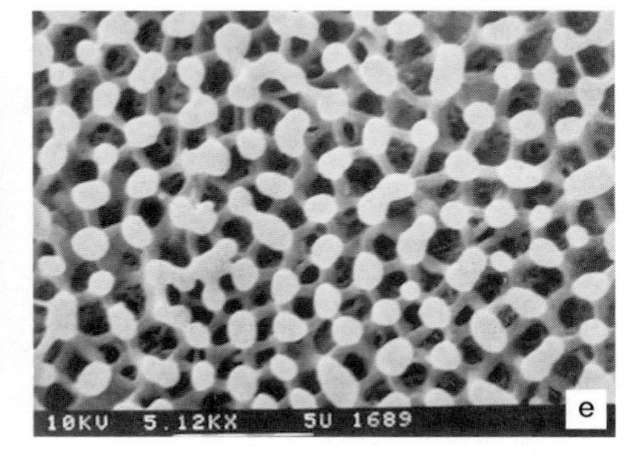
10KV 5.12KX 5U 1689
e

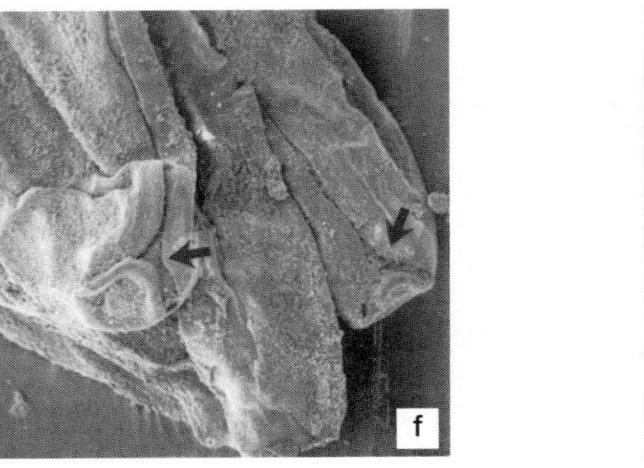

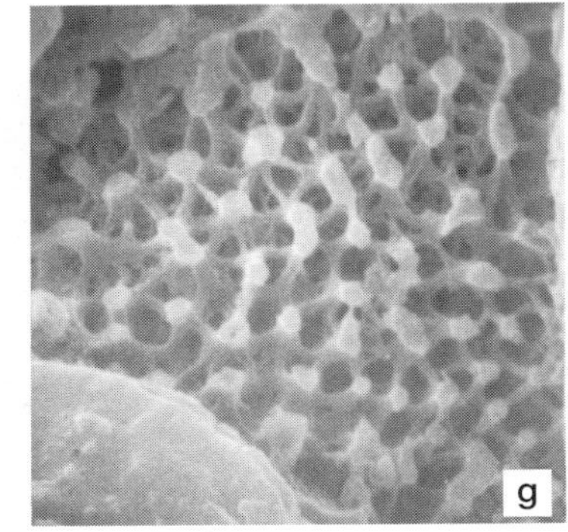

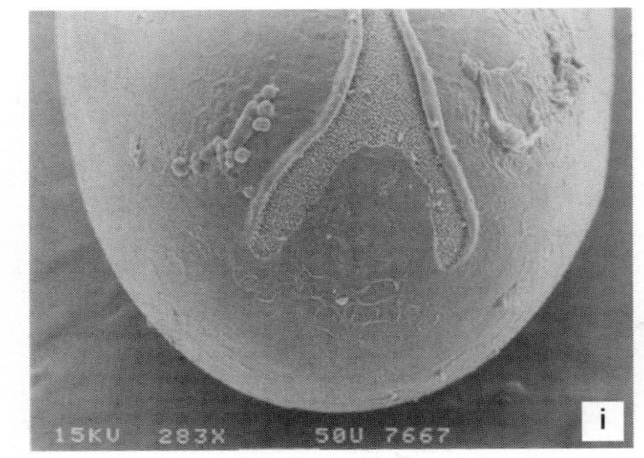

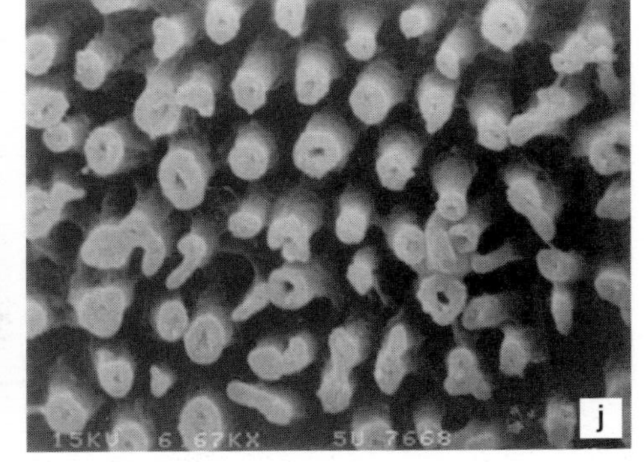

Fig. 3.4. SEM of blowfly eggs. *Cynomyopsis cadaverina:* a, whole egg; b, micropylar end; and c, plastron. *Phaenicia cuprina:* d, micropylar end; and e, plastron. *Phaenicia ibis:* f, micropylar end; and g, plastron. *Chrysomya megacephala:* (Bangalore, India): h, whole egg; i, micropylar end; and j, plastron.

10KV 064X 100U 0112
a
10KV 278X 100U 1287
b
10KV 188X 50U 4537
c
03KV 125X 100U 7394
d
10KV 0.91KX 10U 7391
e
207X 50U
f

Fig. 3.5. SEM of blowfly eggs. *Megaselia scalaris:* a, whole eggs; b, micropylar end and median area; and c, chorion opposite median area. *Piophila casei:* d, whole egg; and e, anterior end showing micropyle, median area, and chorion. *Muscina stabulans:* f, note pronounced flanges, prominent median area, longitudinal folds of the chorion best seen in h, and micropyle. *Muscina assimilis:* g, whole egg; h, anterior end; and i, plastron.

Table 3.1 *Hatching time (hours) of blowfly eggs*

Species	Temperature (°C)												Author(s)
	10	11	12.5	15	17	19	22	23	25	27	29	35	
Protophormia terraenovae			91					17			14	12	Greenberg & Tantawi (1993)
Protophormia terraenovae										12			Kamal (1958)
Phormia regina				49	27						15		Greenberg (unpublished)
Phormia regina							20				18		Greenberg (1991)
Phormia regina										13–16			Cyr (1993)
Phormia regina										10			Kamal (1958)
Calliphora vicina	88		38			19			14				Greenberg (1991)
Calliphora vicina		68		29		23		21					Reiter (1984)
Calliphora vicina									14			13 (33 °C)	Evans (1936)
Calliphora vicina										20			Kamal (1958)
Calliphora vomitoria			65					22			17	12	Greenberg & Tantawi (1993)
Calliphora vomitoria										23			Kamal (1958)
Calliphora terrae-novae										18			Kamal (1958)
Eucalliphora lilaea										14			Kamal (1958)
Cynomyopsis cadaverina										14			Kamal (1958)
Phaenicia (=*Lucilia*) *sericata*									14			9 (33 °C)	Evans (1936)
Phaenicia (=*Lucilia*) *sericata*				42		24 (20°C)						10	Wall *et al.* (1992)
Phaenicia (=*Lucilia*) *sericata*										12			Kamal (1958)
Phaenicia (=*Lucilia*) *sericata*						29				14		10	Ash & Greenberg (1975)
Phaenicia (=*Lucilia*) *sericata*							23				18		Greenberg (1991)
Phaenicia (=*Lucilia*) *sericata*							18				14	12	Greenberg (unpublished)

Phaenicia pallescens	29		13		10	Ash & Greenberg (1975)
Lucilia illustris					10–15 (30 °C)	DasGupta & Roy (1969)
Cochliomyia macellaria					12 (30 °C)	Cunha e Silva and Milward-de-Azevedo (1992)
Cochliomyia macellaria	27	20		12	10	Greenberg (unpublished)
Chrysomya megacephala			12			Wells and Kurahashi (1994)
Chrysomya megacephala			9–10 (24–29 °C)			Wijesundara (1957)
Chrysomya rufifacies		24			12	Greenberg (unpublished)

Table 3.2. *Hatching time of blowfly eggs of the Peruvian Andes*

Species	Hours
Compsomyops boliviana	50
Sarconesia chlorogaster	24
Sarconesia magellanica	25
Sarconesia splendida	35
Sarconesia versicolor	37

Notes:
Average maximum and minimum daily temperatures were 20.3 °C (SD ± 1.5) and 13.7 °C (SD ± 1.5). (Greenberg and Szyska, 1984.)

Table 3.3. *Hatching time of blowfly eggs of the Peruvian Amazon*

Species	Hours
Chrysomya chloropyga putoria	14.5
Cochliomyia macellaria	11
Compsomyops verena	16
Hemilucilia flavifacies	22
Hemilucilia hermanlenti	13
Paralucilia fulvinota	22
Calliphora nigribasis	13
Phaenicia cuprina	12
Phaenicia eximia	12
Phaenicia ibis	22.5
Sarconesia splendida	25.5
Sarconesia versicolor	14

Notes:
Data are from replicate rearings with average maximum and minimum temperatures of 26.0 °C (SD + 3.1) and 21.7 °C (SD + 1.9). (Greenberg and Szyska, 1984.)

a useful range from 10–35° C because the incubation period is temperature dependent. The numbers given in Table 3.1 typically are the average minimum, or average hatching time. Precise times of oviposition and hatching (e.g. <0.5 hours) often are neither realistic nor important in a typical forensic case.

Tables 3.2 and 3.3 provide hatching times of eggs of South American

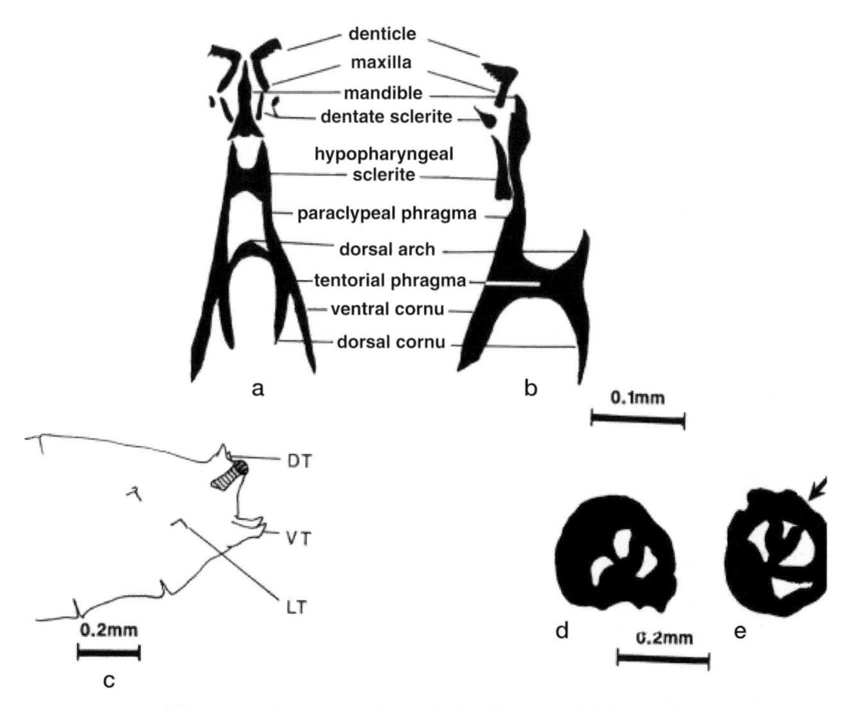

Fig. 3.6a. Larval diagnostic features. a, b, Cephalopharyngeal skeleton first instar calliphorid, dorsal and lateral aspects. c, *Piophila casei* third instar posterior showing dorsal (DT), lateral (LT), and ventral (VT) tubercles. Posterior spiracles, third instar: d, *Muscina assimilis*. e, *Muscina stabulans*. Arrow indicates triangular area.

blowflies. The data summarized in Table 3.2 are of five species that are endemic to the Peruvian Andes. The experiments were conducted in an open but sheltered structure in Montaro Valley, near Jauja, Peru at 3550 m above sea level (a.s.l.), in December 1979. Table 3.3 summarizes data of 12 species that occur at lower elevations in the Peruvian Amazon. These rearings were conducted near San Ramón at 1000 m a.s.l., during June and July 1980, and December 1981, in a room open to the outdoors. The rearing temperatures in both sites approximated outdoor temperatures in shade. Twelve species of flies were reared during several trips and their immature stages were described for the first time (Greenberg and Szyska, 1984). A key to the known third instar blowfly larvae of Peru is based largely on these rearings and is accompanied by Figs. 3.6 to 3.17 of the spinose bands, anterior and posterior spiracles, and mouthparts. The larva offers more diagnostic features than the egg but the common burrowing habit of maggots constrains morphological diversity and accounts for the limited number of species keyed.

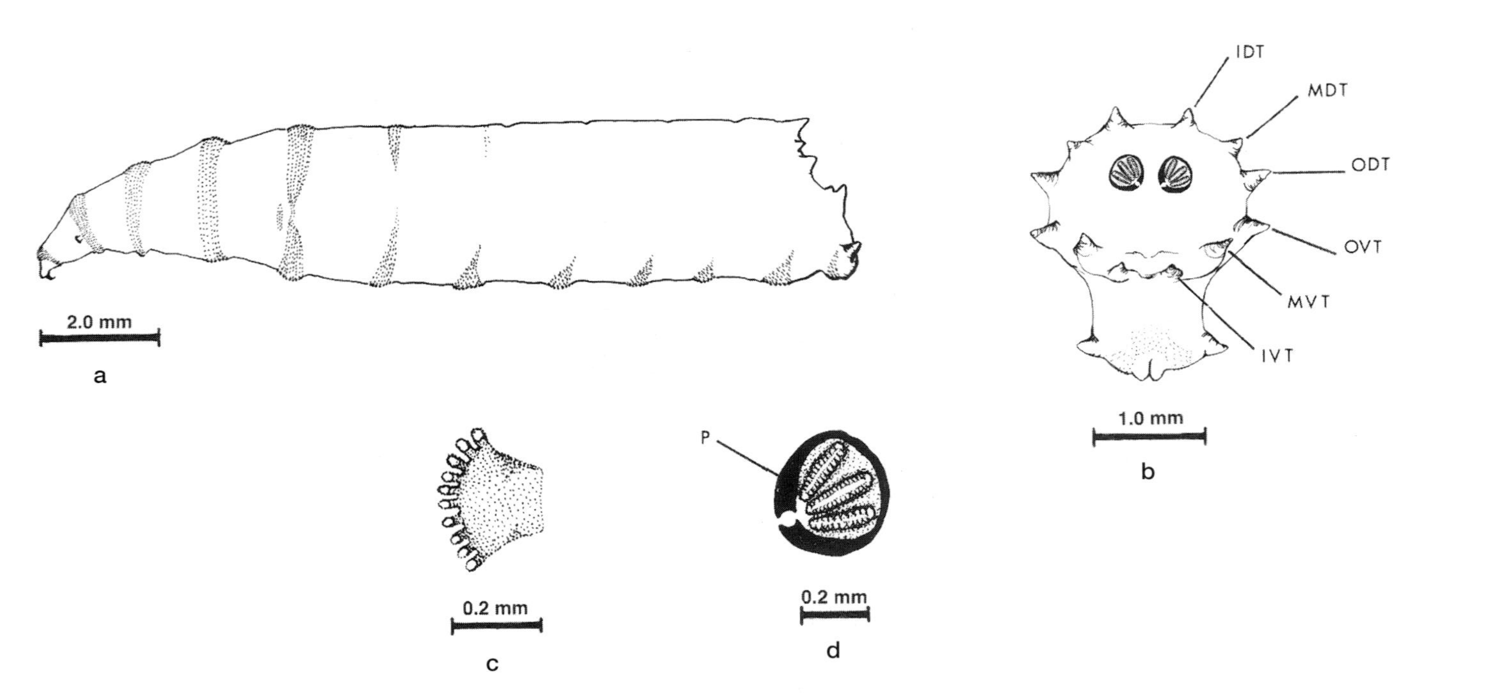
2.0 mm
a
IDT
MDT
ODT
OVT
MVT
IVT
1.0 mm
b
0.2 mm
c
P
0.2 mm
d

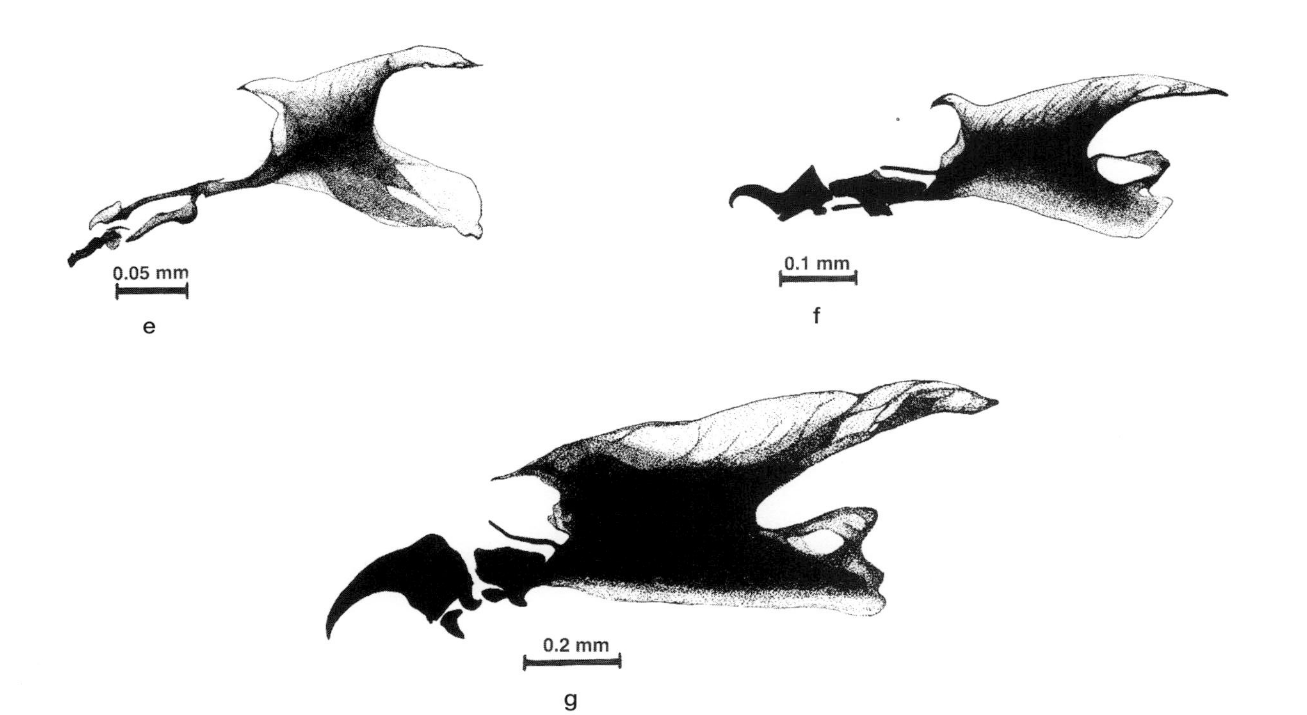

Fig. 3.6b. *Chrysomya chloropyga putoria,* third instar larva. a, Lateral view. b, Posterior view. IDT, Inner dorsal tubercles; IVT, inner ventral tubercles; MDT, median dorsal tubercles; MVT, median ventral tubercles; ODT, outer dorsal tubercles; OVT, outer ventral tubercles. c, Anterior spiracle. d, Right posterior spiracle. P, Peritreme. e, Cephalopharyngeal skeleton, first instar. f, Cephalopharyngeal skeleton, second instar. g, Cephalopharyngeal skeleton, third instar.

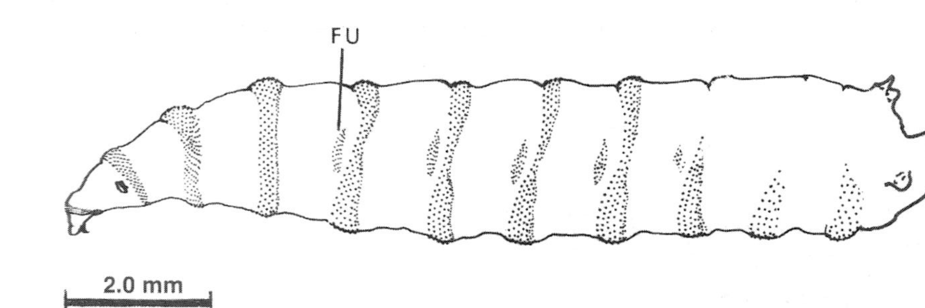
FU
2.0 mm
a

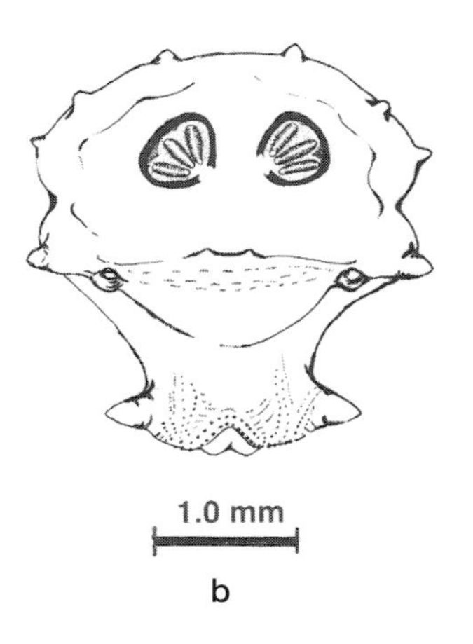
1.0 mm
b

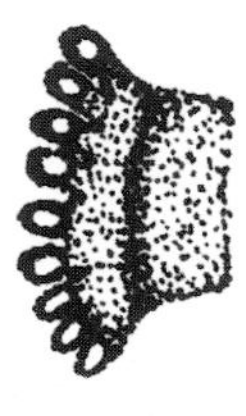
0.2 mm
c

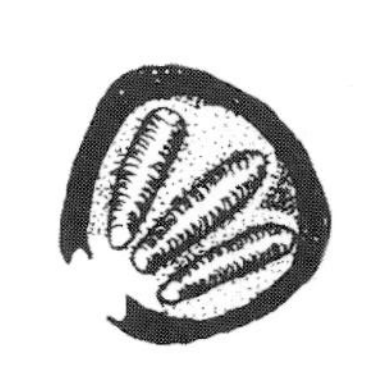
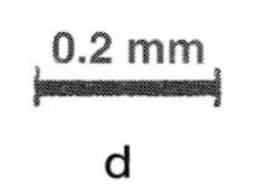
0.2 mm
d

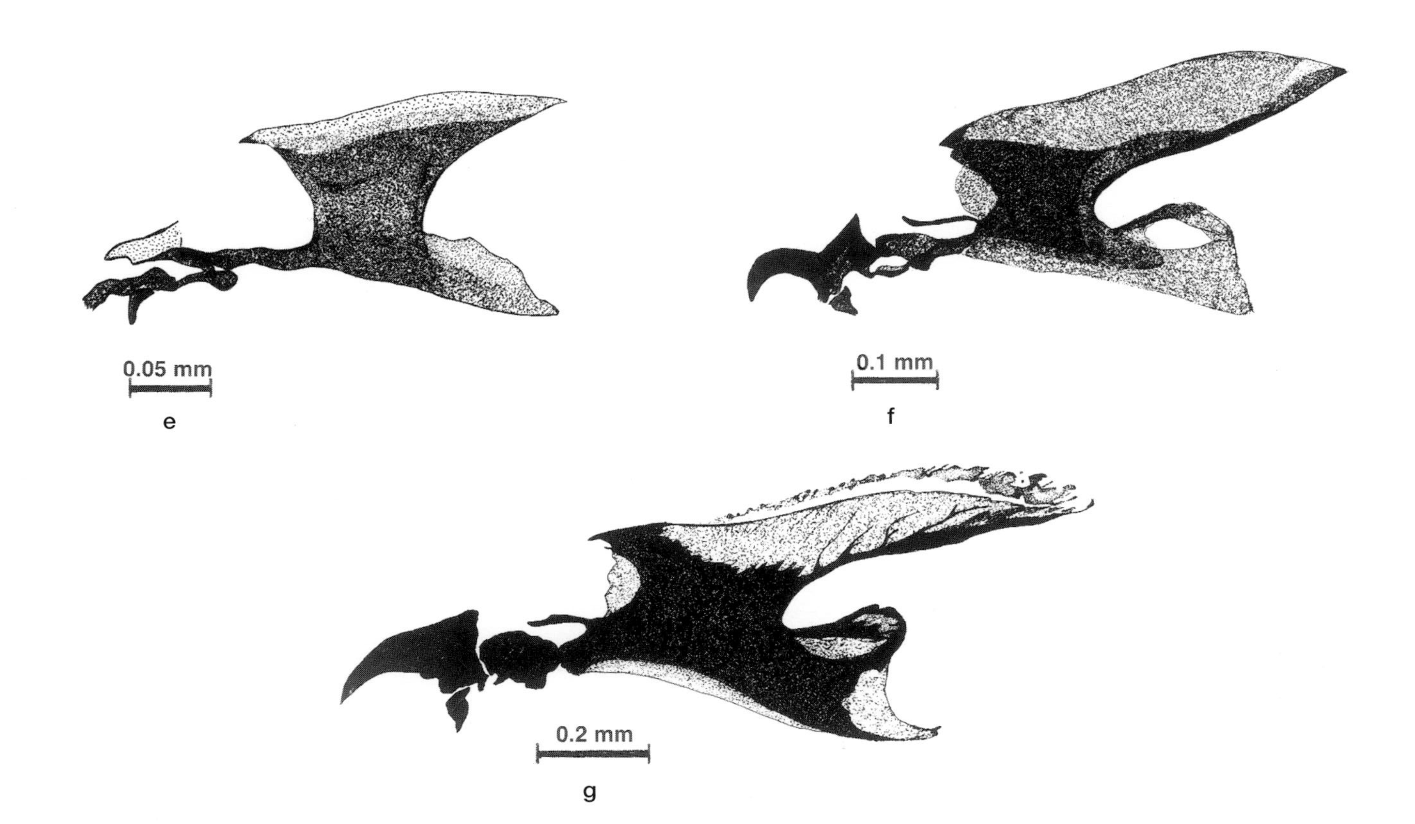

Fig. 3.7. *Cochliomyia macellaria*, third instar larva. a, Lateral view. FU, Fusiform area. b, Posterior view. c, Anterior spiracle. d, Right posterior spiracle. e, Cephalopharyngeal skeleton, first instar. f, Cephalopharyngeal skeleton, second instar. g, Cephalopharyngeal skeleton, third instar.

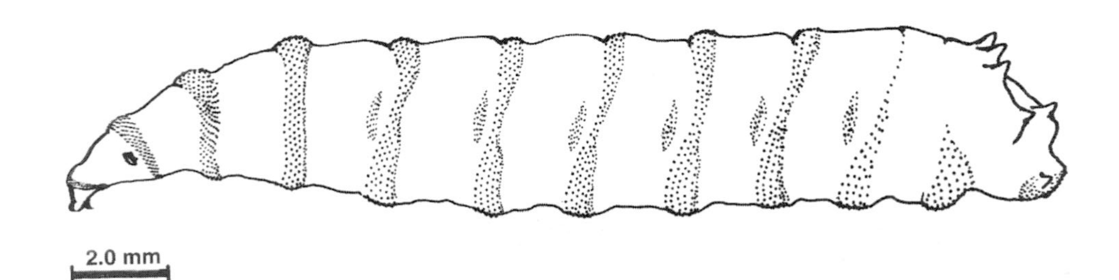
2.0 mm
a

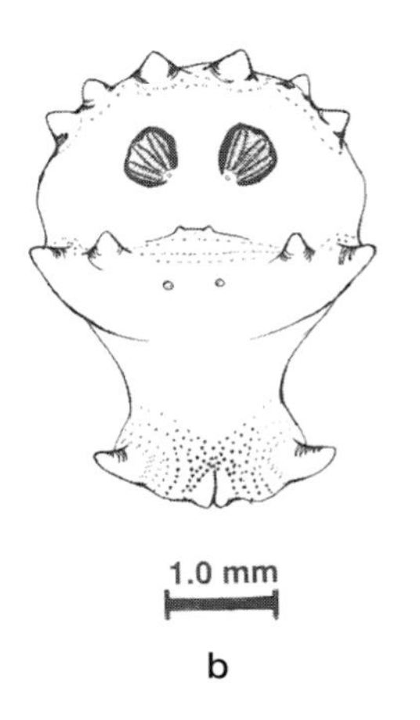
1.0 mm
b

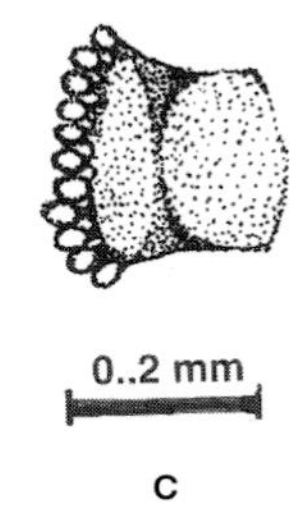
0..2 mm
c

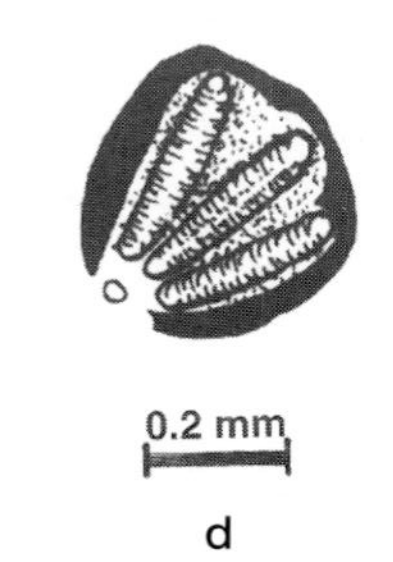
0.2 mm
d

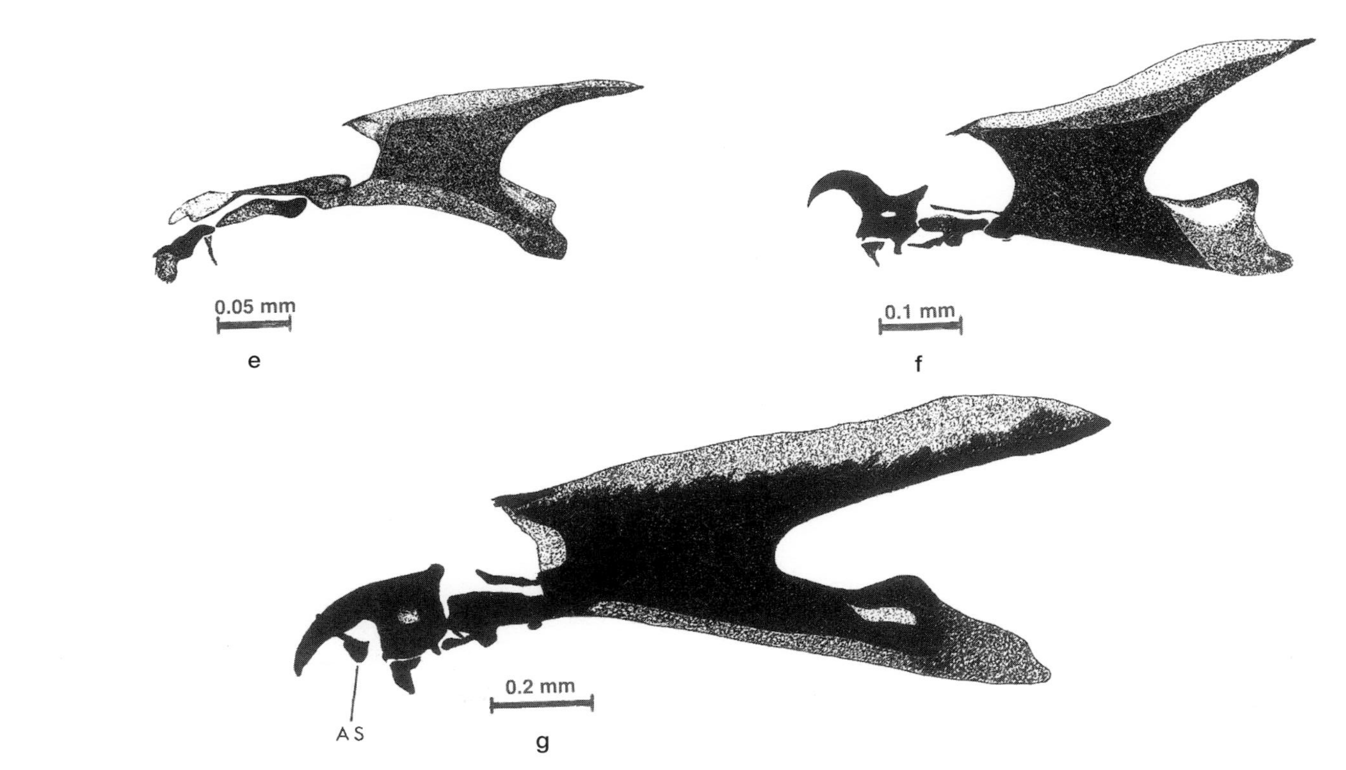

Fig. 3.8. *Compsomyiops boliviana*, third instar larva. a, Lateral view. b, Posterior view. c, Anterior spiracle. d, Right posterior spiracle. e, Cephalopharyngeal skeleton, first instar. f, Cephalopharyngeal skeleton, second instar. g, Cephalopharyngeal skeleton, third instar. AS, Accessory sclerite.

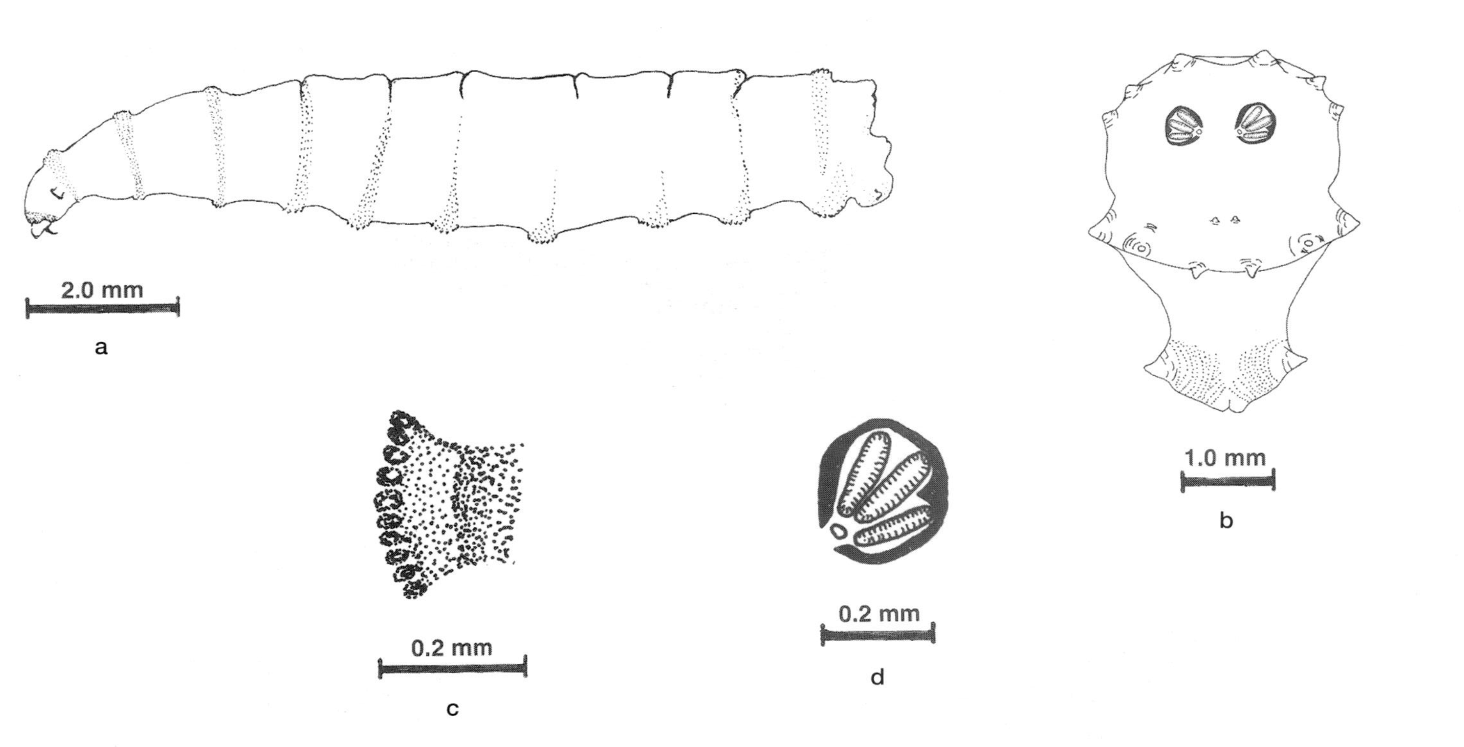
2.0 mm
a
1.0 mm
b
0.2 mm
c
0.2 mm
d

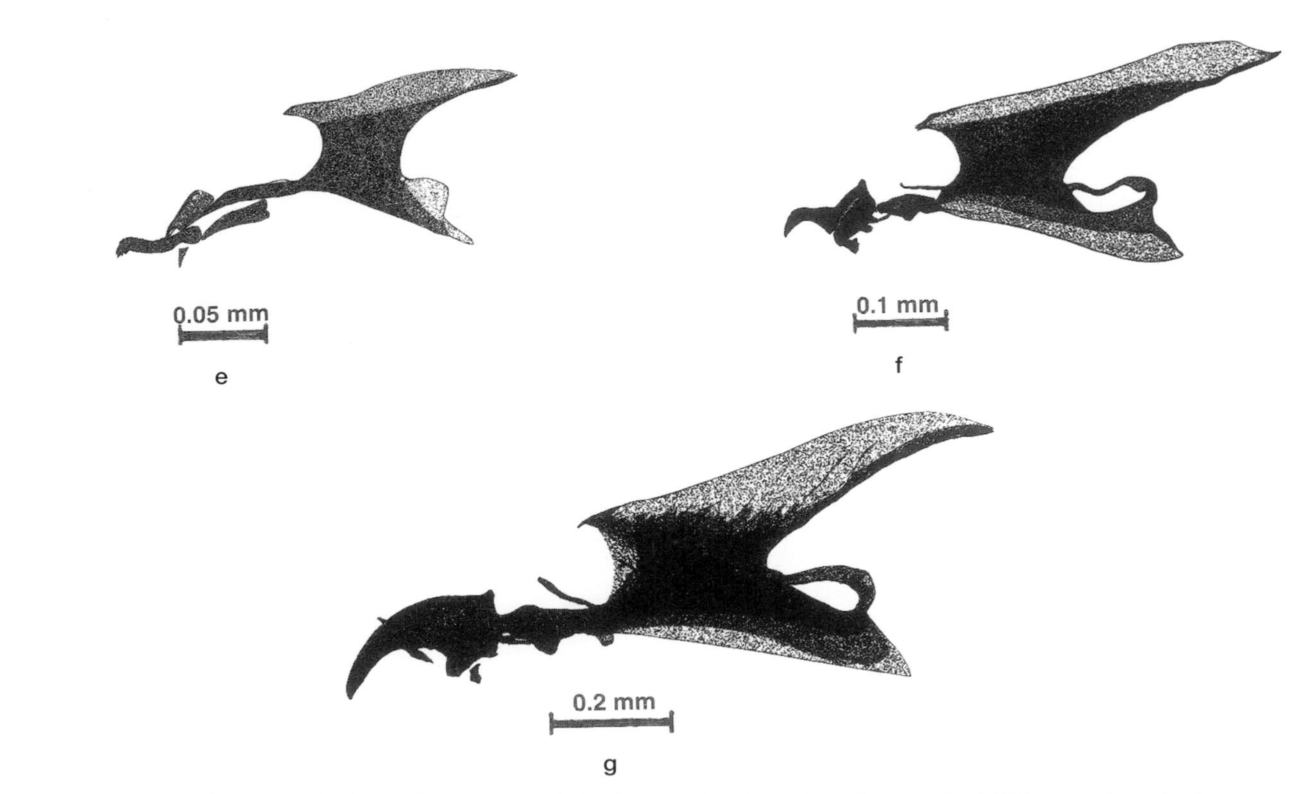

Fig. 3.9. *Hemilucilia flavifacies*, third instar larva. a, Lateral view. b, Posterior view. c, Anterior spiracle. d, Right posterior spiracle. e, Cephalopharyngeal skeleton, first instar. f, Cephalopharyngeal skeleton, second instar. g, Cephalopharyngeal skeleton, third instar.

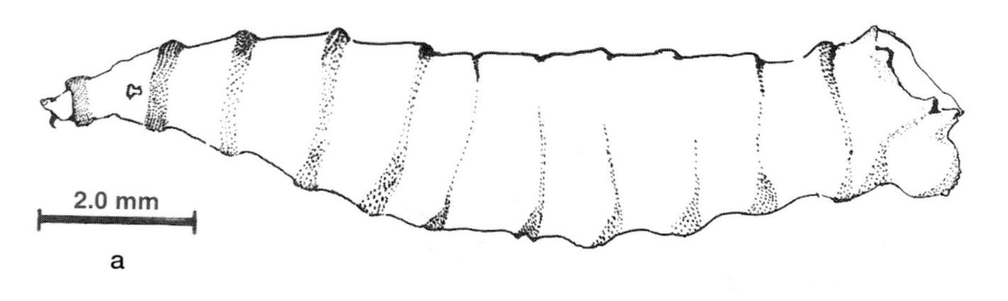
2.0 mm
a

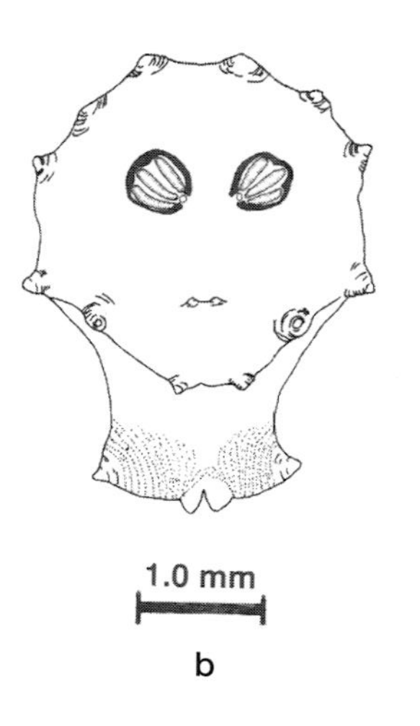
1.0 mm
b

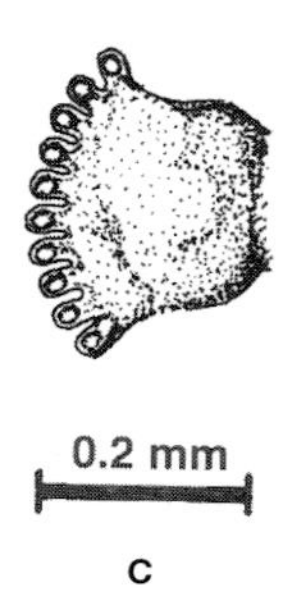
0.2 mm
c

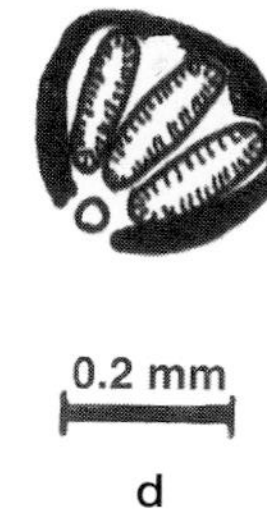
0.2 mm
d

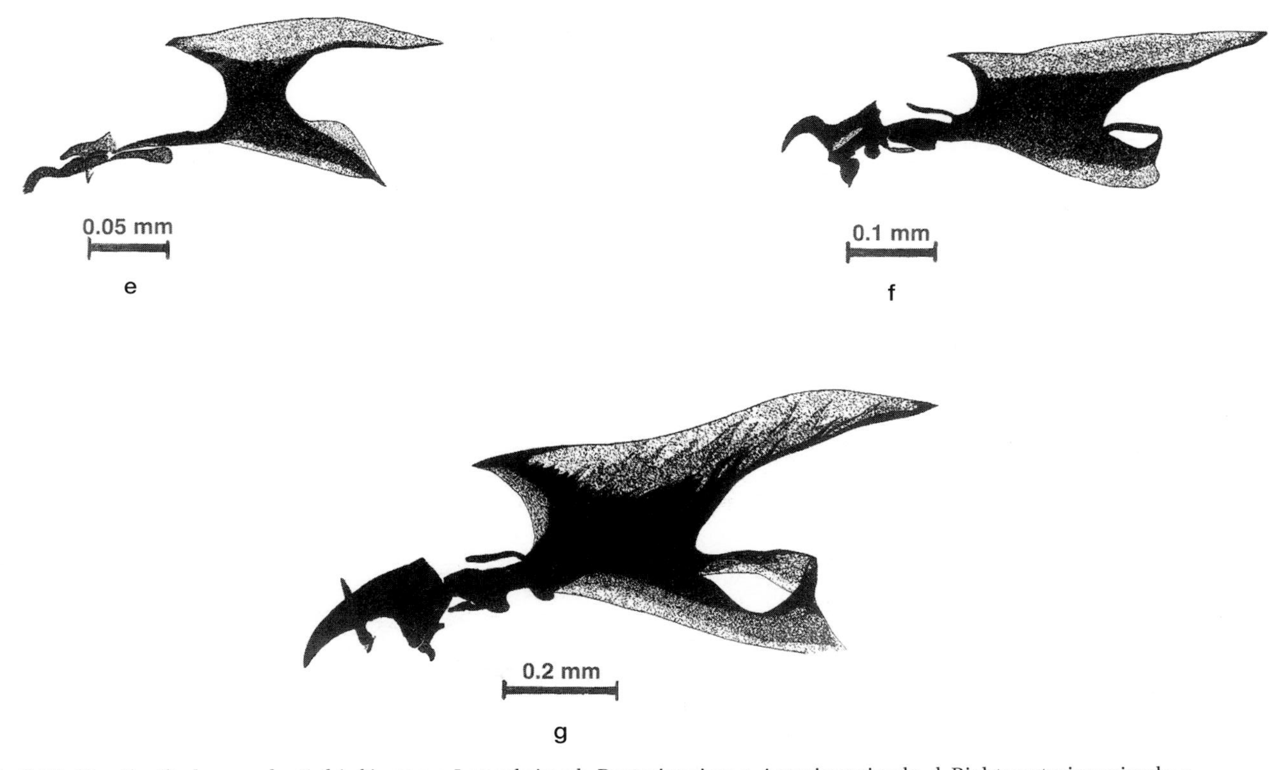

Fig. 3.10. *Hemilucilia hermanlenti*, third instar. a, Lateral view. b, Posterior view. c, Anterior spiracle. d, Right posterior spiracle. e, Cephalopharyngeal skeleton, first instar. f, Cephalopharyngeal skeleton, second instar. g, Cephalopharyngeal skeleton, third instar.

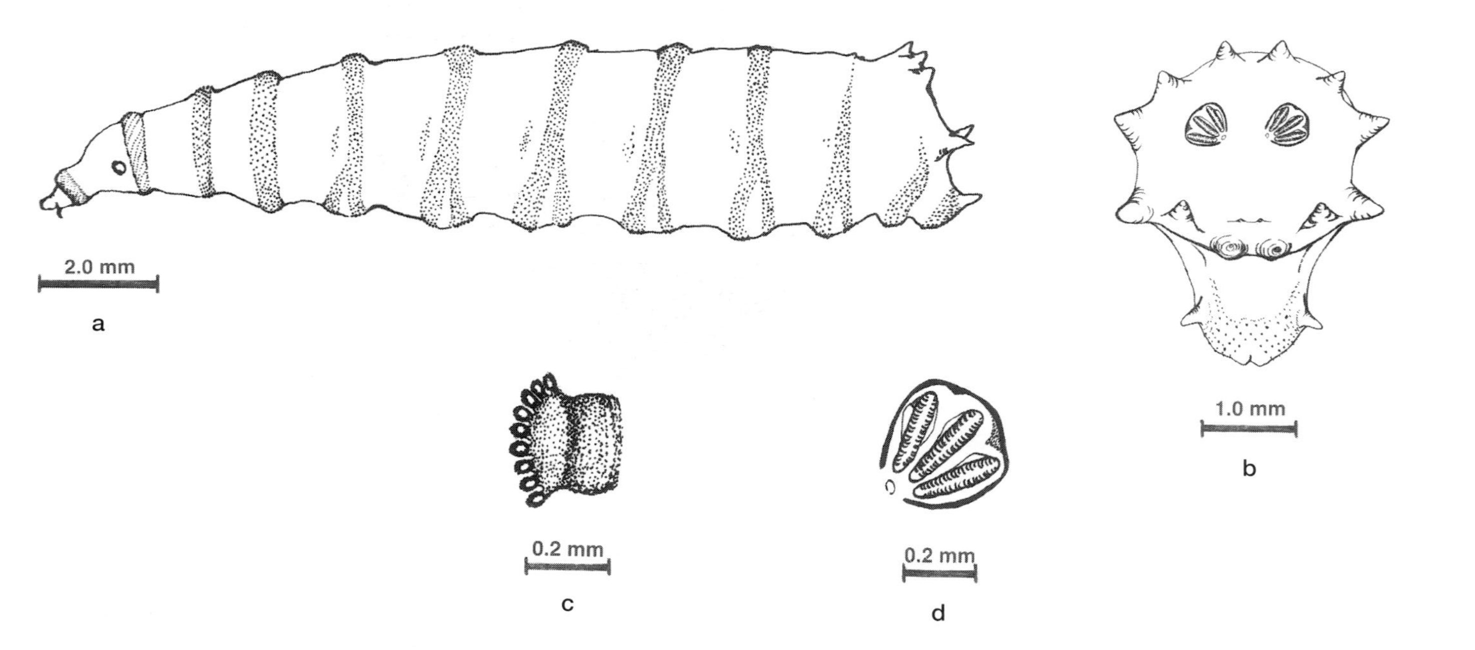
2.0 mm
a
1.0 mm
b
0.2 mm
c
0.2 mm
d

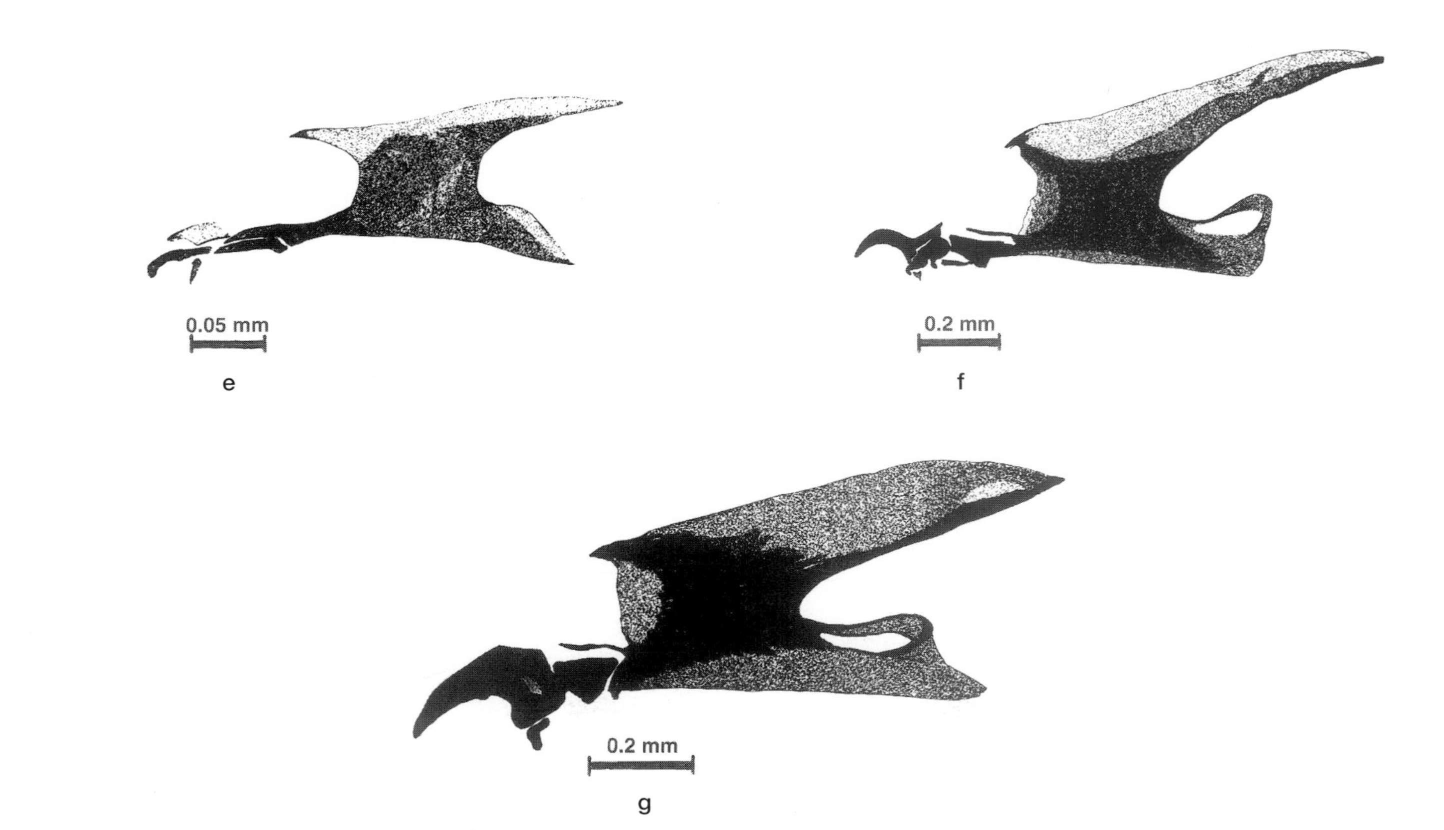

Fig. 3.11. *Paralucilia fulvinota*, third instar larva. a, Lateral view. b, Posterior view. c, Anterior spiracle. d, Right posterior spiracle. e, Cephalopharyngeal skeleton, first instar. f, Cephalopharyngeal skeleton, second instar. g, Cephalopharyngeal skeleton, third instar.

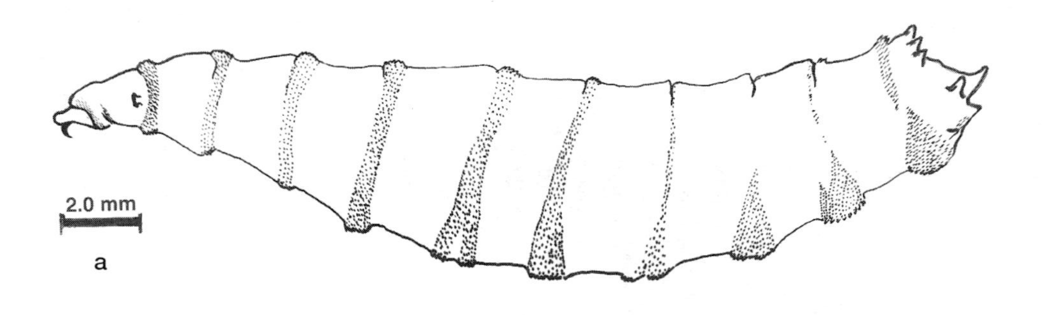
2.0 mm
a

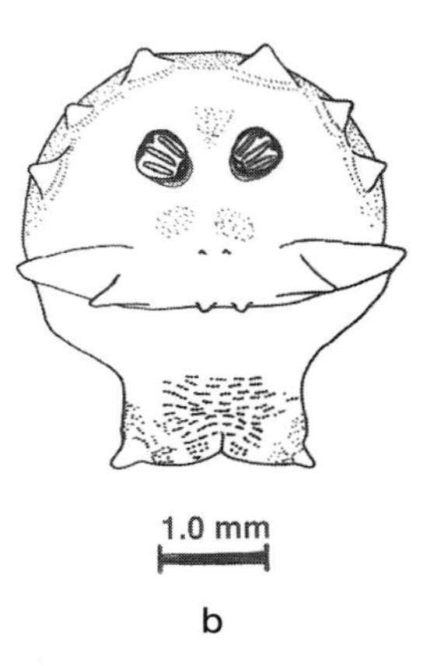
1.0 mm
b

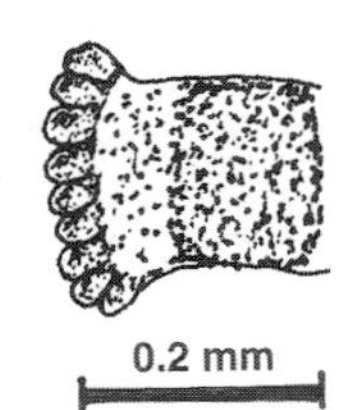
0.2 mm
c

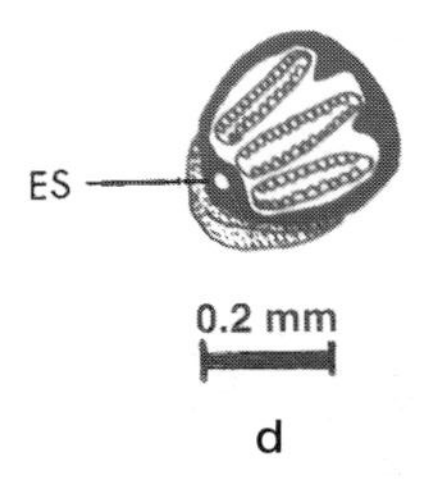
ES
0.2 mm
d

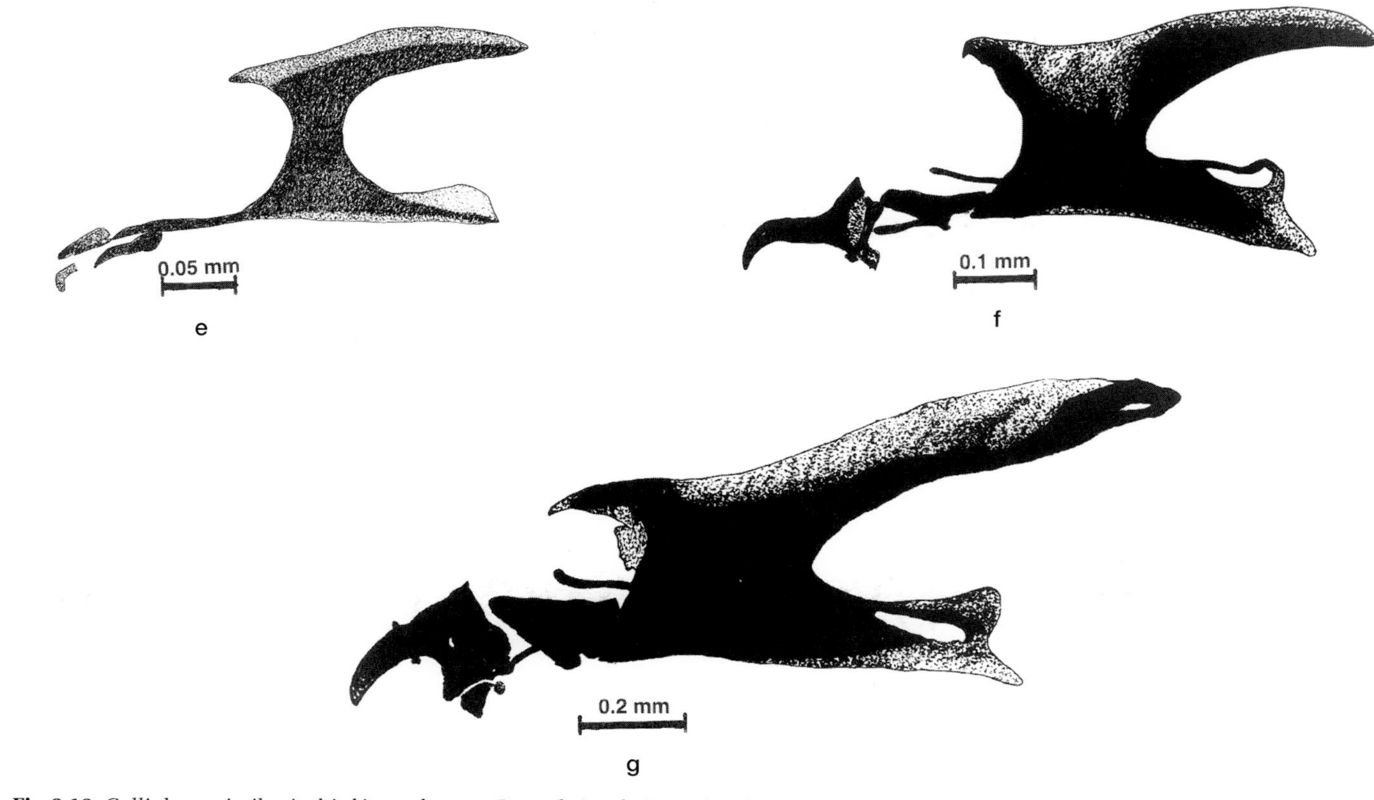

Fig. 3.12. *Calliphora nigribasis,* third instar larva. a, Lateral view. b, Posterior view. c, Anterior spiracle. d, Right posterior spiracle. e, Cephalopharyngeal skeleton, first instar. f, Cephalopharyngeal skeleton, second instar. g, Cephalopharyngeal skeleton, third instar.

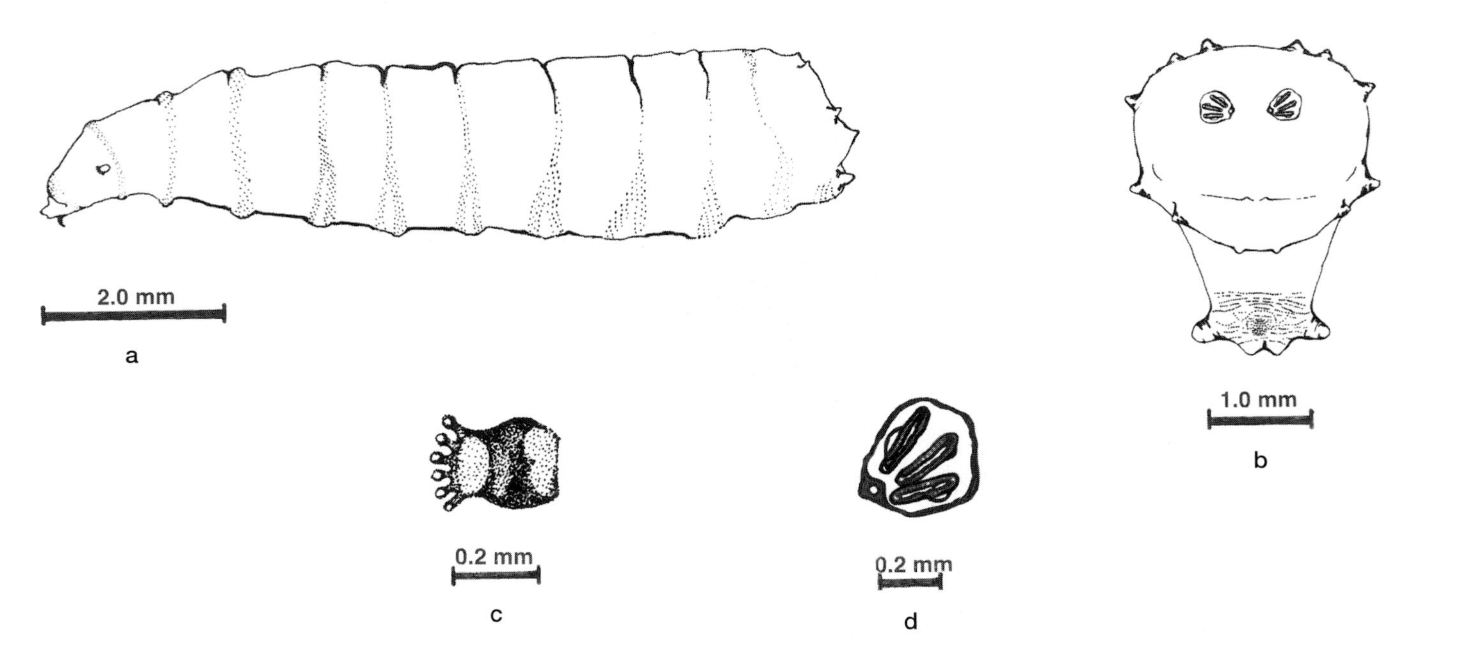
2.0 mm
a
1.0 mm
b
0.2 mm
c
0.2 mm
d

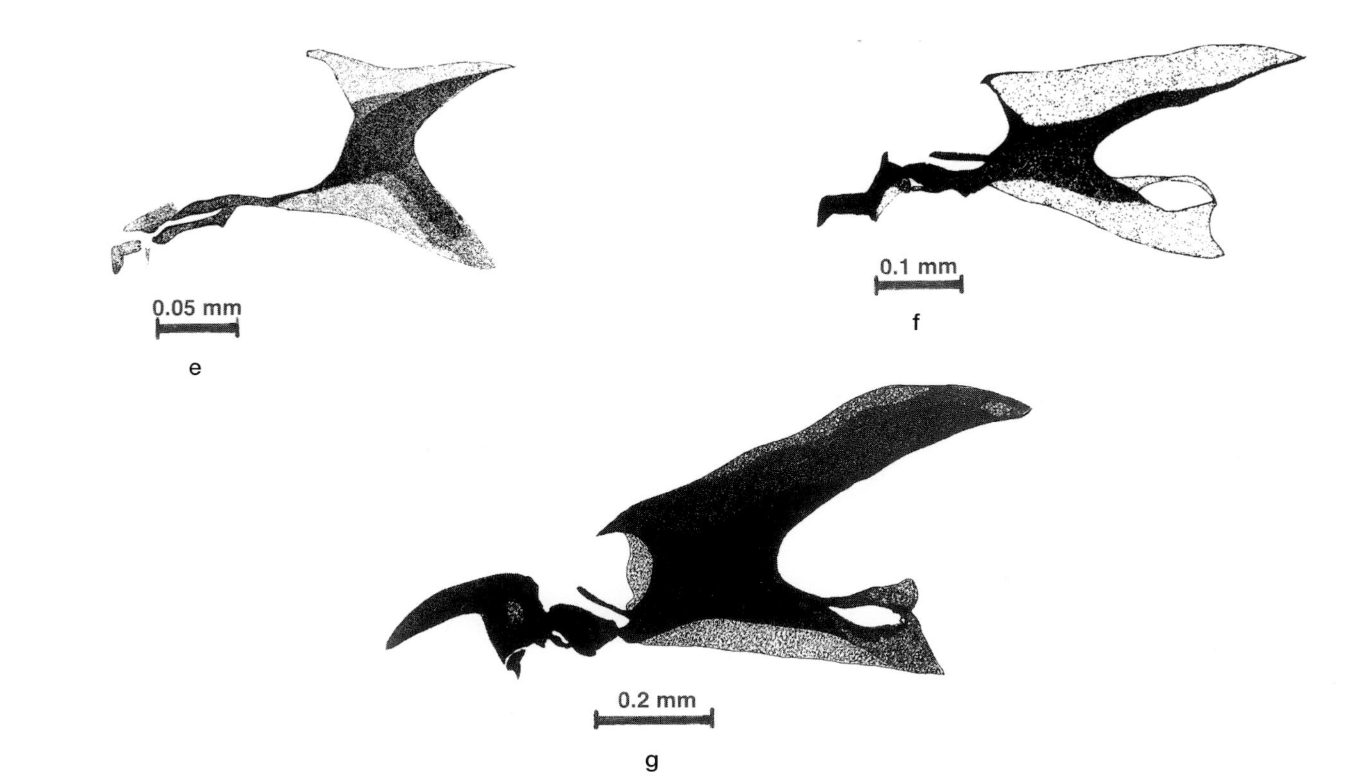

Fig. 3.13. *Phaenicia cuprina*, third instar larva. a, Lateral view. b, Posterior view. c, Anterior spiracle. d, Right posterior spiracle. e, Cephalopharyngeal skeleton, first instar. f, Cephalopharyngeal skeleton, second instar. g, Cephalopharyngeal skeleton, third instar.

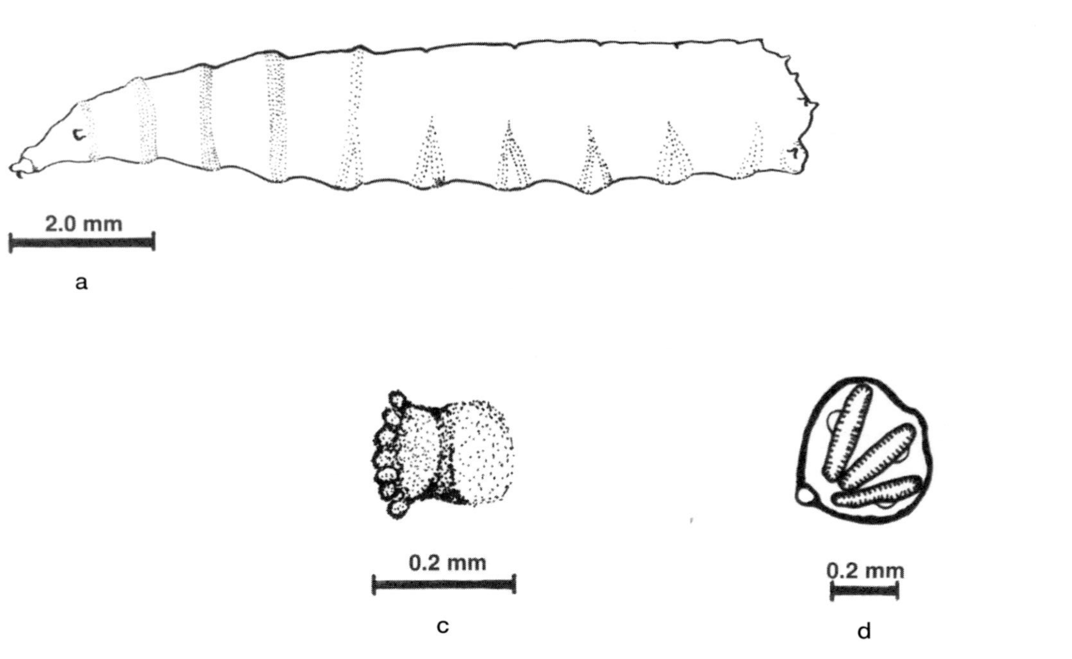
2.0 mm
a
0.2 mm
c
0.2 mm
d

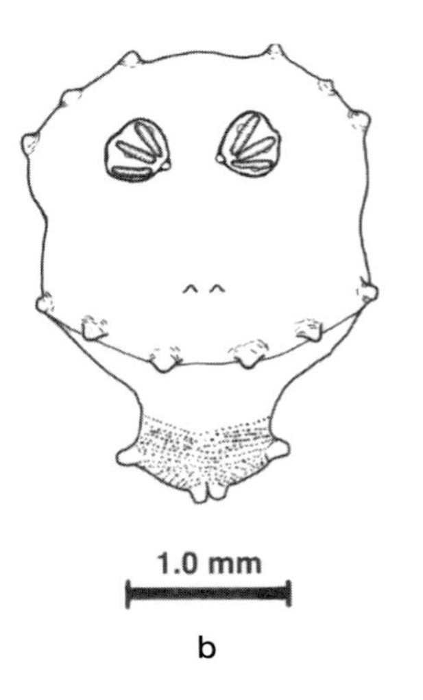
1.0 mm
b

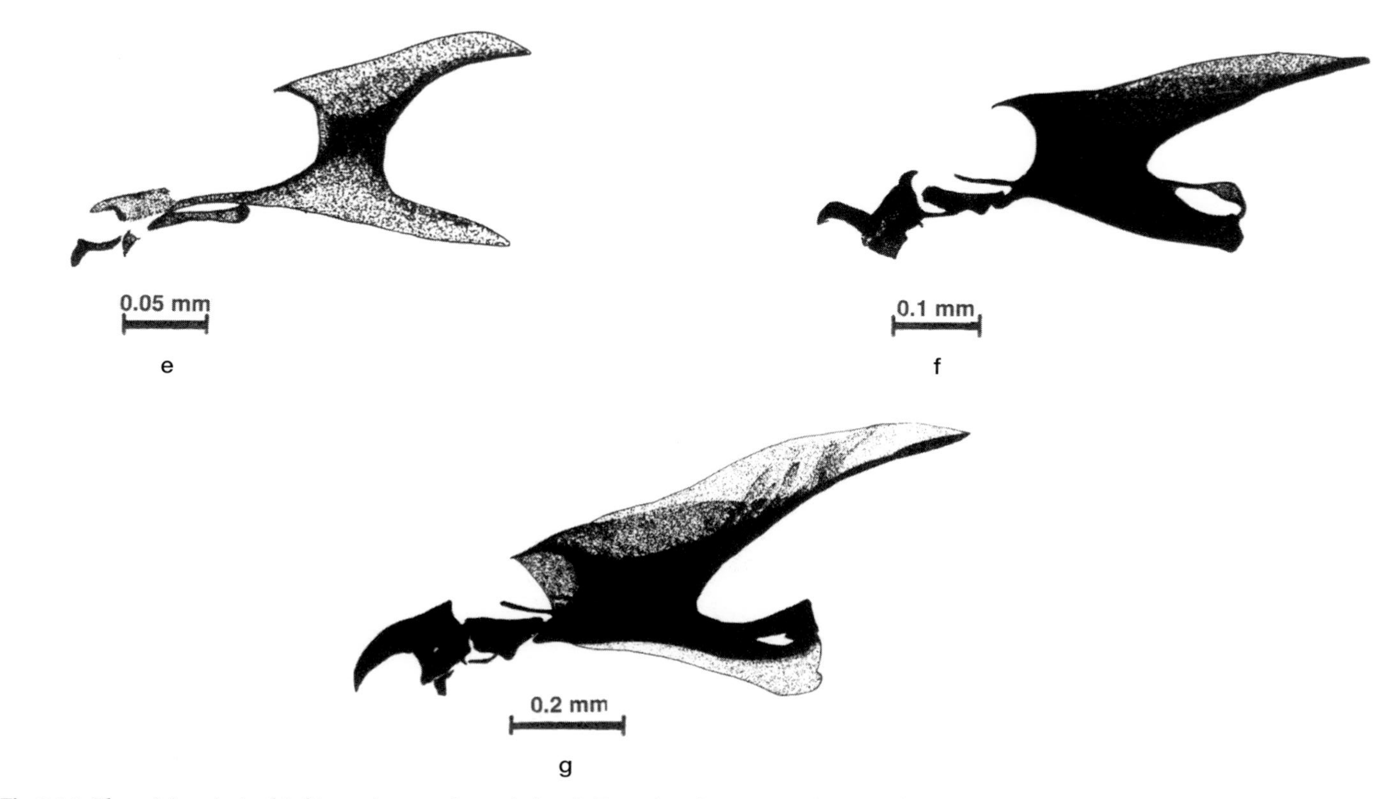

Fig. 3.14. *Phaenicia eximia*, third instar larva. a, Lateral view. b, Posterior view. c, Anterior spiracle. d, Right posterior spiracle. e, Cephalopharyngeal skeleton, first instar. f, Cephalopharyngeal skeleton, second instar. g, Cephalopharyngeal skeleton, third instar.

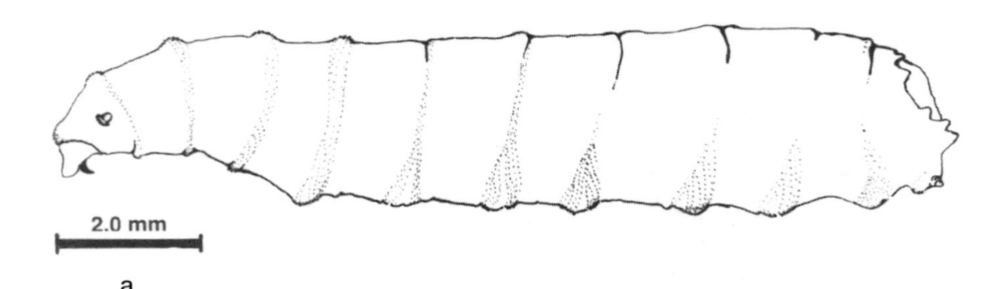
2.0 mm
a

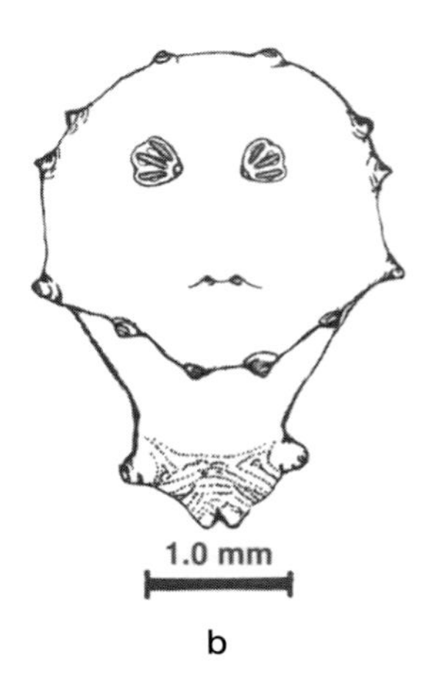
1.0 mm
b

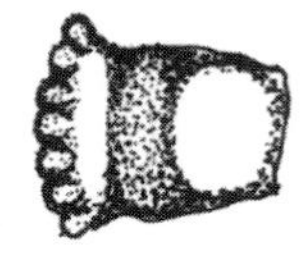
0..2 mm
c

0.2 mm
d

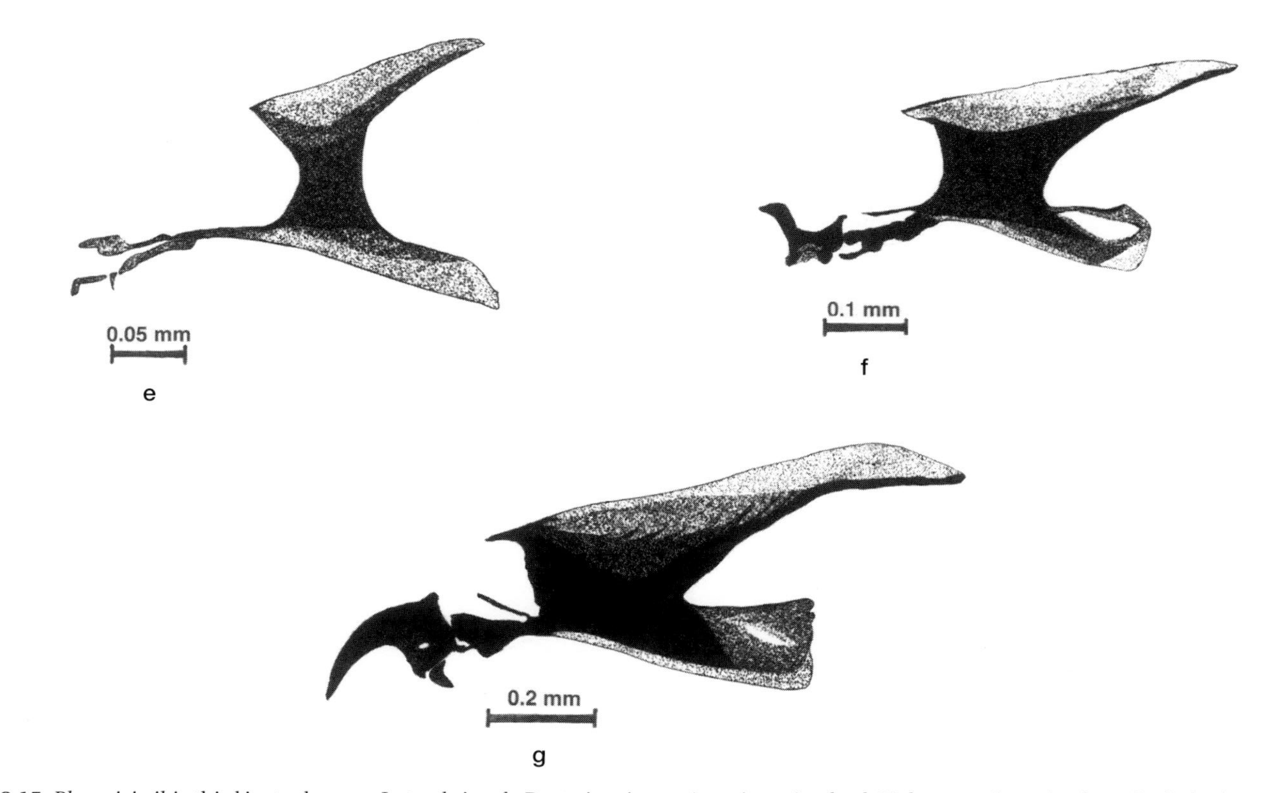

Fig. 3.15. *Phaenicia ibis*, third instar larva. a, Lateral view. b, Posterior view. c, Anterior spiracle. d, Right posterior spiracle. e, Cephalopharyngeal skeleton, first instar. f, Cephalopharyngeal skeleton, second instar. g, Cephalopharyngeal skeleton, third instar.

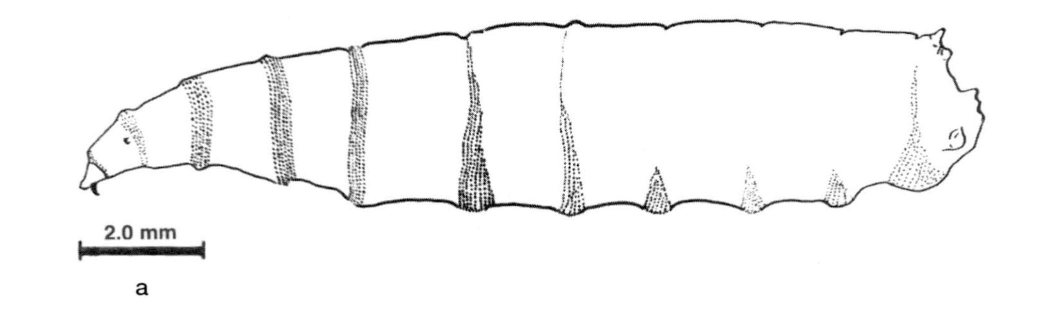

2.0 mm
a

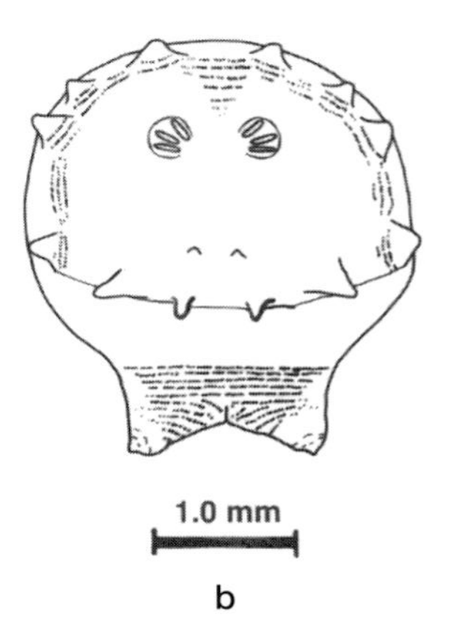

1.0 mm
b

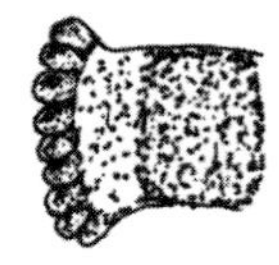

0.2 mm
c

0.2 mm
d

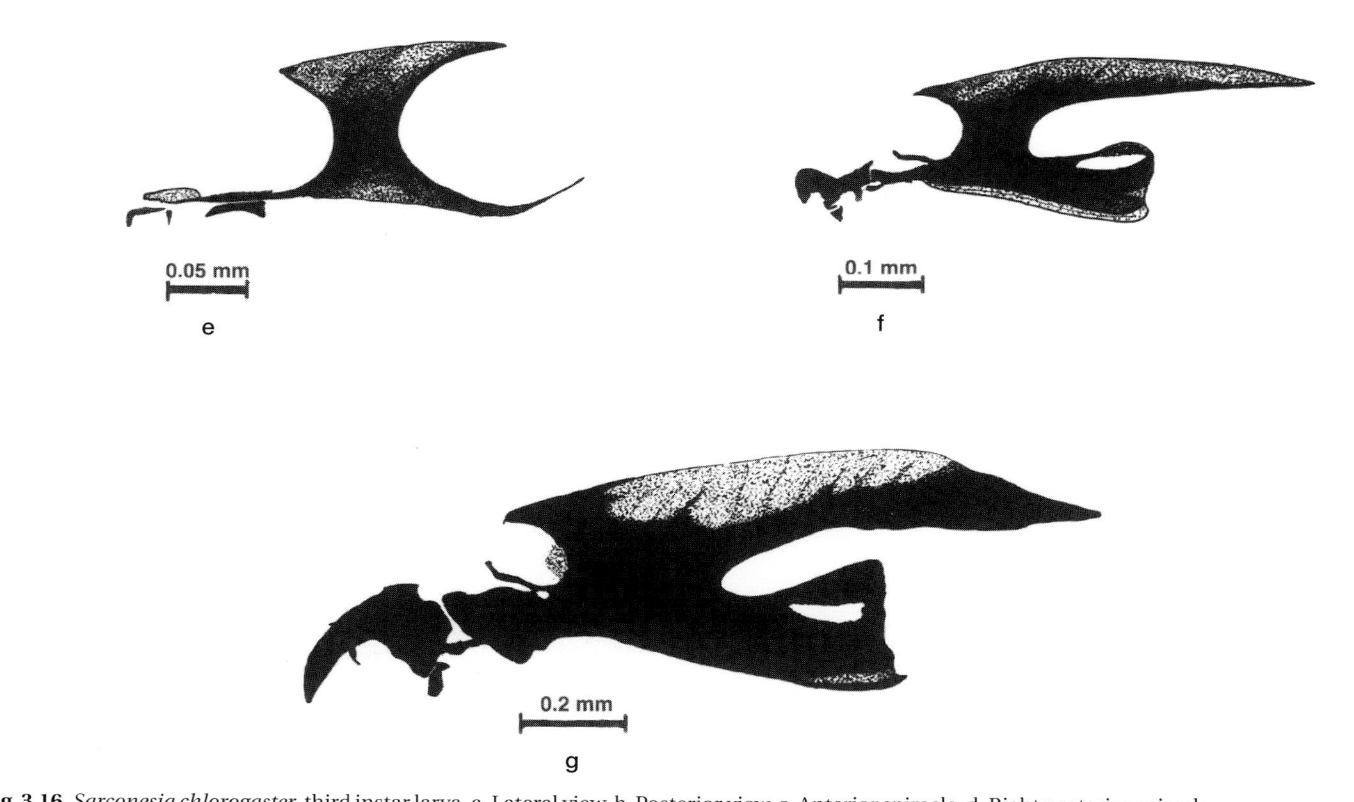

Fig. 3.16. *Sarconesia chlorogaster*, third instar larva. a, Lateral view. b, Posterior view. c, Anterior spiracle. d, Right posterior spiracle. e, Cephalopharyngeal skeleton, first instar. f, Cephalopharyngeal skeleton, second instar. g, Cephalopharyngeal skeleton, third instar.

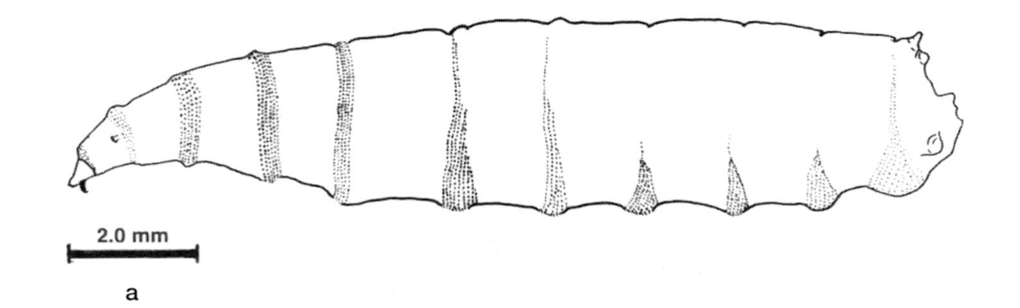
2.0 mm
a

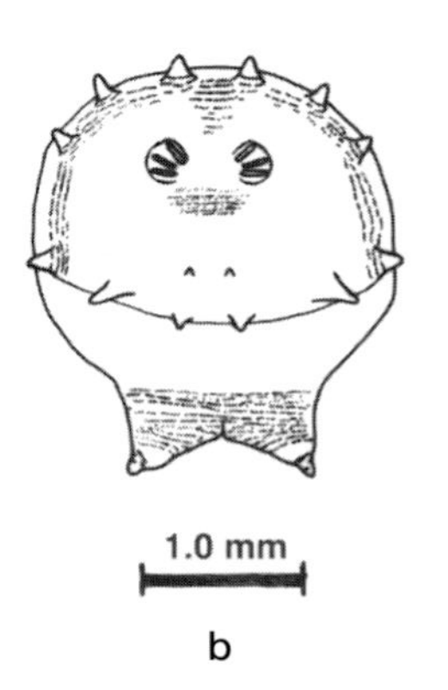
1.0 mm
b

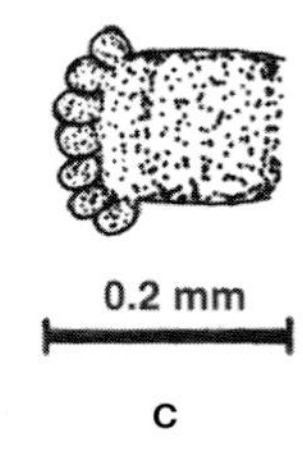
0.2 mm
c

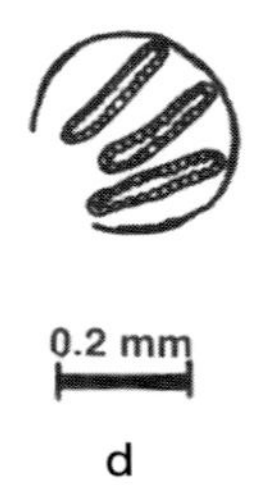
0.2 mm
d

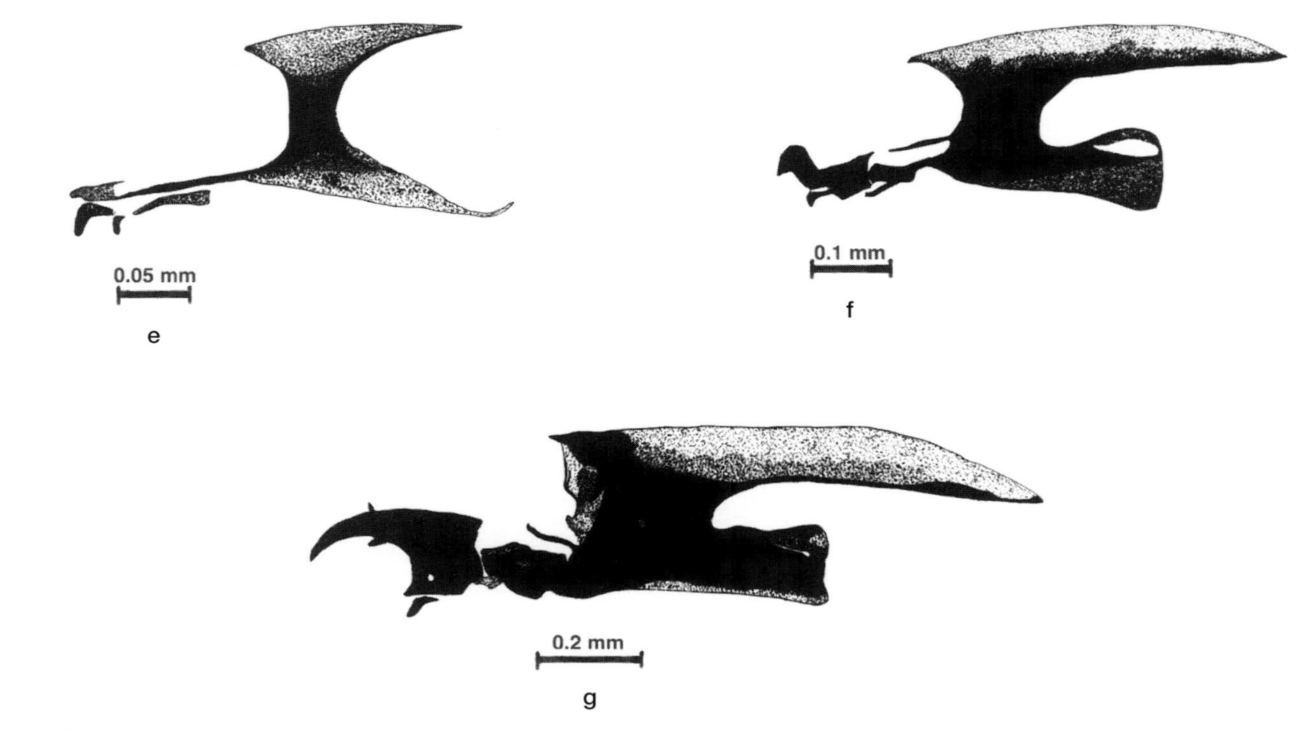

Fig. 3.17. *Sarconesia magellanica*, third instar larva. a, Lateral view. b, Posterior view. c, Anterior spiracle. d, Right posterior spiracle. e, Cephalopharyngeal skeleton, first instar. f, Cephalopharyngeal skeleton, second instar. g, Cephalopharyngeal skeleton, third instar.

Key to Known Third Instar Larvae of Peru
(modified from Greenberg and Szyska, 1984)

1. – Peritreme incomplete (Fig.3.7c) 2
   – Peritreme complete (Fig. 3.14d) 7
2. – Lateral fusiform areas absent. With accessory dental sclerite (Fig.3.8g) 3
   – Lateral fusiform areas present (Fig. 3.7a). With or without accessory dental sclerite 4
3. – All spines single-pointed, small.. *Sarconesia*
   – Spines single-and multi-pointed, especially large and numerous ventrad on segment 12.. *Hemilucilia*
4. – With accessory dental sclerite .... *Compsomyiops*
   – Without accessory dental sclerite (Fig. 3.7g) 5
5. – Spine pattern on anal protuberance convex or bell-shaped (Fig. 3.6b(b)) .... *Chrysomya c. putoria*
   – Spine pattern on anal protuberance U- or V-shaped (Figs. 3.7b; 3.8b) 6
6. – Mature third instar blue-gray .... *Paralucilia fulvinota*
   – Mature third instar nonpigmented.. *Cochliomyia macellaria*
7. – With accessory dental sclerite (Fig. 3.12g).. *Calliphora nigribasis*
   – Without accessory dental sclerite 8
8. Our specimens of *P. cuprina, eximia,* and *ibis* are difficult to separate due to similarity and variability of characters.
   – Distance between inner tubercles on upper stigmal field usually less than distance between inner and outer tubercles (Fig. 3.15b) .... *Phaenicia ibis*
   – Distance between inner tubercles about equal to distance between inner and outer tubercles 9
9. – Median tubercles on upper stigmal field usually closer to inner than to outer tubercles (Fig. 3.13b) .... *Phaenicia cuprina*
   – Median tubercles usually closer to outer tubercles (Fig. 3.14b) ....*Phaenicia eximia*

The key to common adult blowflies of South America includes 27 species in 14 genera. Labeled drawings of the generalized adult head, thorax, wing, and hind leg by Gregor (1971) (Figs. 3.18 to 3.22) facilitate use of the adult keys (see also Guimarães and Papavero, 1999).

Key to Common Adult Blowflies of South America
(Juan Carlos Mariluis,
Administración Nacional de Laboratorios e Institutos de Salud,
"Dr. Carlos G. Malbrán", Buenos Aires, Argentina)

The Calliphoridae of South America was studied by Shannon (1926). The Neotropical Catalogue was made by James (1970). Since then, revision papers of the South American fauna were published by Dear (1979,1985) and Mariluis and Peris (1984).

1. – Stem vein setulose above 2
   – Stem vein entirely bare above 17
2. – Stem vein setulose below 24
   – Stem vein entirely bare below 3

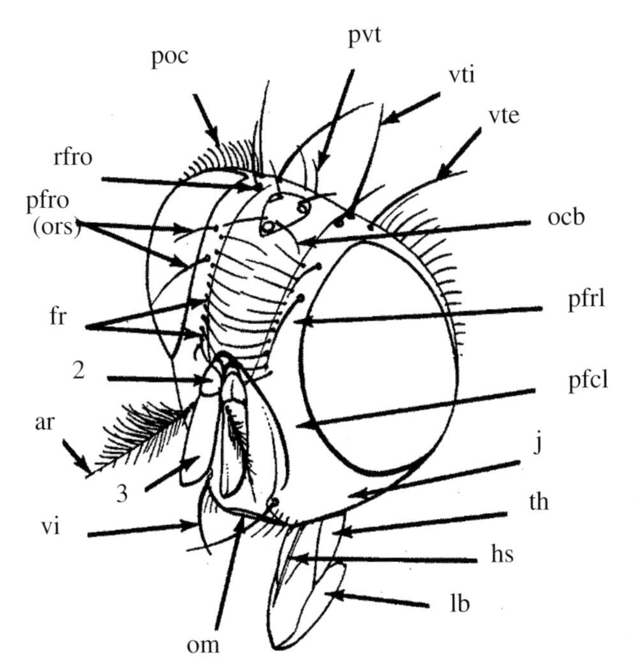

Fig. 3.18. Fronto-lateral view of muscoid head: ar, arista; fr, frontals; hs, haustellum; j, jowls; lb, labella; ocb, ocular (ocellar bristles); om, oral margin; pfcl, parafacial; pfrl, parafrontal; pfro (=ors), proclinate fronto-orbitals; poc, postocular setulae; pvt, postverticals; rfro, reclinate fronto-orbitals; th, theca; vi, vibrissae; vte, external verticals; vti, internal verticals; 2,3, second, third, antennal segments.

3. – Mesonotum flattened on center behind the suture. Lower calyptra bare.......*Protophormia terraenovae*
   – Mesonotum not flattened on center behind the suture. Lower calyptra totally or partially pilose.......4
4. – Palp short, does not reach the oral margin.......5
   – Palp normal, reaches the oral margin.......6
5. – Male and female: parafrontal with pale hairs on the exterior row of frontal bristles; male: yellow to orange basicosta; yellow to orange legs; female: with one or two proclinate fronto-orbital bristles.......*Cochliomyia macellaria*
   – Male and female: parafrontal with black hairs on the exterior row of frontal bristles; male: black basicosta; black legs; female: without proclinate fronto-orbital bristles.......*Cochliomyia hominivorax*
6. – Greater ampulla with long pilosity.......7
   – Greater ampulla with short pilosity.......9
7. – Male and female: white mesothoracic spiracle; white lower calyptra; black antenna; jowls totally or ventrally black; male: eyes with equal-sized facets.......8
   – Male and female: hazel mesothoracic spiracle; hazel lower calyptra; with orange red antenna; with orange red jowls; male: eyes with a definite area of large superior facets and small inferior facets.......*Chrysomya megacephala*

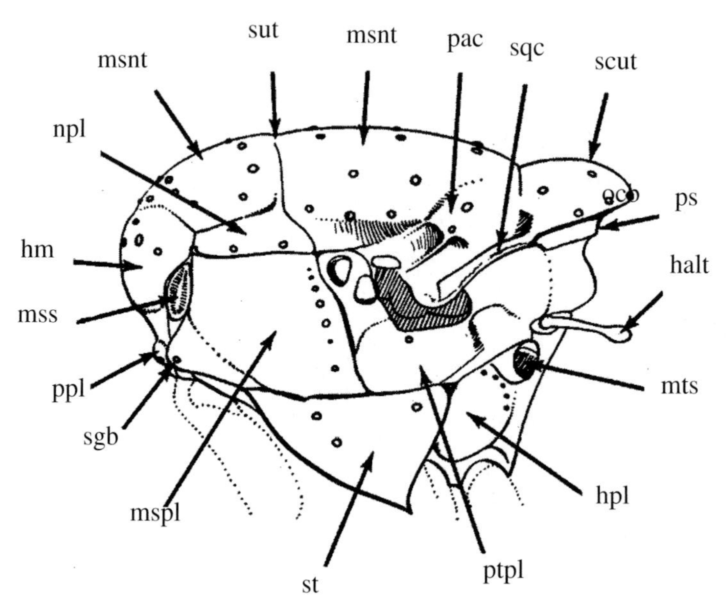

Fig. 3.19. Chaetotaxy of thorax (lateral view): halt, haltere; hm, humerus; hpl, hypopleuron; msnt, mesonotum; mspl, mesopleuron; mss, mesothoracic spiracle (anterior spiracle); mts, metathoracic spiracle (posterior spiracle); npl, notopleuron; pac, postalar callus; ppl, propleuron; ps, postscutellum; ptpl, pteropleuron; scut, scutellum; sgb, stigmatic bristle; sqc, suprasquamal carina; st, sternopleuron (position of sternopleural bristles 2:1); sut, suture.

8. – Male and female: without stigmatic bristle; four to six propleural bristles; yellow or dark orange jowls .... *Chrysomya albiceps*
   – Male and female: with one robust stigmatic bristle; one propleural bristle and sometimes with an additional small bristle; blackish jowls .... *Chrysomya chloropyga*

9. – Parafacial bare 11
   – Parafacial setulose 10

10. – Male and female: jowls with white hairs; white lower calyptra; femurs predominantly dark brown; femur I with a yellowish line on ventral-posterior area; femurs II and III with middle yellow ring; male: eyes subholoptic, the anterior facets usually larger .... *Compsomyiops fulvicrura*
   – Male and female: jowls with golden hairs; smoky lower calyptra; femurs with large middle yellow ring; male: eyes holoptic, the anterior facets greatly enlarged .... *Compsomyiops verena*

11. – Lower calyptra setulose; clear wings 12
   – Lower calyptra bare; smoky wings 13

12. – Male and female: lower area parafrontal with very large black setae, duplicating the upper setae; mesonotum with middle brilliant fringe; jowls with golden to yellow hairs .... *Paralucilia pseudolyrcea*
   – Male and female: lower area parafrontal with short white setae;

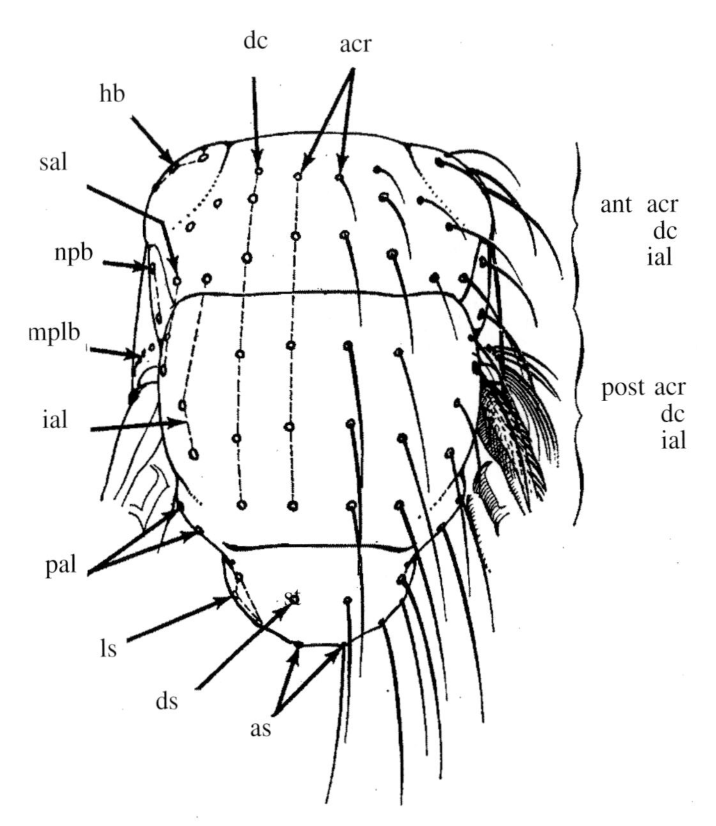

Fig. 3.20. Chaetotaxy of thorax (*Phaenicia sericata*, dorsal view): acr, acrostical bristles; as, apicoscutellar bristles; ds, discoscutellar bristles; dc, dorsocentral bristles; hb, humeral bristles; ial, intraalar bristles; ls, lateroscutellar bristles; mplb, mesopleural bristles; npb, notopleural bristles; pal, postalar bristles; sal, supraalar bristles.

mesonotum without middle brilliant fringe; jowls with yellow whitish hairs .... *Paralucilia fulvinota*

13. – Male and female: wings with smoky spots on costa; basal area of abdomen yellow; prosternum with yellow to golden hairs; male: facets almost equal 14
    – Male and female: wings uniformly smoky, more on costa, without spots; basal area of abdomen brown; prosternum with brown to golden hairs; male: facets unequal,the upper anterior duplicates the lower .... *Chloroprocta idioidea*

14. – Male and female: parafacial in side view wide, prominent at the ocular border, almost equal to the width of the third antennal segment 15
    – Male and female: parafacial, in side view narrow, not prominent at the ocular border, less than the width of the third antennal segment 16

15. – Humerus green metallic .... *Hemilucilia souzalopesi*
    – Humerus yellow .... *Hemilucilia benoisti*

16. – Male and female: dark brown metathoracic spiracle; green or blue metallic

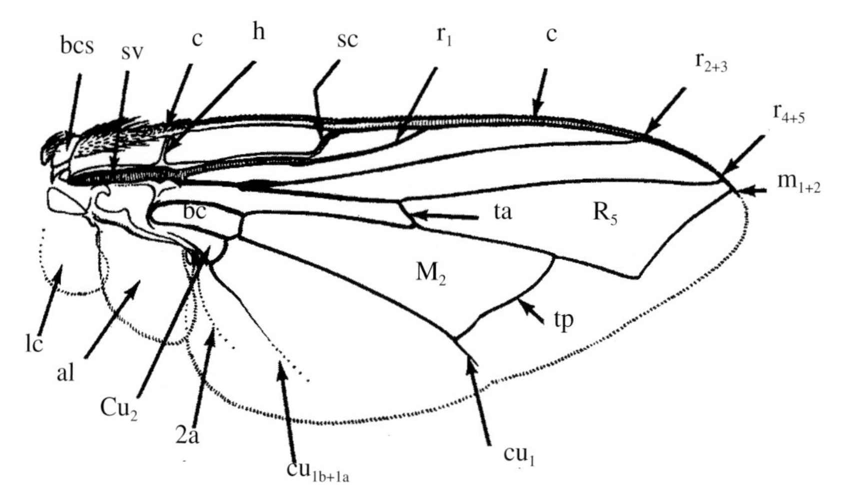

Fig. 3.21. Wing of a muscoid fly: al, allula; bc, basal cell; bcs, basicosta (basicostal scale); c, costa (costal vein); cu_1, cubital vein; Cu_2, cubital vein (hind basal cell or anal cell); Cu_{1b+1a}, anal vein; h, humeral cross vein; lc, lower calyptera (lower squama); m_{1+2}, medial vein; M_2, medial or discal cell; sv, stem vein; R_5, radial or apical cell; sc, subcosta; ta, (r–m) = anterior cross vein; tp, posterior cross vein; 2a, anal vein.

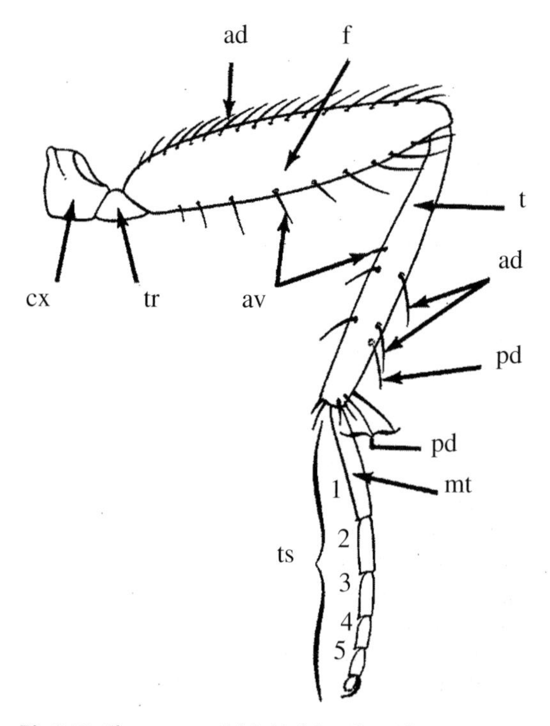

Fig 3.22. Chaetotaxy of third left leg, frontal view (*Muscina*): ad, anterodorsals; av, anteroventrals; cx, coxa; f, femur; mt, metatarsals; pd, posterodorsals; pr, preapicals; t, tibia; tr, trochanter; ts, tarsal segments.

humerus; green or blue metallic propleuron; green metallic mesopleuron; black tegula .... *Hemilucilia semidiaphana*
- Male and female: white yellowish metathoracic spiracle; yellow humerus; yellow propleuron; yellow mesopleuron; yellow tegula .... *Hemilucilia segmentaria*

17. - Parafacial setulose; without tuft of parasquamal setae; lower calyptra pilose 18
- Parafacial bare; with tuft of parasquamal setae; lower calyptra bare 19

18. - Male and female: reddish jowls, at least on half of its anterior; yellow or yellow orange basicosta .... *Calliphora vicina*
- Male and female: black jowls; black basicosta .... *Calliphora nigribasis*

19. - Male and female without anterior acrostical bristles .... *Blepharicnema splendens*
- Male and female with anterior acrostical bristles 20

20. - With three postsutural acrostical bristles 21
- With two postsutural acrostical bristles 22

21. - Male and female: metasternum setulose; central occipital area with five setae on each side, sometimes reduced to two or rarely one; humerus with several very small setae behind humeral bristles; notopleuron with several setae on posterior border; male: abdomen no arch; sternite V shorter than sternite IV, short setulose .... *Phaenicia sericata*
- Male and female: metasternum bare; central occipital area with 1 seta on each side; humerus with two or three very small setae behind humeral bristles; notopleuron with two or three setae on posterior border; male: abdomen rather arch; sternite V as long or longer than the length of sternite IV, dense setulose .... *Phaenicia cuprina*

22. - Male and female: color dark blue, similar to *Calliphora*; wing basal area dark to subcosta and vicinity; lower propleuron fine golden setulose, sometimes some black brown setulose, always thinner than the humeral bristles .... *Phaenicia peruviana*
- Male and female: color variable, green more or less blue with gold or purple reflection, generally very brilliant, not like *Calliphora*; wing basal area not dark or very little; lower propleuron black setulose or more or less with clear or tawny tonality 23

23. - Male and female: anterior area of mesonotum between humeri without whitish pollinosity; white pilosity of the occiput invading some of the jowls .... *Phaenicia eximia*
- Male and female: anterior area of mesonotum between humeri with whitish pollinosity; white pilosity of the occiput not invading the jowls .... *Phaenicia cluvia*

24. - Male and female: large robust flies (12–16 mm); brown allula; male: metatarsus II with enlarged apex; with fused forceps; paralobi reduced; female: tergite V in side view concave, with strong discal bristles .... *Neta chilensis*
- Male and female: small flies (7–11 mm); white or yellowish allula; male: metatarsus II without enlarged apex; forceps not fused; paralobi not reduced; female: tergite V in side view not concave without strong discal bristles 25

25. - Male and female: without postsutural acrostical bristles; without anterior intraalar bristles near the suture .... *Sarconesia chlorogaster*

Table 3.4. *Blowfly Rearings in Peruvian Andes, December 1979*

	Median duration of immatures (h)						
	Egg batch	Larva I	Larva II	Larva III	Larval duration	Pupa (days)	Egg to adult (days)
Chrysomyinae							
Compsomyops boliviana	57	23	49	144*	216	9	20.4
	50	23	96	214*	333	10	26.0
Toxotarsinae							
Sarconesia chlorogaster	36	24	60	203	287	12*	25.5
	27	49	49	191	289	11*	24.2
	24	24	–	189*	266	11	23.1
Sarconesiopsis	26	24	38	265*	315	13	27.2
magellanica	25	35	38	208*	281	12	24.8
Sarconesia splendida	35	20	–	186*	291	11	24.6
	–	20	22	182*	224	9	–
Sarconesia versicolor	37	38	22	179*	239	11	22.5

Notes:
Data are replicate rearings in most cases. Data preceding the number marked with an asterisk are from rearings near Jauja, in Montaro Valley, December 1979; average maximum and minimum daily temperatures: 20.3°C (SD ± 1.5) and 13.7°C (SD ± 1.5): developmental data subsequent to that stage were obtained in Lima: average maximum and minimum daily temperatures: 26.7°C (SD ± 1.7) and 24.8°C (SD ± 1.6).
Source: From Greenberg and Szyska, 1984.

– Male and female: with postsutural acrostical bristles; with anterior intraalar bristles near the suture....................26

26. – Male and female: eyes with scarce pilosity, at 60× very light cilia, more evident in anterior-lower area; arista plumose along half or more of its length, the longer hairs exceeding the width of third antennal segment; blue brilliant abdomen, without changing spots under the incidence of light; brown lower calyptra; postsutural acrostical bristles located at same level as the postsutural dorsocentral bristles....................*Sarconesiopsis magellanica*

– Male and female: eyes not pilose; arista plumose less than half its length, the longer hairs not exceeding the width of third antennal segment; golden abdomen with dense dusting, with changing spots under the incidence of light; white or yellowish lower calyptra; postsutural acrostical bristles located at different level than the postsutural dorsocentral bristles ....*Chlorobrachycoma versicolor*

Table 3.4 summarizes the duration of the pre-adult stages of six blowflies in the Peruvian Andes. Table 3.5 does the same for 12 species found at two lower elevations in the rain forests of Peru. Figure 3.23 completes the Neotropical

Table 3.5. *Blowfly rearings in Peruvian upper rain forest, June 1980*

	Median duration of immatures (h)						
	Egg batch	Larva I	Larva II	Larva III	Larval duration	Pupa (days)	Egg to adult (days)
Chrysomyinae							
Chrysomya chloropyga	14.5	18	24	76	118	4	9.5
putoria	16.5	14.5*	23	96	137	4	10.4
Cochliomyia macellaria	14	11	24	71	106	5	10.0
	22.5	18	24	71	113	6	11.6
	21	12.5	24	91	127.5	5	11.2
Compsomyops verena	16	27.5	24	190	241.5	9	19.7
	16	25	15	136	176	8	16.0
	23.5	18	22.5	133.5	174	7	15.2
Hemilucilia flavifacies	22	14	23	72	109	4.5	10.0
Hemilucilia hermanlenti	25	16.5	24	78	118.5	5	11.0
	14	23	24	117	164	6	13.4
	–	15*	12	92	119	5	–
	13	22	24	122	168	6.5	14.0
	13.5	18	23	72	113	6	11.3
Paralucilia fulvinota	28	18	20.5	99	137.5	5	12.0
	22	26	12	183	221	5	15.1
Calliphorinae							
Calliphora nigribasis	13	37	23	323	383	12	28.5
	17	26	29	246	301	12	25.3
Phaenicia cuprina	12	17	20*	96	133	8	14.0
	20	23	24	120	167	7	14.8
	13	11.5	14*	96	121.5	10	15.6
	15.5	23	24	96	143	8.5	15.1
Phaenicia eximia	14	13	34	144	191	15	23.5
	12	13	13	144	170	15	22.6
	12	13	12	144	169	15	22.5
Phaenicia ibis	28	23	12	279	314	12.5	26.8
	22.5	12	12.5	279	303	15	28.6
Toxotarsinae							
Sarconesia splendida	25.5	13	21	231	265	9	21.1
Sarconesia versicolor	14	23	11	170	204	11	20.1
	16	28	20	184	232	11	21.3

Notes:
Data are from replicate rearings at San Ramón, elevation 1000 m, : average maximum and minimum daily temperatures, 26.0 °C (SD ± 3.1) and 21.7 °C (SD ± 1.9). Data preceding the number with an asterisk are rearings at Puerto Bermudez, elevation 500 m, : average maximum and minimum daily temperatures: 29.5 °C (SD ± 1.0) and 22.9 °C (SD ± 1.1).
Source: From Greenberg and Szyska, 1984.

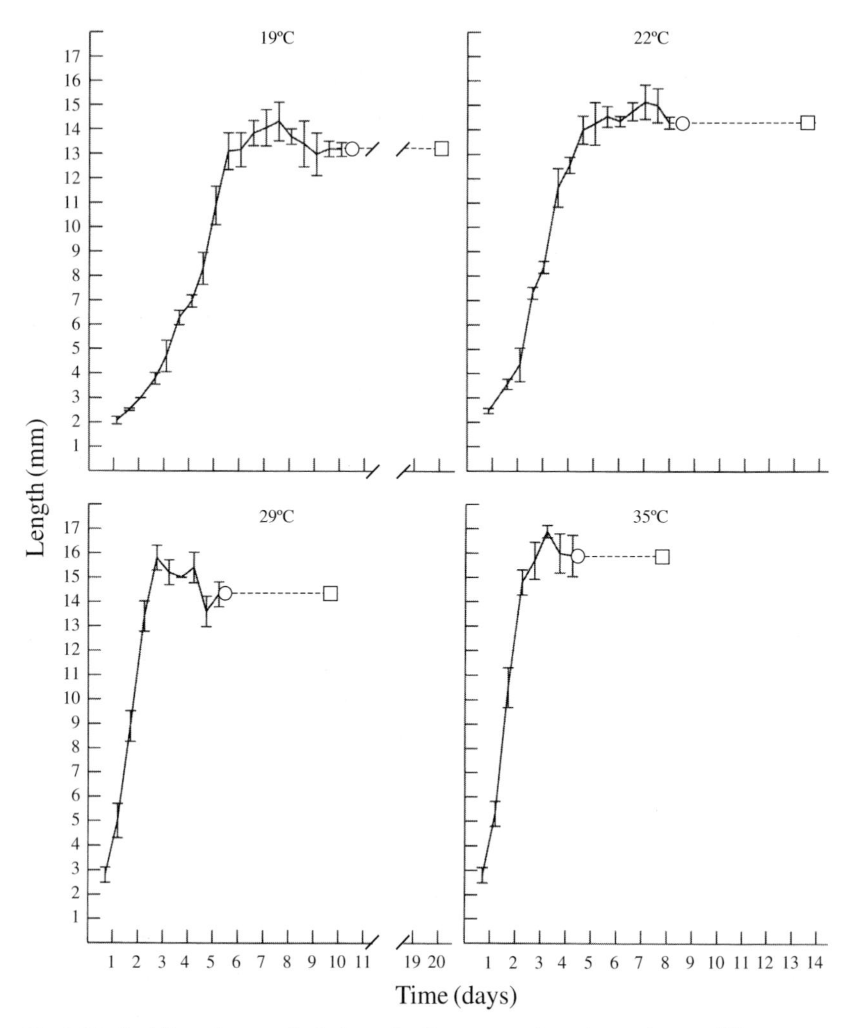

Fig. 3.23. *Cochliomyia macellaria.* Length of larvae monitored over time at different rearing temperatures. Plotted with standard deviation calculated from $n = 5$ specimens; ○, pupariation; □, adult emergence.

section with schedules of development of *Cochliomyia macellaria* at four temperatures (see also Cunha e Silva and Milward-de-Azevedo, 1992; Byrd and Butler, 1996). This is one of the most widespread of the forensically important flies in the New World.

Until the latter part of the twentieth century certain species of the genus *Chrysomya* were confined to the tropics and subtropics of the Old World and Australasia. In 1978, Guimarães *et al.* reported *C. albiceps* and *C. megacephala* in the State of São Paolo, Brazil. Along with *C. putoria* (primarily coprophagous in Africa), these flies were probably introduced a few years earlier with the influx of Portuguese refugees from Africa. In the same year in Costa Rica, Jirón (1979), reared *C. rufifacies* from a human cadaver (Jirón and Cartin, 1981). This fly and *C. megacephala* are endemic in the Australasian region and are now established in the southern tier of states in the United States and in Puerto Rico (Baumgartner and Greenberg, 1984; Greenberg, 1988; Tantawi and Greenberg, 1993b). *Chrysomya rufifacies* has become the dominant blowfly on human cadavers in north and central Florida, outnumbered only by *Chrysomya megacephala* in southern Florida (Byrd and Butler, 1997). These authors provide rearing data and temperature preferences for *Chrysomya rufifacies.* Like *C. macellaria,* the cold-intolerance of both *C. rufifacies* and *C. megacephala* will probably confine them to the southern parts of North America, with summer forays northward. *Chrysomya rufifacies* has not dispersed into South America where its sibling, *C. albiceps* is now established, nor vice versa. Morphologically and behaviorally, the two species are similar but distinguishable as larvae and adults. It would be interesting to monitor their populations if and wherever the two meet.

There are numerous treatments of North American flies, including many of forensic importance, but Hall's (1948) *The Blowflies of North America* is the definitive work on the group. It contains keys to genera and species of three subfamilies of North American adults and to some of their larvae, including species the forensic entomologist is likely to encounter. The following publications are also worth consulting. Wells *et al.* (1999) provide a key to third instar larvae of nine species belonging to the Chrysomyinae of continental United States, including four species of *Chrysomya.* Liu and Greenberg (1989) have published keys and diagnostic descriptions of the eggs, larval instars, and puparia of common carrion flies, including several South American species (see also Greenberg and Singh, 1995; Wells *et al.*, 1999). With the closing circle of trade and tourism, *C. megacephala* has attained global status and is now found in the two Americas, Africa, and the Australasian region. For its schedule of development the reader is referred to Wijesundara (1957), Subramanian and Mohan (1980), O'Flynn (1983) and Wells and Kurahashi (1994).

North America and Europe share a number of blowfly species, more than do North and South America. The latter has been intermittently isolated biologically, and until recently, was outside the brisk trade routes of the North Atlantic that dumped numerous insects onto the shores of North America. For larval keys with a European focus the reader is referred to Erzinçlioglu (1985,1987a, b, 1988, 1989a,b, 1990b). Gregor's (1971) key to adult blowflies and fleshflies of the Holarctic region is reproduced below; Figs. 3.24 to 3.32 by Gregor, accompany his key. Rognes (1991) provides a definitive treatment of the calliphorids of the Scandanavian region. Zumpt (1965) and James (1947) have written excellent treatises on myiasis with keys and illustrations to many of the common forensic flies.

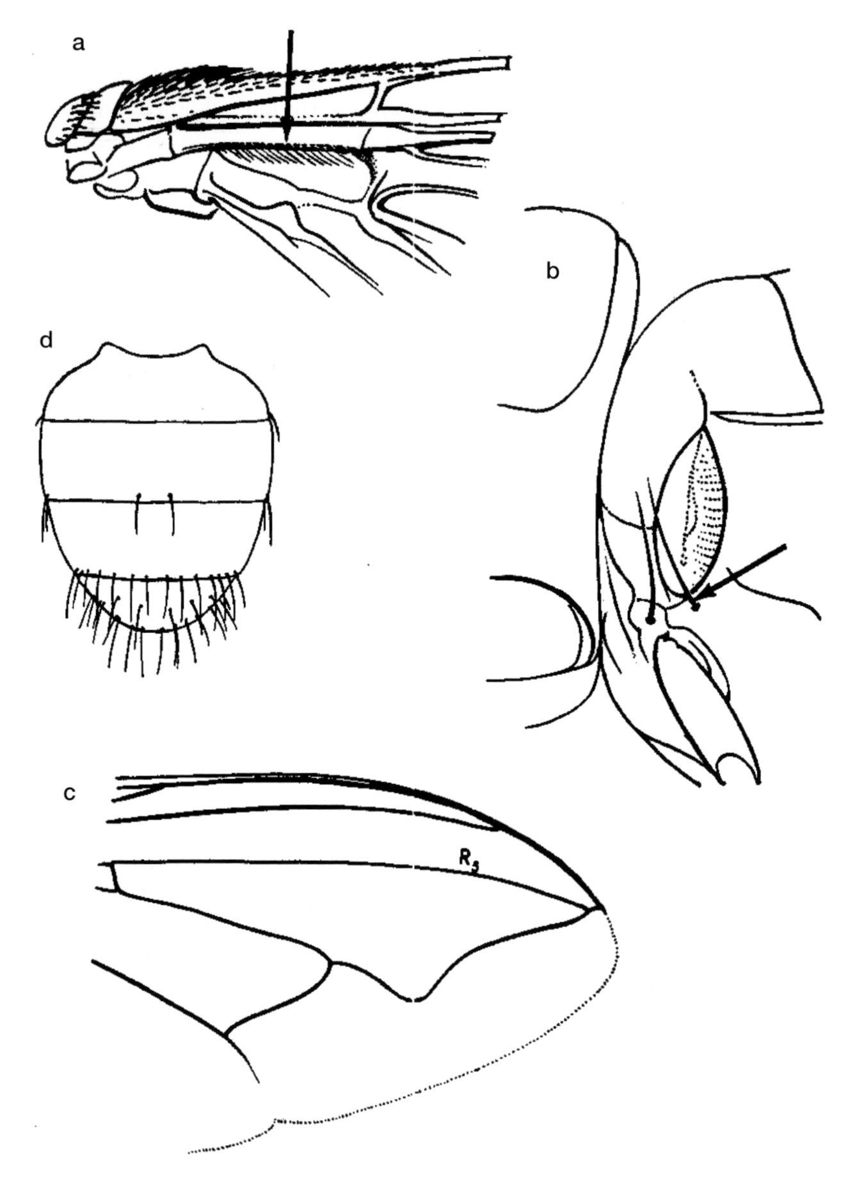

Fig. 3.24. a, Upper surface of wing base of *Chrysomya albiceps* (Wied.). b, Area surrounding anterior spiracle of *Chrysomya albiceps.* c, Apex of wing of *Pollenia atramentaria* (Meig.). d, Macrochaetae on abdominal tergites of *Bufolucilia silvarum* (Meig,). (From Gregor, 1971.)

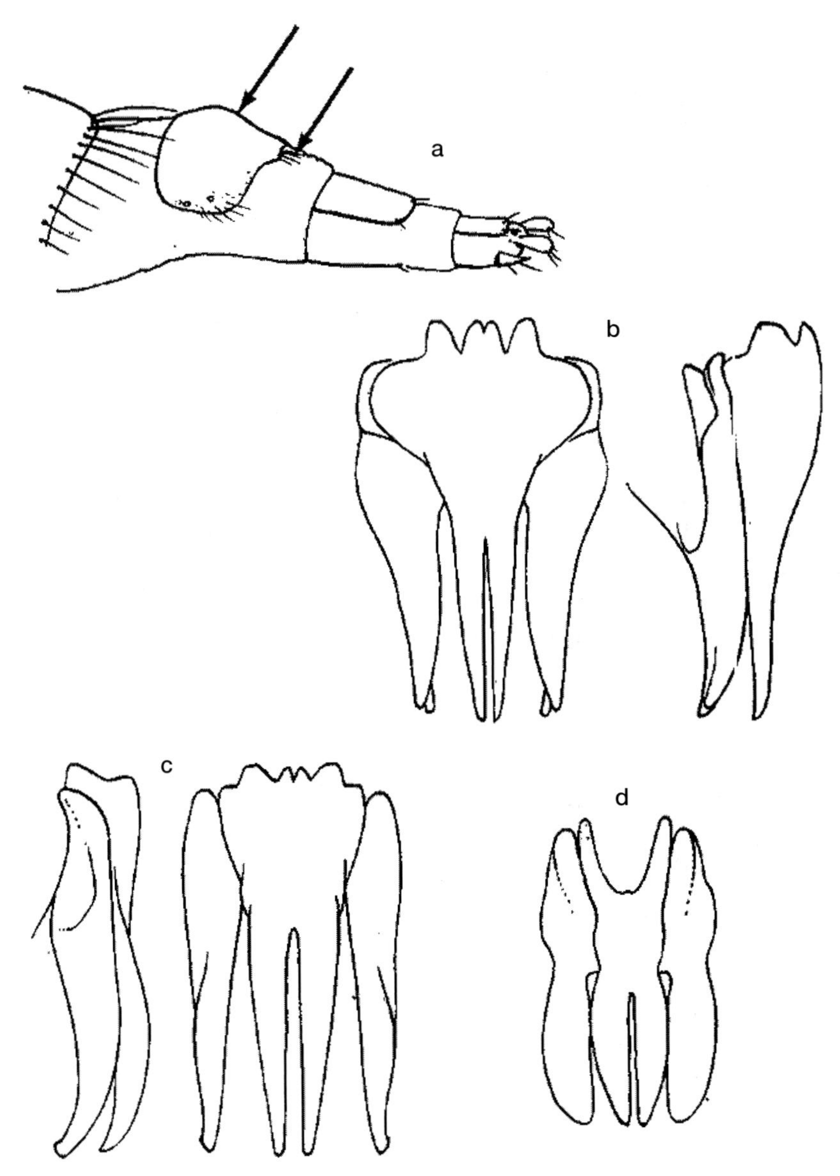

Fig. 3.25. a, Ovipositor of *Lucilia caesar* (L.). b, Cerci and surstyli in dorsoventral (left) and lateral (right) view in *Lucilia caesar*. c, Cerci and surstyli in lateral (left) and dorsoventral (right) view in *Lucilia illustris* (Meig.). d, Cerci and surstyli in dorsoventral view in *Lucilia ampullacea* Vill. (From Gregor, 1971.)

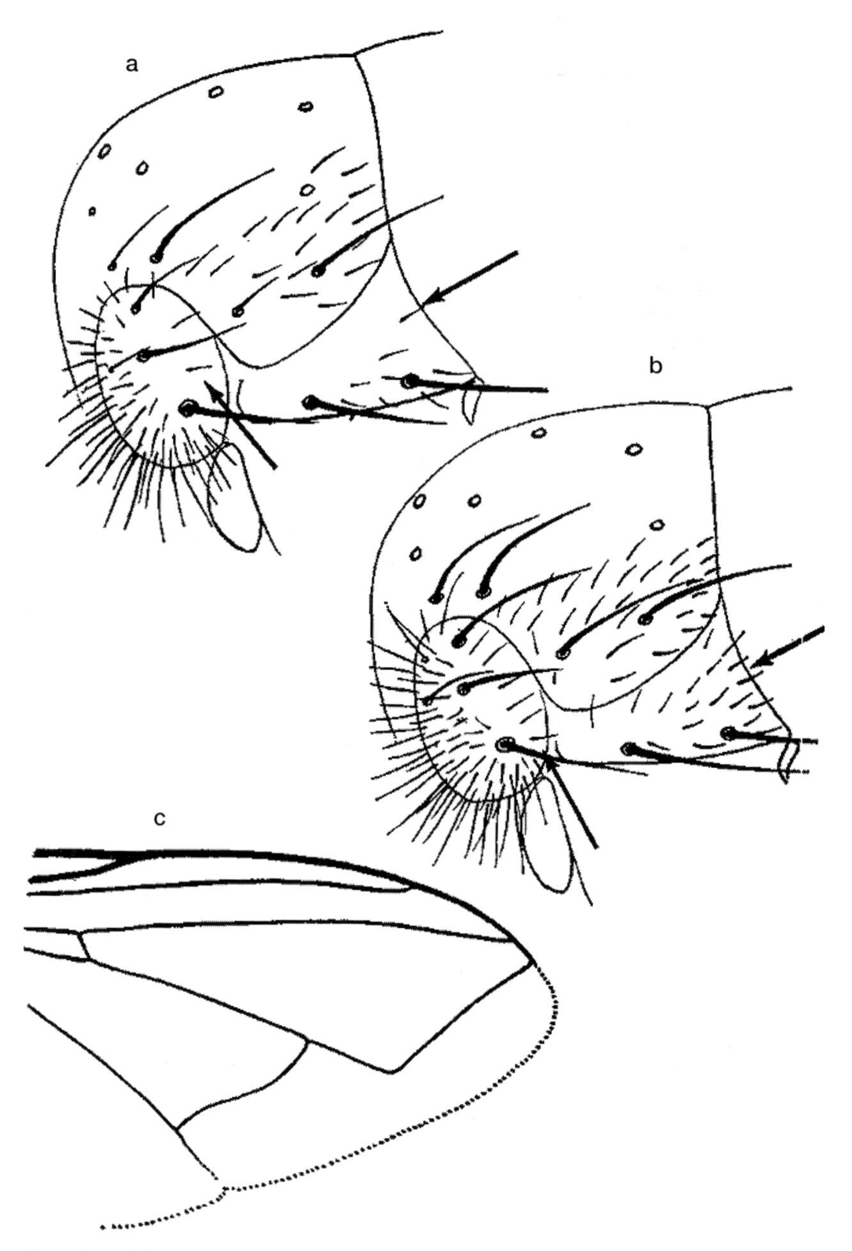

Fig. 3.26. a. Microsetae on humerus and notopleura in *Phaenicia cuprina* (Weid.). b, Microsetae on humerus and notopleura in *Phaenicia sericata* (Meig.). c, Apex of wing of *Phaenicia sericata*. (From Gregor, 1971.)

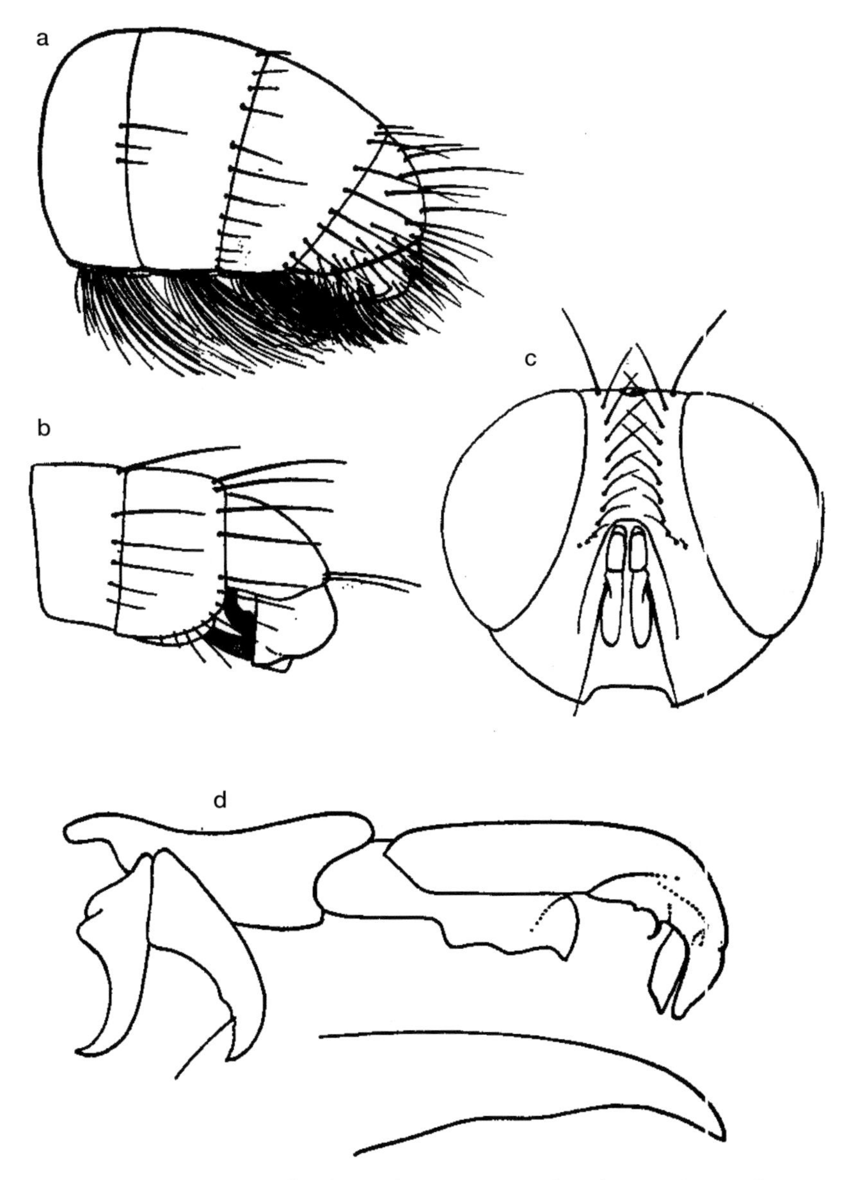

Fig. 3.27. a, Macrochaetae of abdominal sternites in *Lucilia pilosiventris* Kram. b, Distribution of macrochaetae on the apex of abdomen in male *Sarcophaga melanura* (Meig.). c, Head of female *Sarcophaga melanura*. d, Phallosoma and apex of cerci of *Sarcophaga melanura*. (From Gregor, 1971.)

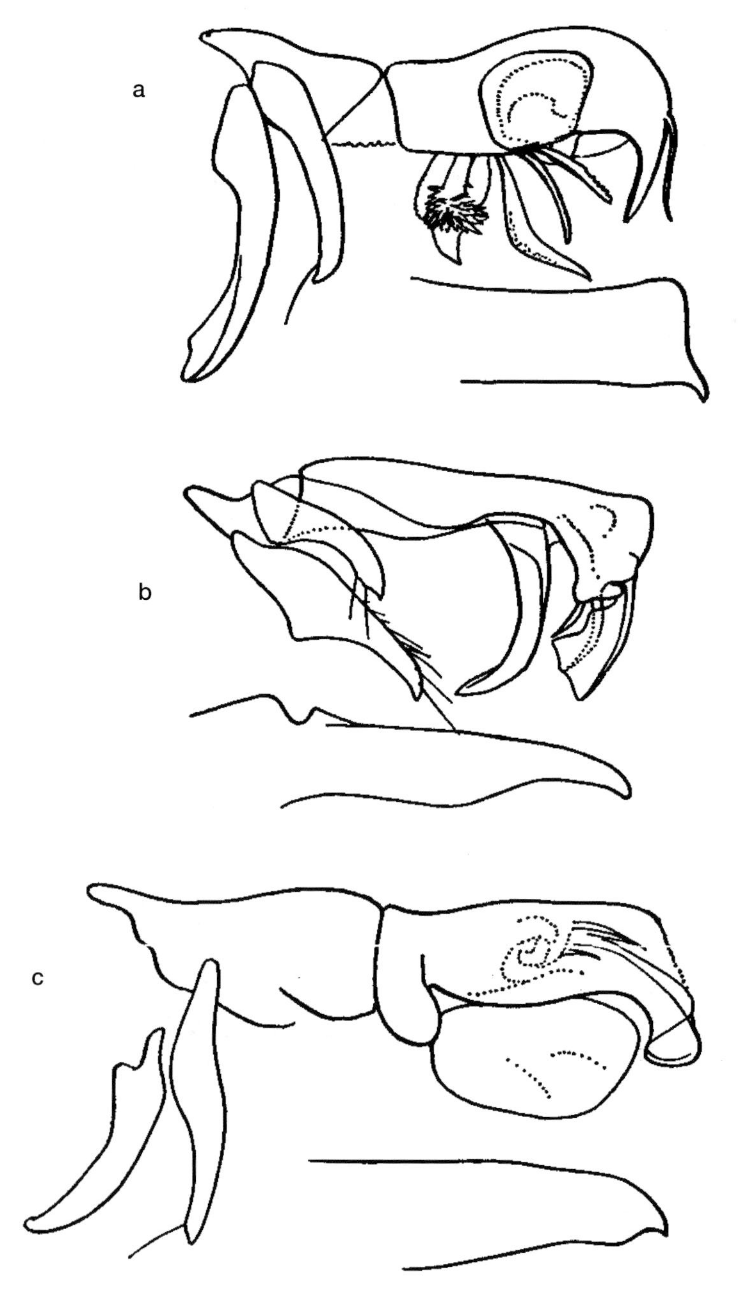

Fig. 3.28. a, Phallosoma and apex of cerci of *Thyrsocnema incisilobata* (Pand.). b, Phallosoma and apex of cerci of *Sarcophaga haemorrhoidalis* (Fall.). c, Phallosoma and apex of cerci of *Sarcophaga subvicina* Rohd. (From Gregor, 1971.)

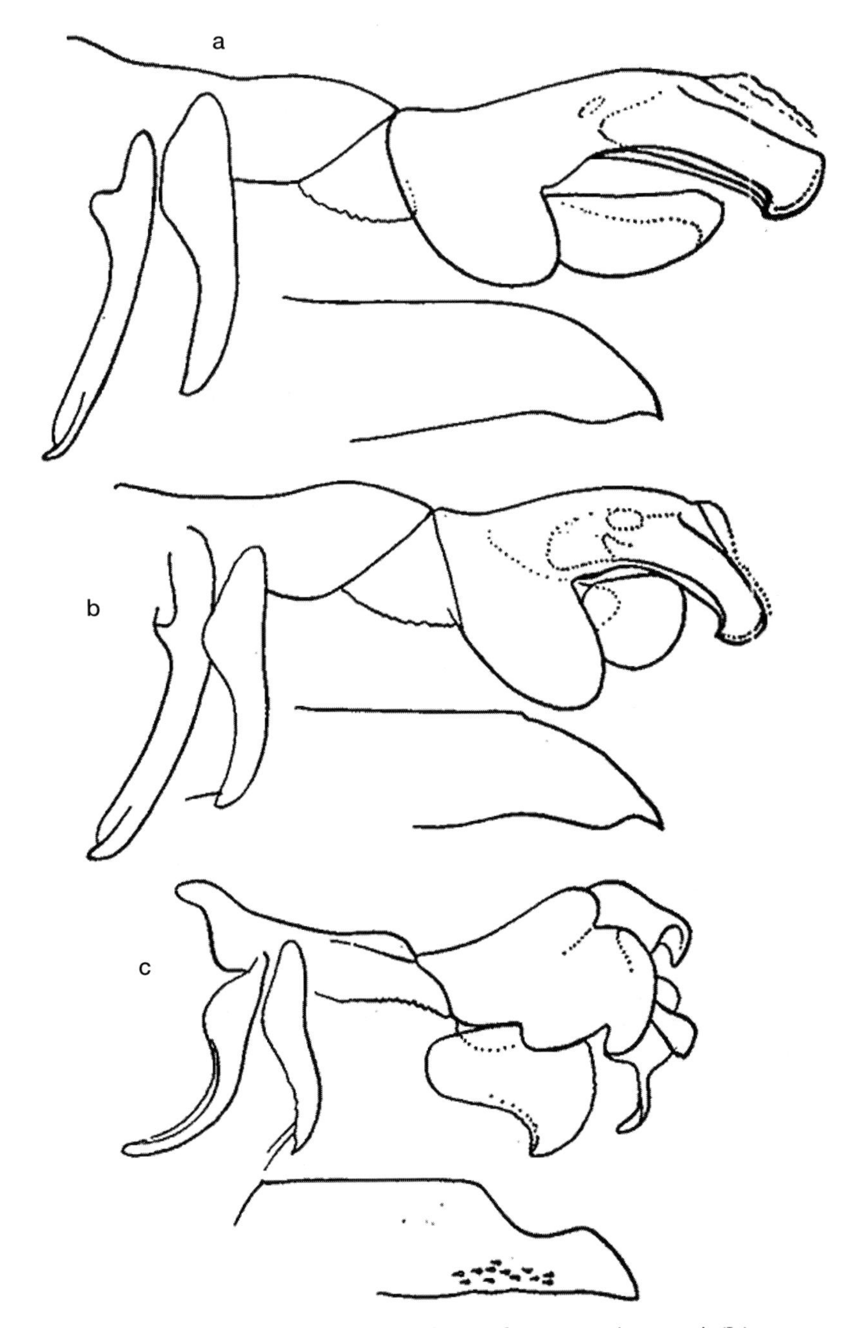

Fig. 3.29. a, Phallosoma and apex of cerci of *Sarcophaga carnaria carnaria* (L.). b, Phallosoma and apex of cerci of *Sarcophaga carnaria lehmanni* Mull. c, Phallosoma and apex of cerci of *Parasarcophaga barbata* Thom. (From Gregor, 1971.)

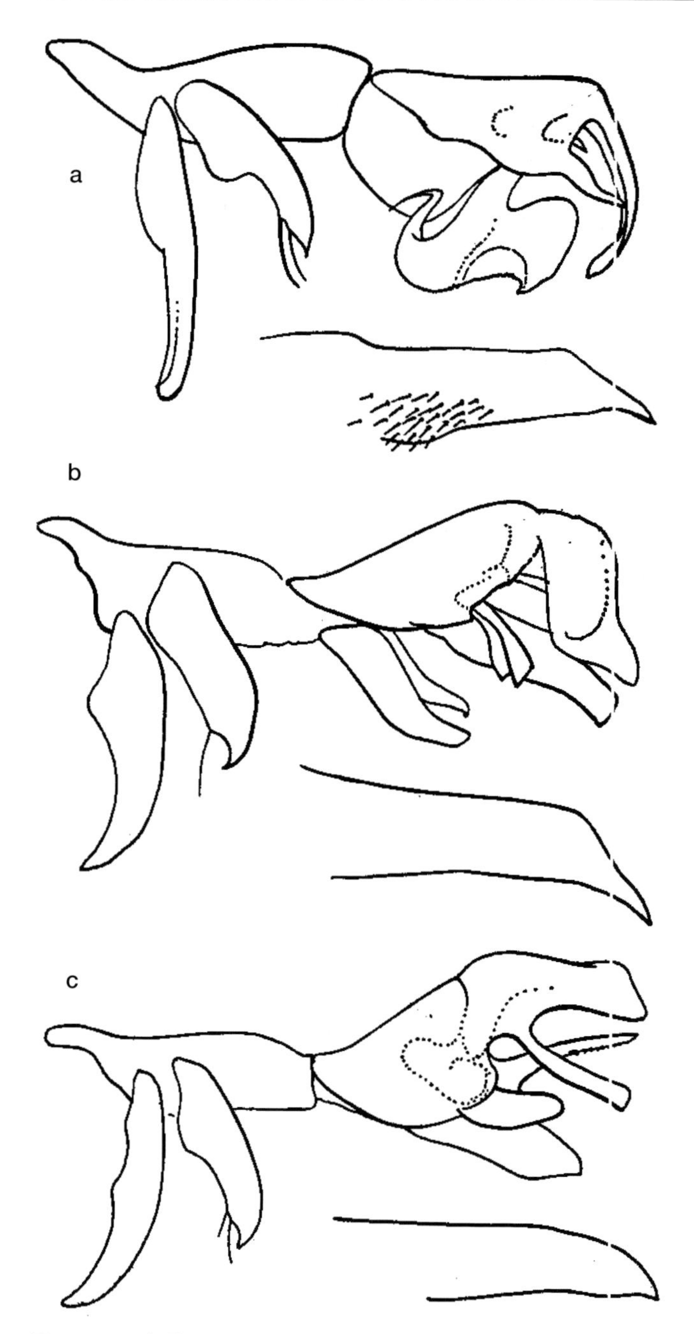

Fig. 3.30. a, Phallosoma and apex of cerci of *Parasarcophaga albiceps* (Meig.). b, Phallosoma and apex of cerci of *Parasarcophaga aratrix* Pand. c, Phallosoma and apex of cerci of *Parasarcophaga teretirostris* Pand. (From Gregor, 1971.)

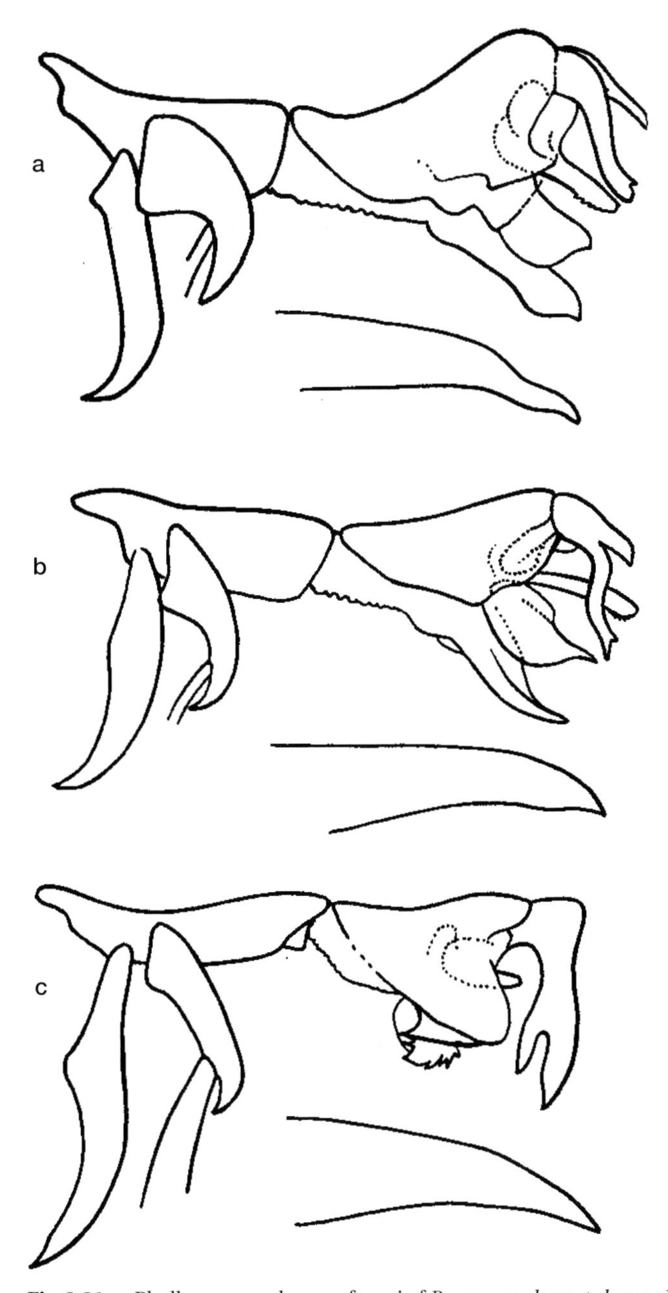

Fig. 3.31. a, Phallosoma and apex of cerci of *Parasarcophaga tuberosa* Pand. b, Phallosoma and apex of cerci of *Parasarcophaga jacobsoni* Rond. c, Phallosoma and apex of cerci of *Parasarcophaga scopia* Pand. (From Gregor, 1971.)

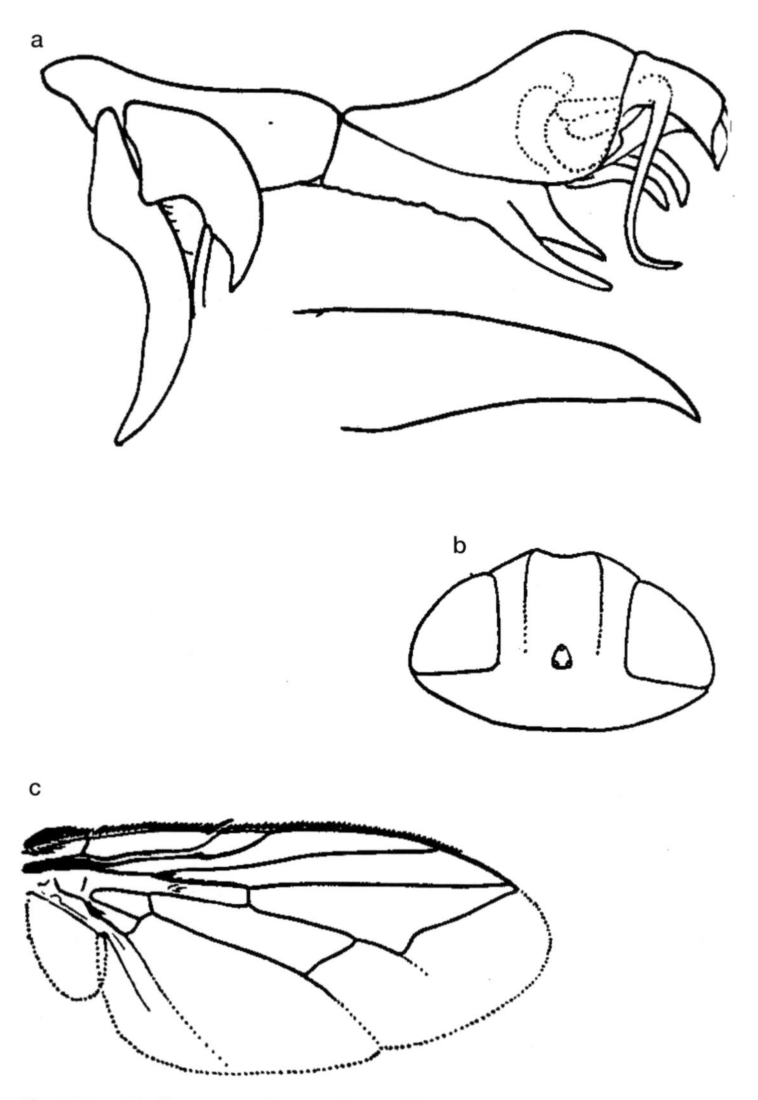

Fig. 3.32. a, Phallosoma and apex of cerci of *Parasarcophaga similis* Pand. b, Head of male *Agria latifrons* Fall. c, Wing of *Agria latifrons* Fall. (From Gregor, 1971.)

Key to Adult Holarctic Blowflies and Fleshflies of Forensic Importance.

(František Gregor 1971, Czech Academy of Sciences, Brno)

Thorax with pale and dark vittae, abdomen with pearly maculae or dark round maculae. Body densely pollinose, never metallic blue or green. As a rule 3 or more notopleural bristles present. Vestigial m2 vein or veinlike fold in place of bifurcation of m1and m2 present (Fig. 3.32c): Sarcophagidae.Body metallic blue, green or purple. Two notopleural bristles, vein m2 entirely absent and not suggested by a fold of wing membrane (Fig. 3.26c): Calliphoridae. The abbreviations in the key refer to structures illustrated in Figs. 3.24 to 3.32.

CALLIPHORIDAE

1. – Base of the radius before the humeral cross vein ciliated posteriorly above (Fig. 3.24a) (Chrysomyinae) 2
   – Base of radius bare posteriorly above 8
2. – Hind-coxae pilose posteriorly, lower part of head yellow to reddish orange 3
   – Hind coxae bare posteriorly, lower part of head predominantly dark 6
3. – Lower squama pilose above. Body compact, head large, distinctly concave posteriorly. Body metallic, with more or less uniform but feeble pollinosity 4
   – Lower squama bare above. Head not distinctly concave posteriorly. Thorax with dense pollinosity and with three dark bands; abdomen with conspicuous shifting pattern of pollinosity (non-palearctic) 5
4. – Anterior spiracles white, stigmatic bristle (Fig. 3.24b) absent normally; body metallic green, posterior margins of abdominal segments dark; 6–12 mm .... *Chrysomya albiceps* (Wied.)
   – Anterior spiracles brown to black, stigmatic bristle invariably present. Body metallic bluish-green; 6–9 mm .... *Chrysomya megacephala* (Fabr.)
5. – Parafrontal with black hair anteriorly, outside frontal row of bristles. Body deep bluish-black with partial green or purple luster; 8–10 mm .... *Cochliomyia hominivorax* (Cocquerel)
   – Parafrontal with pale hair anteriorly, outside frontal row of bristles. Body normally deep metallic green; 6–9 mm .... *Cochliomyia macellaria* (Fabr.)
6. – Arista long-haired 7
   – Arista short, pubescent. Thorax flattened, mesothoracic spiracle much enlarged. Body black with bluish luster; 6–9 mm .... *Boreellus atriceps* (Zett.)
7. – Mesonotum flattened on disc, preacrostical bristles reduced. Body black with bluish or greenish-blue luster, front of male broad; 8–12 mm .... *Protophormia terraenovae* (R.-D.)
   – Mesonotum convex, preacrostical bristles well developed. Body dark green metallic, yellowish-green or purple in places. Eyes of male subcontiguous; 8–10 mm .... *Phormia regina* (Meig.)
8. – Head conspicuously golden yellow, front wide, protruding. One pair of acr in front of the scutellum as a rule; 9–15 mm .... *Cynomya mortuorum* (Linn.)
   – No such combination of characters 9
9. – Body metallic blue, shining, with pale pollinosity forming stripes on thorax and more or less distinct mother-of-pearl blotches on abdomen. Lower squama normally haired (*Calliphora*) 10
   – Body metallic green or blue, without distinct pollinosity. Lower squama bare (*Lucilia, Phaenicia*) 14
10. – Basicosta yellow to yellow-brown, anterior thoracic spiracle orange. Bucca reddish except for posterior third of occipital part; 5–12 mm .... *Calliphora vicina* R.-D.
    Basicosta black 11
11. – Lower calyptra more or less broadly infuscate (less intensely in female). Male: cerci and paralobi covered with fine setulae only 12
    – Lower calyptra greyish-white, with the part adjacent to the scutellum somewhat infuscate.Male: cerci and paralobi covered with stout bristles, lobes of fifth sternite very large and erect 13

12. – Hair on lower part of jowls and of occiput fulvous to golden; 8–14 mm .... *Calliphora vomitoria* (Linn.)
    – Hair on jowls and occiput black. Anterior thoracic spiracle strongly infuscate .... *Calliphora uralensis* Villen.

13. – Scutellum with four to five pairs of strong marginals, lateral seta well developed; 7–13 mm .... *Calliphora subalpina* (Ringd.)
    – Scutellum with only three pairs of strong marginals, lateral seta absent; 8–11 mm .... *Calliphora alpina* (Zett.)

14. – Basicosta black or brown 15
    – Basicosta yellowish, at most with infumated margin 18

15. – Third tergite with conspicuous marginals (Fig. 3.24d). Subcostal sclerite with microscopic pile only .... *Bufolucilia spp.* not attracted to carrion.
    – Third tergite with outstanding marginal setae. Subcostal sclerite in addition to the microscopic pile with some blackish setulae near apex 16

16. – Male: hypopyg large, protruding, glossy bluish-green. Inferior forceps bifid at apex (Fig. 3.25b), eyes separated by about half the width of the third antennal segment. Female: Sixth tergite feebly convex, its posterior margin in middle with only one or two pairs of small setae (Fig. 3.25a). 6–11 mm .... *Lucilia caesar* (Linn.)
    – Male: hypopyg small, inconspicuous, black or green, paralobi never bifid at apex. Female: sixth tergite flattened, with a complete row of long marginal bristles 17

17. – Male: eyes separated by almost the width of third antennal segment. Inferior forceps with a slight apical knob which is curved forward (Fig. 3.25c). Female: underside of arista with 9–12 rays, third antennal segment less long than width of interfrontalia plus one parafrontale, 2.5 times as long as wide; 5–9 mm .... *Lucilia illustris* (Meig.)
    – Male: eyes separated by about half the width of the third antennal segment. Inferior forceps parallel-sided with the apex broadly rounded (Fig. 3.25d). Female: underside of arista with 18–22 rays; 6–11 mm .... *Lucilia ampullacea* Villen.

18. – Third tergite with two or more long lateral setae, front of male one-third as wide in the narrowest place as the diameter of eye; 5–9 mm .... *Lucilia regalis* (Meig.)
    – Third tergite without conspicuous marginal setae 19

19. – Mid-tibia with one ad seta only 20
    – Mid-tibia with two strong ad setae 21

20. – Posterior part of humeral callus with two to four setulae (Fig. 3.26a). Abdomen usually strongly coppery. Male: abdominal sternites long-haired, two pairs of ocellar bristles, front about one-fifth head width. Female: bucca less than one-third eye height; 5–10 mm .... *Phaenicia cuprina* (Wied.)
    – Posterior part of humeral callus with six to eight setulae (Fig. 3.26b). Abdomen usually bright green. Male: abdominal sternite with ordinary pile, accessory pair of ocellars absent, front about one-eighth head width. Female: bucca almost two-fifths eye height; 5–10 mm .... *Phaenicia sericata* (Meig.)

21. – Male: eyes separated by not much more than width of third antennal segment, abdominal sternites of male and female normally haired; 5–11 mm .... *Lucilia richardsi* Collin.

- Male: eyes separated by one-third eye width, abdominal sternites of male with dense tufts of long hairs (Fig. 3.27a), those of female with strong setae; 7–10 mm .... *Lucilia pilosiventris* Kramer

SARCOPHAGIDAE

1. – Abdominal tergites with irregular pearly maculae 2
 – Abdominal tergites with symmetrical pattern of dark, dull, oval maculae on sides and medial stripe 9
2. – Three strong postsutural dc 3
 – Four or more postsutural dc, or at least the series distinctly spaced for four or more 5
3. – Second genital segment and hypopygium largely red, frontal bristles straight or very slightly curved outward from end; 4–8.5 mm .... *Ravinia striata* Fabr.
 – Second genital segment entirely black, frontal bristles opposite the antennae diverging outward (Fig. 3.27c) 4
4. – Male: first genital segment without marginal setae, hypopygium as in Fig. 3.28a. Female: frons just over one-third head width; 6–13 mm .... *Thyrsocnema incisilobata* (Pand.).
 – Male: first genital segment with distinct marginal setae (Fig. 3.27b), hypopygium as in Fig. 3.27d. Female: frons distinctly broader than eye; 6.5–13 mm .... *Sarcophaga melanura* Meig.
5. – Prescutellar ac absent, genital segments mostly red. Hypopygium of male as in Fig. 3.28b; 8–14 mm .... *Sarcophaga haemorrhoidalis* (Fall.)
 – Prescutellar ac present 6
6. – Third tergite with strong marginals, normally one pair present; genital tergite of male with a series of marginal setae 7
 – Third tergite without long marginals, genital tergite of male without a series of marginal setae as a rule 8
7. – Male: hypopygium as in Fig. 3.28c, first genital segment dusted except on basal part. Female: eighth tergite present; 8–15 mm .... *Sarcophaga subvicina* Rohd.
 – Male: hypopygium as in Fig. 3.29a, subsp. *carnaria* Linn. Or in Fig. 3.29b, subsp. *lehmanni* Mull. First genital segment glossy for the most part. Female: eighth tergite present; 8–16 mm .... *Sarcophaga carnaria* (Linn.)
8. – Genital segments red, hypopygium as in Fig. 3.29c .... *Parasarcophaga barbata* Thom.
 – Genital segments black. This group includes a number of *Parasarcophaga* spp., reliably distinguishable only by their hypopygia: *P. albiceps* (Meig.) (Fig. 3.30a), *P. aratrix* Pand. (Fig. 3.30b), *P. teretirostris* Pand. (Fig. 3.30c), *P. tuberosa* Pand. (Fig. 3.31a), *P. jacobsoni* Rohd. (Fig. 3.31b), *P. scoparia* Pand. (Fig. 3.31c), *P. similis* Pand. (Fig. 3.32a). (Many species of *Parasarcophaga*, e.g., *alibiceps*, *argyrostoma*, *barbata*, *hirtipes*, and *parkeri*, are placed in *Sarcophaga* by some specialists.)
9. – Lateral maculae nearer to anterior margin of tergites, three to four notopleural bristles, three sternopleural bristles more or less in 1 row (1:1:1); 6–16 mm .... *Bellieria maculata* (Meig.)
 – Lateral maculae nearer to posterior margin of tergites 10
10. – Sternopleural bristles in a position 2 (or 3):1, frons in either sex relatively very broad (Fig. 3.32b); 4.5–9 mm .... *Agria latifrons* Fall.

Table 3.6. *The development of body length (in mm) of some fly species during their metamorphosis*

	Species (L = larva, P = puparium, A = adult fly)				
Day	*Musca domestica*	*Calliphora vomitoria*	*Lucilia caesar*	*Sarcophaga carnaria*	*Piophila nigriceps*
2	L 2	L 3–4	L 2	L 3–4	L 1
3	L 2–3	L 5–6	L 2–3	L 5–6	L 2–3
4	L 4–5	L 7–8	L 3–4	L 7–9	L 4–5
5	L 6–7	L 10–12	L 5–6	L 10–12	L 5–6
6	L 7–8	L 13–14	L 7–8	L 13–14	pupariation
7	L 8	pupariation	L 8–9	L 15–16	P 3–4
8	pupariation	P 9–10	pupariation	L 16–18	P 3–4
9	P 5–6	P 9–10	P 6–7	L 19–20	P 3–4
10	P 5–6	P 9–10	P 6–7	pupariation	P 3–4
11	P 5–6	P 9–10	P 6–7	P 10–12	P 3–4
12	P 5–6	P 9–10	P 6–7	P 10–12	A 4–5
13	P 5–6	P 9–10	P 6–7	P 10–12	
14	A 7–8	A 12–13	A 7–9	P 10–12	
15				P 10–12	
16				P 10–12	
17				P 10–12	
18				A 16–18	

Notes:

[a] Modified from Schranz. In B. Muller, 1953, *Gerichtliche Medizin*. Springer-Verlag. Berlin-Gottingen-Heidelberg.

Source: From Nuorteva, 1977.

The section on carrion flies common to North America and Europe concludes with rearing schedules of *Lucilia caesar* (Table 3.6), *Phaenicia sericata* (Table 3.7), *Phormia regina* (Tables 3.8, 3.9, 3.10, 3.11), *Protophormia terraenovae* (Table 3.12), *Calliphora vomitoria* (Table 3.13), and *Calliphora vicina* (Tables 3.14, 3.15). Davies and Ratcliffe (1994) provide useful information on the development rates of several of these blowflies at low temperature. The sarcophagids are represented by *Sarcophaga carnaria* (Table 3.6), *Musca domestica*, the housefly, and one piophilid – *Piophila nigripes* – are also included in Table 3.6. The housefly is an infrequent visitor to carrion, preferring fecal matter. The development of *Piophila casei* at three temperature regimens is shown in Fig. 3.33. Piophilids are typically part of the clean-up fauna in the late stages of decomposition. See also Figs. 3.34 to 3.42a,b for development details of *Phaenicia sericata*, *Phormia regina* and *C. vicina*. This sample of 10 species reflects the writer's own forensic experience. Byrd and Butler (1998) provide

Table 3.7. *Average minimum duration (days) of pre-adult stages of* Phaenicia sericata

°C	Egg	First instar	Second instar	Third instar	Pupa	Total
19	1.5	1.3	0.4	6.0	7.1	16.3
22	0.9	1.0	0.9	5.4	6.0	14.2
29	0.6	0.7	0.5	4.8	5.5	12.1
35	0.6	0.7	0.3	[a]	6.0	

Note:

[a] At 35 °C many postfeeding larvae entered diapause.

Table 3.8. *Average minimum duration (days) of pre-adult stages of* Phormia regina

°C	Egg	First instar	Second instar	Third instar	Pupa	Total
19	1.1	1.8	2.1	3.8	6.8	15.6
22	0.8	1.0	0.7	6.3	5.8	14.6
29	0.8	0.5	0.6	5.6	4.1	11.6
35	0.5	0.3	0.4	5.0	3.8	10.0

Table 3.9. *Chronology of development within the puparium of* Phormia regina

	Hours after pupariation when stage/event first occurs at:	
Stage/Event	22 °C	29 °C
Prepupa	0	0
Cryptocephalic pupa	9	7
Phanerocephalic pupa	23.5	16
Onset of pupal-adult molt: membrane around appendages	42	25
Molt completed: pharate adult completely enclosed in membrane	52	28
"Segmented" abdomen	60	30
Posterior part of eye pigmented	81	47
Entire eye pigmented	88	52
Setae tanned on head and thorax	120	70
Setae tanned on abdomen	134	77
Eclosion	149	90

Source: Greenberg, 1991.

Table 3.10. *Color code for changes in eye color of pharate adult* Phormia regina

Age(h)	Color code[a] 22°C	Color code 29°C
52		B00Y60M10
58		B00Y80M20
59		**B00Y80M20**
60		**B00Y80M40**
62		**B00Y90M40**
65		**B00Y99M50**
72		**B00Y99M50**
78		**B10Y90M50**
88	B00Y70M10	**B50Y90M60**
93	[c]	**B60Y60M50**
96	**B00Y80M30**	**B60Y99M60**[b]
98	B00Y80M30	
99	B00Y80M30	
102	B00Y80M30	
108	B00Y90M50	
116.5	**B00Y99M50**	
120	**B50Y60M60**	
132	B50Y90M60	
134	**B60Y99M60**	
140	**B60Y99M60**	
144	**B60Y99M70**	
156	**B60Y99M70**	
168	**B70Y80M50**	

Notes:

[a] Kueppers (1982).

[b] Eclosion.

[c] Not observed.

Source: Greenberg, 1991.

Table 3.11. *Number and size of fat bodies in post-emergent* Phormia regina

Age (h)	Number (+ SD)	Size μm (+ SD)
0	5549 (696)	98.5 (35)
24	4365 (372)	–
48	2194 (158)	74.2 (17)
72	1371 (211)	–
96	641 (270)	68.5 (21)

Source: Greenberg, 1991.

Table 3.12. Protophormia terraenovae: *average minimum duration of developmental stages in hours*

	Rearing temperature							
	12.5 °C		23 °C		29 °C		35 °C	
Stage	Hours	%[b]	Hours	%	Hours	%	Hours	%
Egg	91.2	4	16.8	5	14.4	6	12.0	5
First instar	290.4	13	26.4	8	13.2	5	9.6	4
Second instar	240.0 } 1108.8	11 } 51	27.6 } 98.4	8 } 29	18.0 } 80.4	7 } 31	16.6 } 72.0	8 } 33
Third instar	578.4	27	44.4	13	49.2	19	45.8	21
Postfeeding third instar	254.4	12	74.4	22	55.2	21	36.0	16
Pupa[a]	722.4	33	144.0	43	110.4	42	100.8	46
Total (h)	2176.8		333.6		260.4		220.8	

Notes:

[a] Pupariation to eclosion.

[b] Percentage of total developmental time.

Source: From Greenberg and Tantawi, 1993.

Table 3.13. Calliphora vomitoria: *average minimum duration of developmental stages in hours*

	Rearing temperature											
	12.5 °C				23 °C				29 °C		35 °C	
Stage	Hours		%[b]		Hours		%		Hours		Hours	
Egg	64.8		5		21.6		4		16.8		12.0	
First instar	55.2		4		25.2		5		10.8		18.0	
Second instar	60.0	393.6	5	30	19.2	172.8	4	33	18.0	117.6	12.0	74.4
Third instar	278.4		21		128.4		24		88.8		44.4	
Postfeeding third instar	156.0		12		86.4		16		74.4		–	
Pupa[a]	717.6		54		247.2		47		–		–	
Total (h)	1332.0				528.0				208.8		86.4	

Notes:

[a] Pupariation to eclosion.

[b] Percentage of total developmental time.

Source: Adapted from Greenberg and Tantawi, 1993.

data on the development of *Sarcophaga haemorrhoidalis* under laboratory conditions. This is a thermophilous fly with an almost worldwide distribution. It is common in and around dwellings and is often present on cadavers.

Our attention now shifts to the forensic flies of the Australasian region. Among the outstanding taxonomic contributions in this region are the following. *The Flies of China,* edited by Xue and Chao (1996), treats 30 families and over 4000 species of adult Cyclorrhapha, and includes illustrated keys to calliphorids and sarcophagids. The *Key to the Common Flies of China* (1992), edited by Fan, deals with 1546 species or subspecies of calyptrate Diptera, in addition to a key to more than 100 species of synanthropic larvae. Unfortunately, the text in both works is in Chinese but the illustrations are in English (as my former professor would say). Kano and Shinonaga have produced two excellent works on adult Sarcophagidae (Kano *et al.*, 1967) and Calliphoridae (Kano and Shinonaga, 1968) of Japan; the texts are in English and the illustrations are well done and detailed. The following keys were specifically designed for the forensic entomologist working in the Australasian region.

Key to Third Instar Larvae of Flies of Forensic Importance in Malaysia
(Baharudin Omar, Department of Biomedical Sciences, National University of Malaysia, Kuala Lumpur, Malaysia)

1. – Cephalopharyngeal skeleton with conspicuous and longish accessory sclerite running horizontally below denticle (hook), dorsal arch with mesh-like sclerotization, and ventral cornu with horizontally striated sclerotization at base .... Family Muscidae (2)

Table 3.14. *Length and age of* Calliphora vicina *larvae developing at various temperatures*

Length	6.5 °C		10–12 °C		14–16 °C		18–19 °C		22–23 °C		30 °C		35 °C	
(mm)	$\bar{x}$ Tag	S	$\bar{x}$ Tag	S	$\bar{x}$ Tag	S	$\bar{x}$ Tag	S	$\bar{x}$ Tag	S	$\bar{x}$ Tag	S	$\bar{x}$ Tag	S
2.00	0.0		0.0		0.0		0.0		0.0		0.0		0.0	
3.00	2.0	0.7	0.9	0.2	0.8	0.1	0.5	0.2	0.4	0.2	0.3	0.1	0.2	0.1
4.00	4.2	0.8	2.2	0.2	1.5	0.2	1.0	0.2	0.7	0.2	0.5	0.1	0.4	0.1
6.00	9.6	0.6	3.7	0.5	2.2	0.2	1.3	0.2	1.0	0.2	0.8	0.3	0.7	0.1
8.50	14.5	0.7	5.3	0.3	3.2	0.1	1.8	0.3	1.5	0.3	1.1	0.3	1.0	0.2
10.00	16.5	1.2	5.7	0.4	3.4	0.1	2.1	0.3	1.6	0.3	1.2	0.3	1.2	0.2
12.00	18.3	1.0	6.2	0.5	3.8	0.1	2.3	0.4	1.9	0.3	1.5	0.4	1.4	0.3
14.00	20.2	1.0	6.7	0.5	4.1	0.2	2.6	0.4	2.1	0.4	1.7	0.4	1.7	0.2
15.50	21.6	1.1	7.5	0.7	4.4	0.2	2.9	0.4	2.4	0.4	1.9	0.4	1.8	0.1
15.75	21.9	0.9	7.7	1.0	4.5	0.2	2.9	0.4	2.4	0.4	2.0	0.4		
17.00	24.0	1.4	8.4	1.0	5.1	0.2	3.3	0.5	2.8	0.4	2.3	0.4		
17.50	24.9	1.3	9.1	1.0	5.3	0.4	3.7	0.7	3.2	0.4	2.6	0.3		
17.90	25.9	1.2	9.6	1.0	5.6	0.5	3.9	0.6						
18.00	26.0	1.3	9.9	0.8	6.0	0.8	4.2	0.4						
17.50	38.2	2.2	12.8	2.2	7.6	1.0	5.8	1.7	4.7	0.7	3.6	0.6		
17.00	46.7	2.2	15.0	2.2	8.8	1.3	6.3	1.6	5.0	0.9	4.0	0.5		
16.50	56.9	4.8	18.6	2.1	10.3	1.4	7.1	1.5	5.3	0.9	4.2	0.6		
15.50	92.6	4.9	28.3	2.0	12.5	0.6	8.4	0.8	6.7	1.0	5.6	1.2	4.2	1.3
15.30	102.2	4.8	29.8	1.9	13.4	0.6	8.6	0.7	7.0	0.8	5.8	1.2	4.3	1.3
15.10	112.4	4.3	31.2	2.0	14.4	0.7	8.9	0.7	7.3	0.7	6.1	1.4	4.4	1.3

Source: Adapted from Reiter, 1984.

Table 3.15. *Average minimum duration (days) of pre-adult stages of* Calliphora vicina

°C	Egg	First instar	Second instar	Third instar	Pupa	Total
12.5	1.6	2.0	2.4	11.0	27.5	44.5
19	0.8	0.8	1.2	7.5	12.5	22.8
22	0.8	0.3	1.0	6.3	11.0	19.4
25	0.6	0.4	1.0	6.6	11.0	19.6

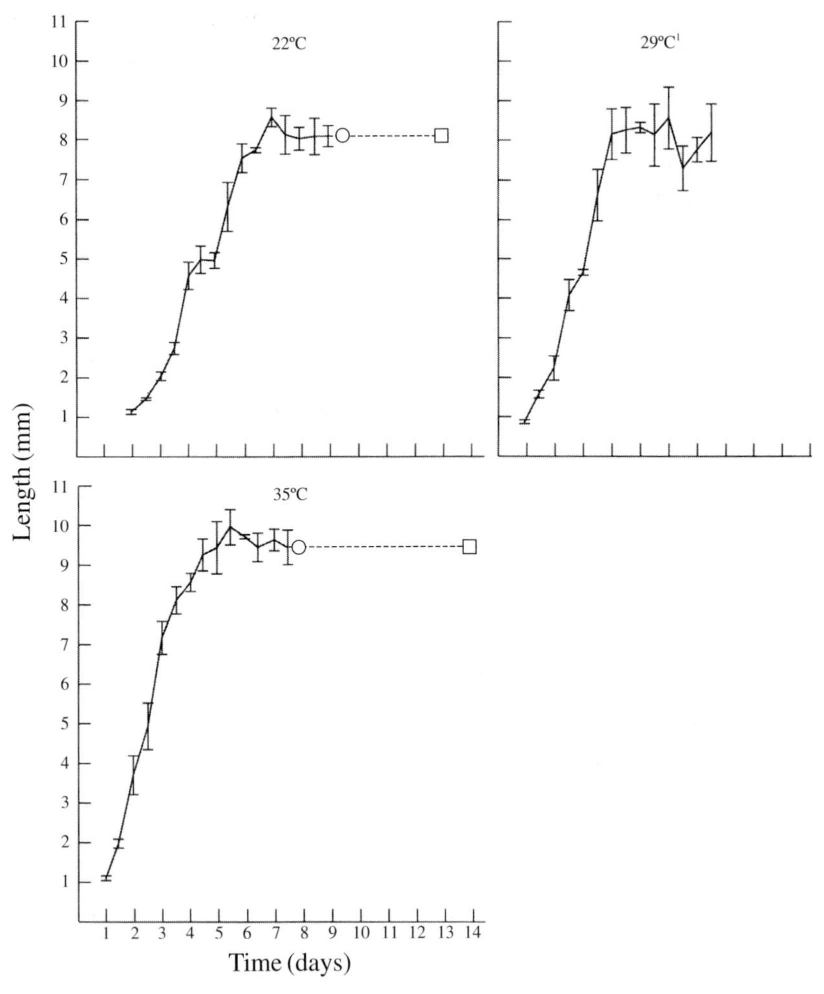

Fig. 3.33. *Piophila casei.* Length of larvae monitored over time at different rearing temperatures. Plotted with standard deviation calculated from $n = 5$ specimens; ○, pupariation; □, adult emergence. [1]Data on pupariation and emergence not collected at 29 °C.

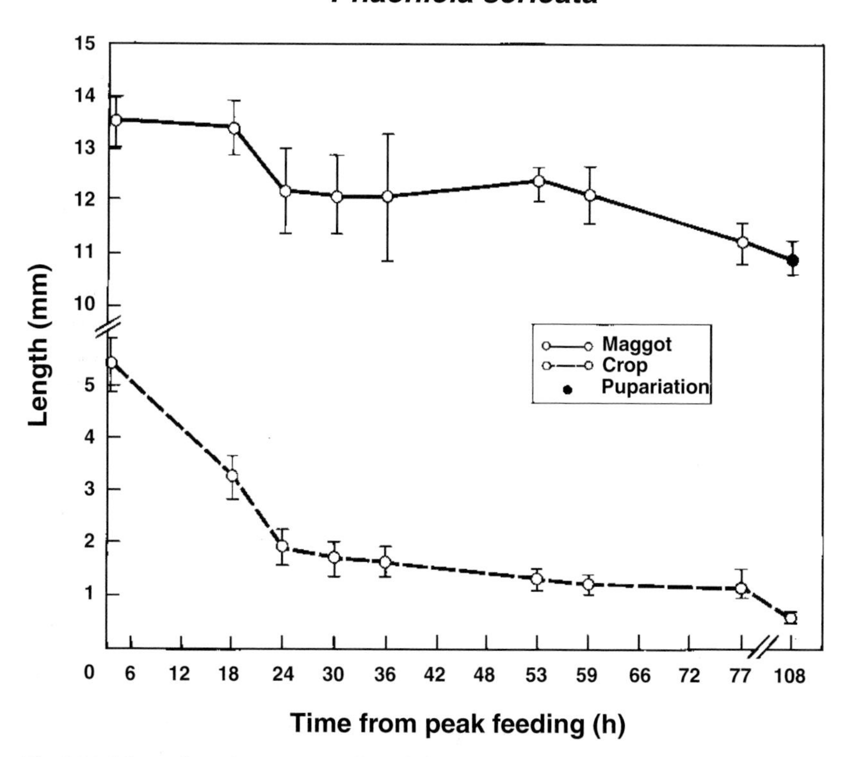

Fig. 3.34. Maggot length versus crop length from peak of feeding until pupariation in *Phaenicia sericata* at 22 °C.

- Cephalopharyngeal skeleton with or without accessory sclerite, if latter is present, comma-, dot-shaped or short streak below denticle; dorsal arch without mesh-like sclerotization; ventral cornu without striated sclerotization 3

2. – Posterior spiracles with S-shaped slits; peritreme complete and highly chitinized; spiracular buttons visible in early third instar but not in late one; anterior spiracles with five to seven papillae each .... *Synthesiomyia nudiseta*
- Posterior spiracles with cigar shaped slits radiating from button, slits subparallel to each other; peritreme complete but weak; anterior spiracles with seven to eight papilllae each .... *Ophyra spinigera*

3. – Dorsal cornu of cephalopharyngeal skeleton split into two; posterior spiracles with delicate, lightly sclerotized and incomplete peritreme, spiracular slits cigar-shaped, delicate, and not radiating from opening of peritreme .... Family Sarcophagidae
- Dorsal cornu of cephalopharyngeal skeleton intact; posterior spiracles well-sclerotized, peritreme complete or incomplete, button distinct or indistinct, spiracular slits cigar-shaped radiating from opening of peritreme .... Family Calliphoridae(4)

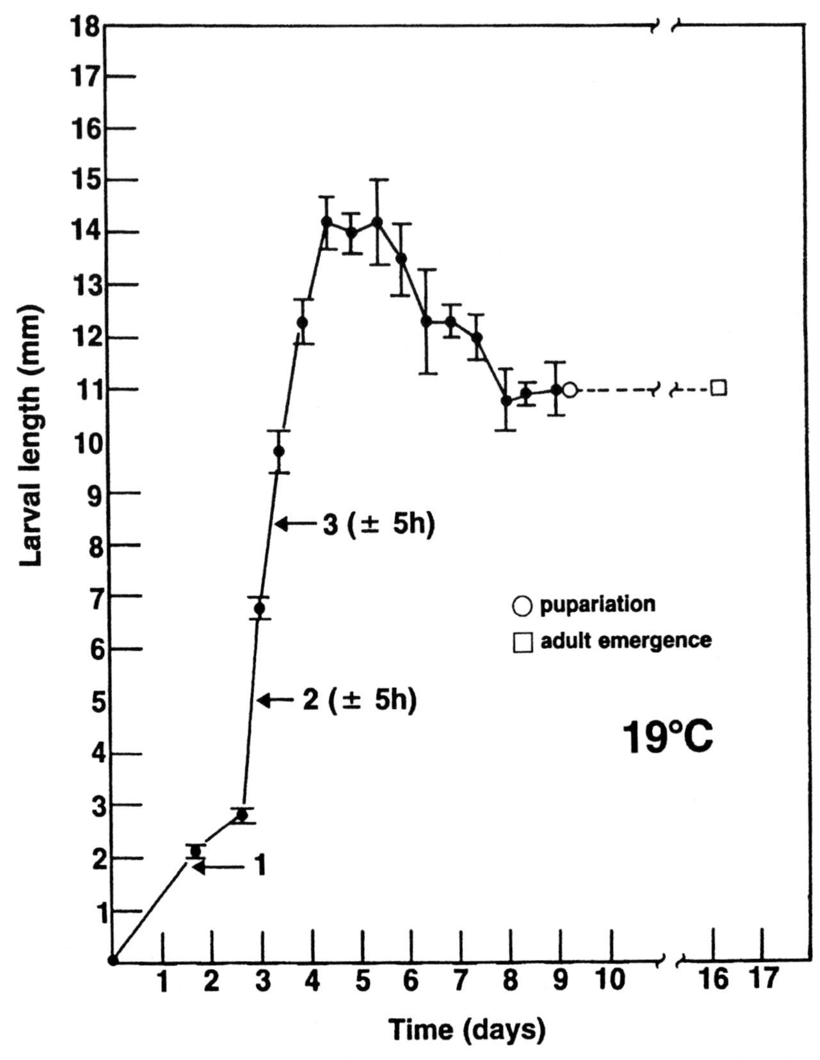

Fig. 3.35. *Phaenicia sericata,* schedule of development at 19 °C. Arrows indicate larval instar.

4. – Posterior spiracles with complete peritremes 5
   – Posterior spiracles with incomplete peritremes...6

5. – Sharp and single-pointed dorsolateral spines found on segments 1 to 7; buttons of posterior spiracles complete; accessory sclerite of cephalopharyngeal skeleton comma-shaped; anterior spiracles with six to nine papillae each..*Hemipyrellia ligurriens*
   – Unicuspid, bicuspid, tricuspid or blunt dorsolateral spines found on segments 2 to 8; stitch-like spines found on segments 3 to 12; midsegmental minute fusiform tuberculations present on segments 5 to 12; accessory sclerite of cephalopharyngeal skeleton, if present, dot-like in shape; posterior

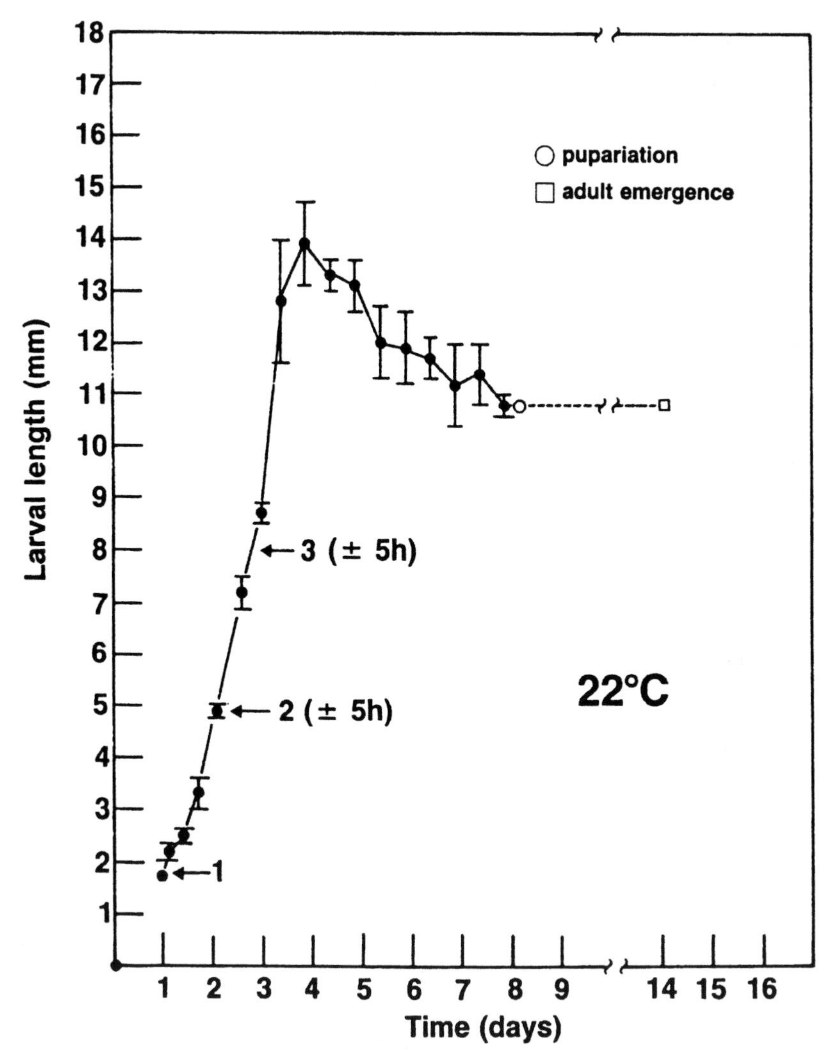

Fig. 3.36. *Phaenicia sericata*, schedule of development at 22 °C. Arrows indicate larval instar.

spiracle weak with incomplete button, peritreme surounded by a thick sclerotized ring; anterior spiracles with 10 papillae each; gross shape of larva highly blunted at posterior end .... *Chrysomya chani*

6. – A row of conical tubercles present on segments 4 to 12 ("hairy" larvae); cephalopharyngeal skeleton without distinct dorsal arch; posterior spiracles with heavily sclerotized peritreme 7
   – Conical tubercles absent on body segments; if present, usually small or inapparent, can be found only on terminal segment of larvae ("smooth" larvae); cephalopharynegeal skeleton with dorsal arch; posterior spiracles with mildly sclerotized peritreme 8

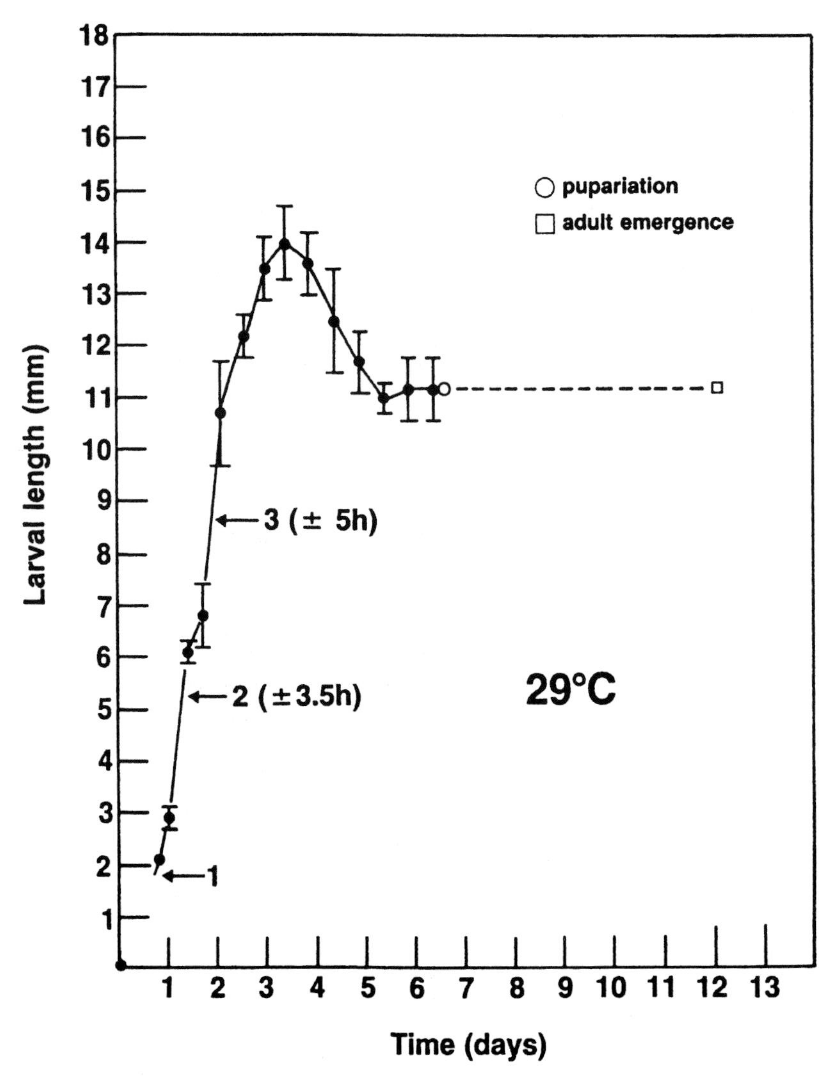

Fig. 3.37. *Phaenicia sericata*, schedule of development at 29 °C. Arrows indicate larval instar.

7. – Conical tubercles randomly covered with blunt spines, crown of each tubercle with sharp spines numbering up to five; anterior spiracles with 13–14 papillae each .... *Chrysomya villeneuvi*
   – Conical tubercles not covered with spines, crown of each tubercle with sharp spines numbering up to 30; anterior spiracles with 10–12 papillae each .... *Chrysomya rufifacies*

8. – Accessory sclerite of cephalopharyngeal skeleton present 9
   – Accessory sclerite of cephalopharyngeal skeleton absent...................................10

9. – Cephalopharyngeal skeleton with very apparent comma or club shaped accessory sclerite; dorsolateral spines unicuspid or bicuspid, mostly unicuspid and blunt, present on all segments except segments 9 and/or 10; posterior spiracles with lightly sclerotized peritreme; anterior spiracles with 9–11 papillae each .... *Chrysomya pinguis*
   – Cephalopharyngeal skeleton with dot or club shaped accessory sclerite; dorsolateral spines unicuspid, bicuspid or tricuspid, mostly blunt in shape, present on all segments except segment 11; posterior spiracles with lightly sclerotized peritreme; anterior spiracles with 11–13 papillae each .... *Chrysomya megacephala*

10. – Laterodorsal spines absent on segments 6 to 10; late third instar larvae with highly pigmented dorsal plate on each segment; cephalopharyngeal skeleton with denticle (hook) strongly bent downward; anterior spiracles with 10 papillae .... *Chrysomya nigripes*
   – Dorsolateral spines absent on segment 11; late third instar larvae without dorsal plate; cephalopharyngeal skeleton with denticle not strongly bent downward; anterior spiracles with 11 to 13 papillae each .... *Chrysomya megacephala*

Key to the calliphorid adults of forensic importance in the Oriental Region

(H. Kurahashi – Department of Medical Entomology, National Institute of Infectious Diseases, Tokyo 162, Japan)

1. – Stem vein of wing without setulae on dorsal side of basal section 2
   – Stem vein of wing with distinct setulae on postero-dorsal side of basal section .... Subfamily Chrysomyinae, 3

2. – Posterior part of suprasquamal carina with posterior parasquamal tuft of black erect hairs on small well-defined black sclerite; thoracic squama quite bare on upper surface; body mostly metallic green to blue, sometimes coppery .... Subfamily Luciliinae, 19
   – Posterior parasquamal tuft absent; thoracic squama hairy on upper surface; body blackish, usually with abdomen metallic blue and more or less dusted .... Subfamily Calliphorinae 36

3. – Wing broadly infuscated; face and antennae bright yellowish orange. Pakistan (Baluchistan) .... *Chrysomya regalis* Robineau-Desvoidy (= *marginalis* (Wiedemann)
   – Wing entirely hyaline 4

4. – External vertical bristle (vte) well developed in both sexes; female tergite 5 with median posterior cleft 5
   – No vte in male; female tergite 5 without median posterior cleft 9

5. – Face, epistome, parafacialia, vibrissarium and mediana testaceous yellow, face and parafacialia sometimes darkened in male; tergite 3 without row of marginal bristles 6
   – Face, epistome blackish; tergite 3 with row of fine marginal bristles 8

6. – Tergites 3 to 4 without distinct marginal band; tergites mainly clothed with black hairs; mesothoracic spiracle fuscous brown; femora remarkably swollen, metallic blue to purple in both sexes; male metatarsus with brush of short spines on entire length .... *C. villeneuvi* Patton
   – Tergites 3 to 4 with broad marginal band; tergites largely clothed with pale

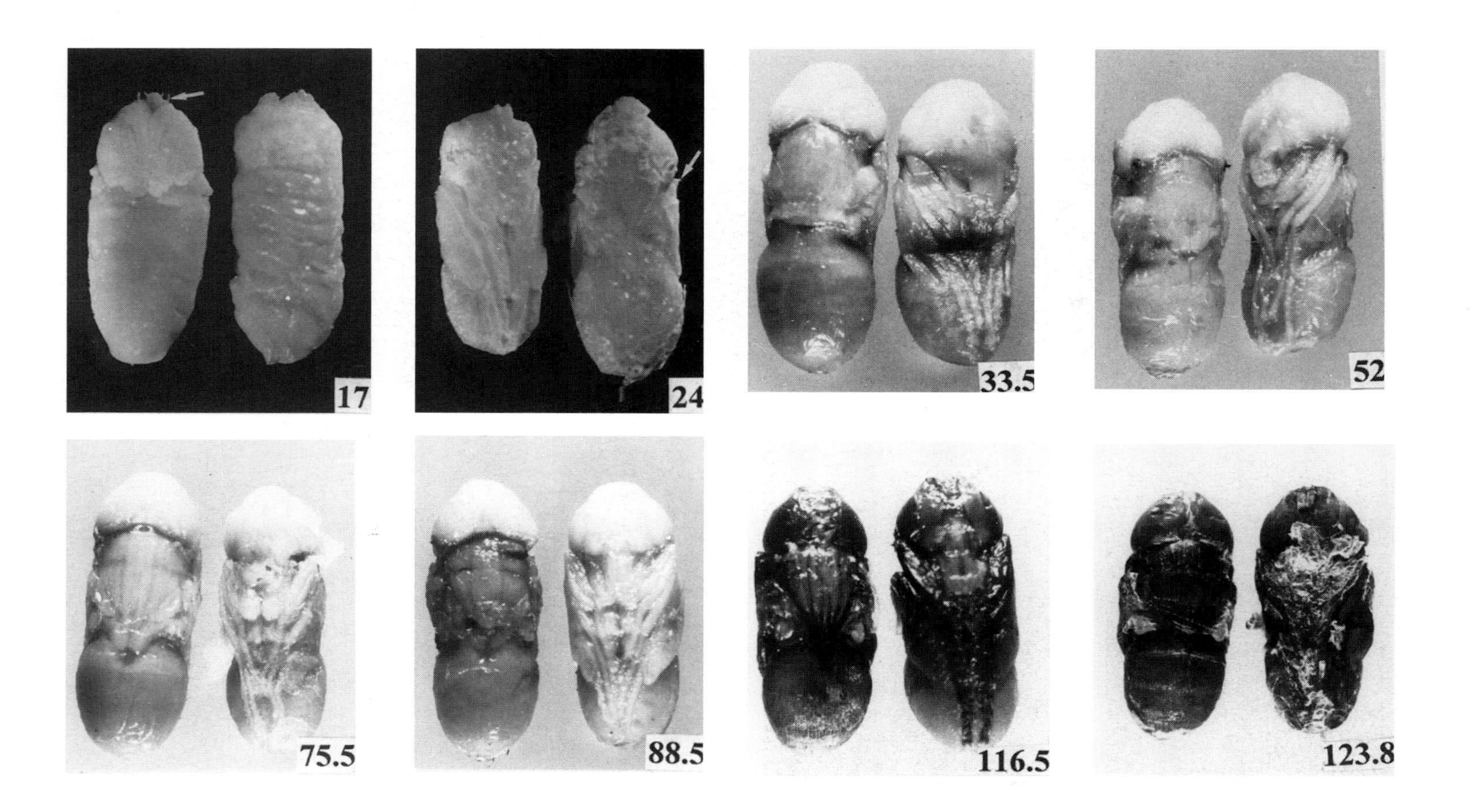
17
24
33.5
52
75.5
88.5
116.5
123.8

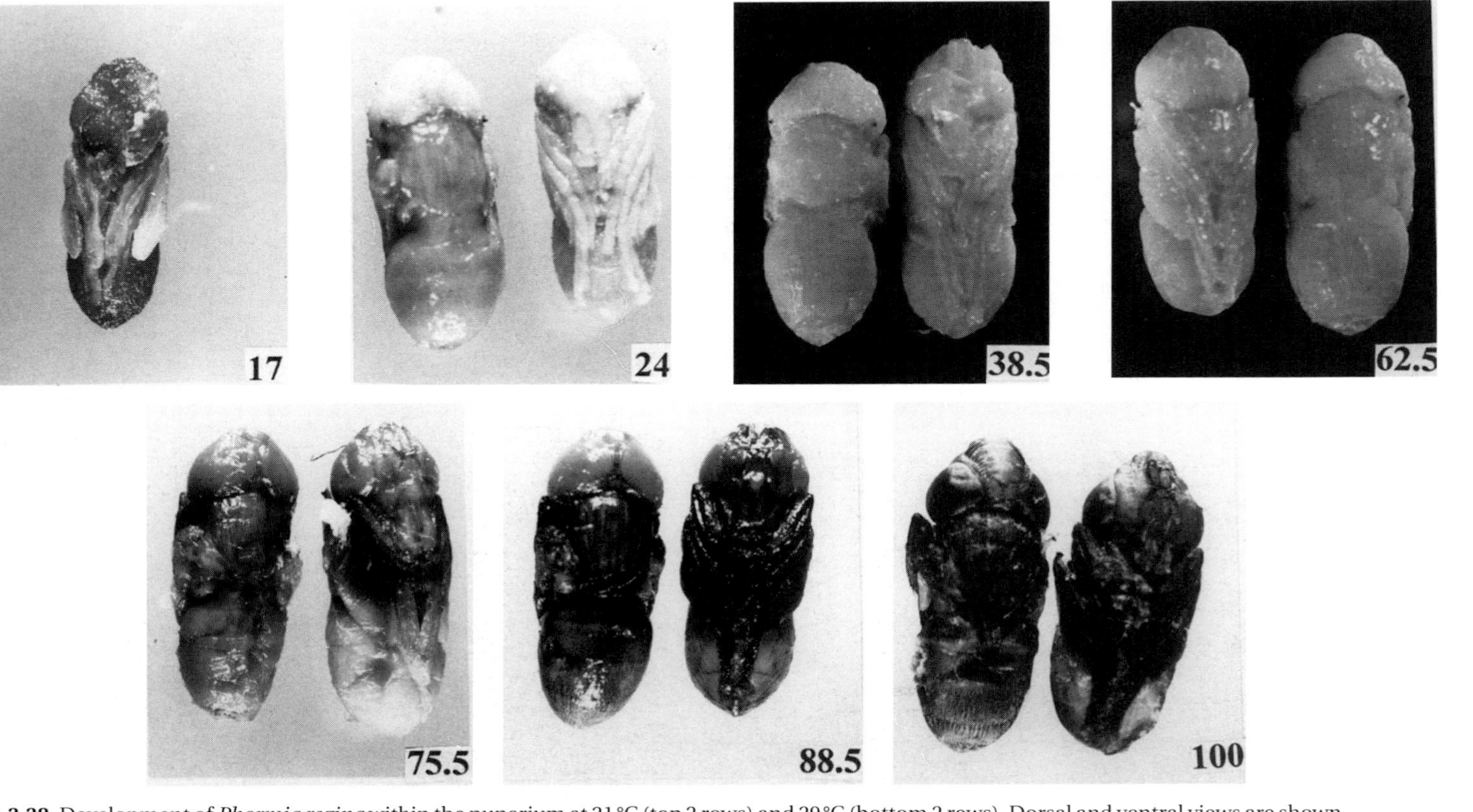

Fig. 3.38. Development of *Phormia regina* within the puparium at 21 °C (top 2 rows) and 29 °C (bottom 2 rows). Dorsal and ventral views are shown. Numbers indicate hours after formation of the white puparium. Arrows indicate the respiratory horns. The premature dark colors, e.g. 88.5 hours at 21 °C and 24 hours at 29 °C, are artifacts of preservation.

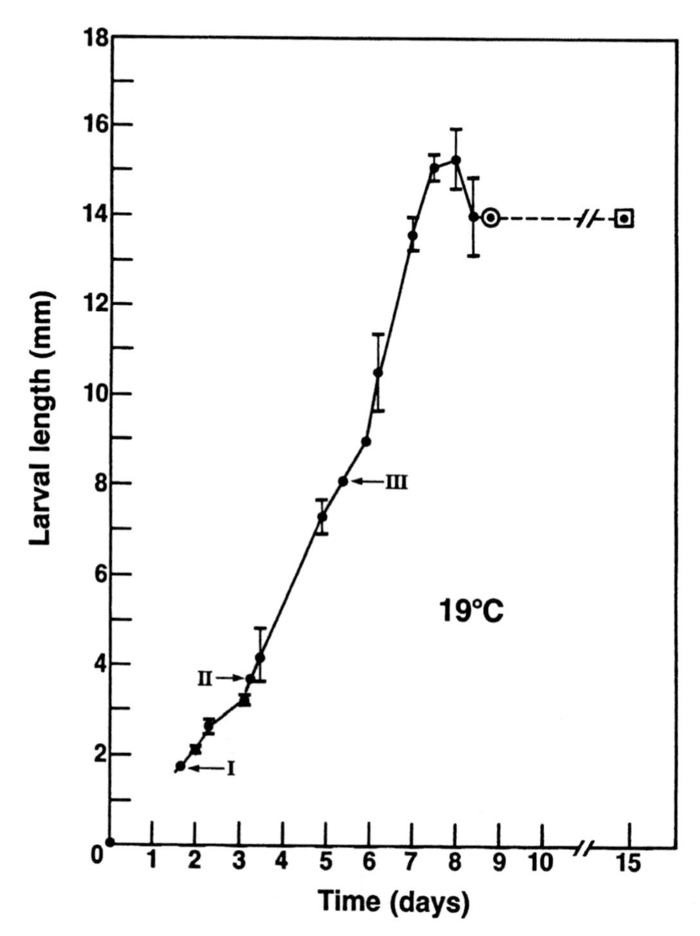

Fig. 3.39. *Phormia regina*, schedule of development at 19 °C. ○, pupariation; □, eclosion. Roman numerals indicate larval instar.

hairs, especially on venter; mesothoracic spiracle yellowish white; femora black, submetallic, but not remarkably swollen; metatarsus without brush of short spines in male 7

7. – Prostigmatal bristle absent; gena reddish anteriorly or entirely. Baluchistan, Pakistan .... *C. albiceps* (Wiedemann)
   – Prostigmatal bristle present .... *C. rufifacies* (Macquart)

8. – Tergites 3 to 4 with distinct broad marginal band in both sexes; tergite 5 with several discal bristles; sternite clothed with yellow hairs; parafacialia and vibrissarium brownish. Mindanao, Philippines .... *C. schoenigi* Kurahashi and Magpayo
   – Tergites 3 to 4 without marginal band in male, in female with distinct band; tergite 5 without discals; sternite 2 largely covered with black hairs; parafacialia and vibrissarium blackish. Indonesia (Sulawesi) .... *C. yayukae* Kurahashi and Magpayo

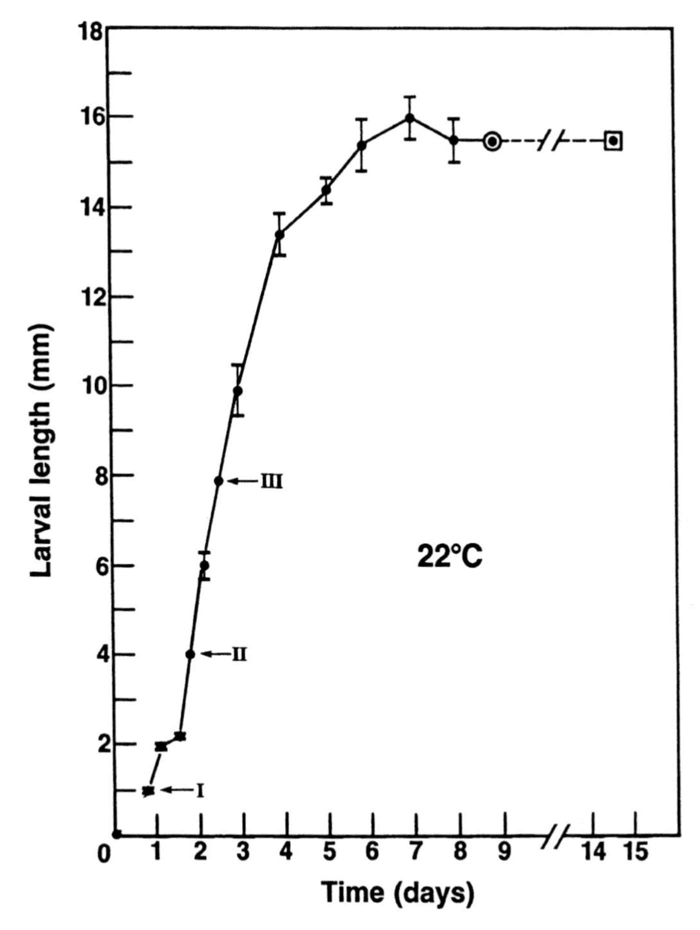

Fig. 3.40. *Phormia regina*, schedule of development at 22 °C. ○, pupariation; □, eclosion. Roman numerals indicate larval instar.

9. – Gena yellow or orange 10
   – Gena fuscous 11

10. – Alar and thoracic squamae entirely white; female frontal stripe parallel-sided (larva myiasis producer, may not have forensic importance) .... *C. bezziana* Villeneuve
    – Only base of squamae white; frontal stripe broadest in middle (Synanthropic species) .... *C. megacephala* (Fabricius)

11. – Mesothoracic spiracle white 12
    – Mesothoracic spiracle fuscous 13

12. – Presutural ial present; st 1 + 1. Philippines (Samar) .... *C. samarensis* Kurahashi
    – Presutural ial absent; st 0 + 1 .... *C. nigripes* Aubertin

13. – At least basal part of alar squama white 14
    – Alar and thoracic squamae entirely fuscous 16

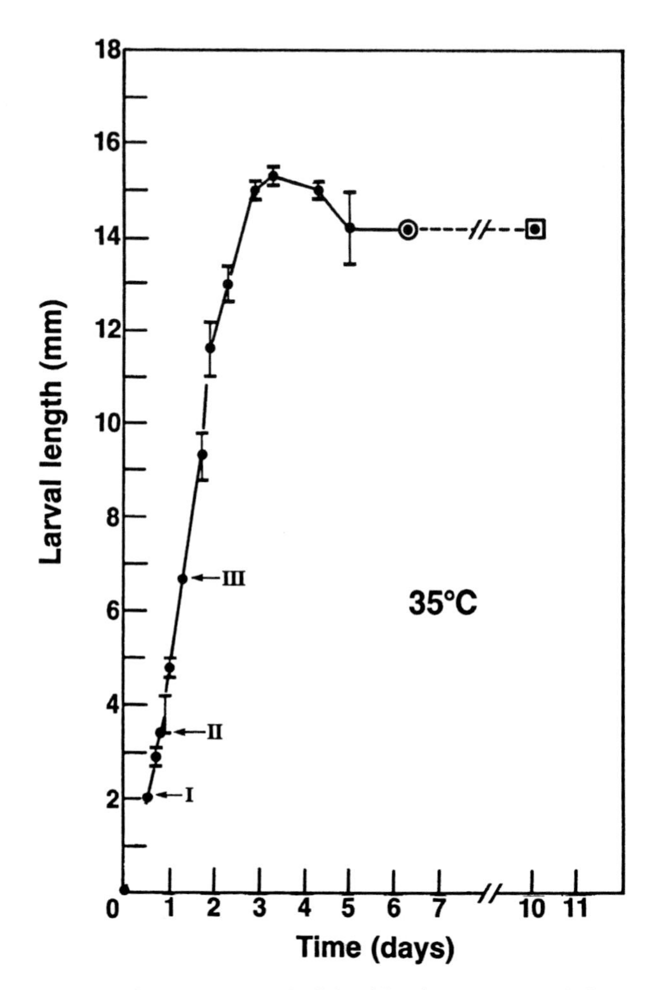

Fig. 3.41. *Phormia regina*, schedule of development at 35 °C. ○, pupariation; □, eclosion. Roman numerals indicate larval instar.

14. – Head dichoptic in male, with widely separated eyes; female tergite 6 split into lateral plates (Montane species, Himalayas) .... *C. phaonis* (Séguy)
    – Head holoptic in male, with eyes almost meeting in center 15
15. – Mesothoracic spiracle small, long axis shorter than third antennal segment; opaque white part of alar squama bare ventrally except for brownish fringe; male head hemispherical in profile; female sternite 5 subequal to 4 in length [Lowland forest species] .... *C. chani* Kurahashi
    – Mesothoracic spiracle large, longer and much broader than third antennal segment; opaque white part of alar squama hairy and with whitish fringe; male head distinctly flattened in profile; female sternite 5 longer than 4. Indonesia (Sulawesi) (Montane species) .... *C. greenbergi* Wells and Kurahashi
16. – Body purple; length usually more than 11 mm; gena reddish, clothed with

golden hairs; tergite 5 often with white dusting; no posthumeral bristle (ph), sometimes weakly developed in female. [Montane species] .... *C. thanomthini* Kurahashi

– Body blue or green, sometimes purple, length usually less than 11 mm: gena fuscous with blackish hairs; tergite 5 metallic, without white dusting; ph developed.. 17

17. – Length of head at epistome less than or equal to length at base of antenna. Wallacea (Lowland forest species) .... *C. cabrerai* Kurahashi and Salazar

– Length of head at epistome greater than length at base of antenna..................... 18

18. – Body blue to purple; postgena usually covered with yellowish hairs; length of gena in profile more than that of eye, height of gena compared to total head about three-tenths in male, about four-tenths in female; male cercus elongate, approximately three times length of surstylus (Montane forest species) *C. pinguis* (Walker)

– Body dark green; postgena usually covered with black and brown hairs; gena in profile same length as eye; height of gena compared to total head about two-tenths in male, about three-tenths in female; male cerci stout, not more than two times length of surstylus. (Lowland forest species) ...... *C. defixa* (Walker)

19. – Supraspiracular convexity clothed with long, upstanding, fine hairs.................. 20

– Supraspiracular convexity bare or pubescent .... *Lucilia*, 27

20. – Legs in male more or less fringed; hypopygium strongly developed; generally large flies, more than 15 mm in length .... *Hypopygiopsis*, 21

– Legs in male not fringed; hypopygium normal; medium and small flies, less than 10 mm in length .... *Hemipyrellia*, 24

21. – Antennae yellowish orange; facial tomentum golden yellow; tarsi in male without fringe 23

– Antennae dark reddish brown; facial tomentum silver white; tarsi in male with long fringes.. 22

22. – Alar and thoracic squamae fuscous brown; hind tibia with two long and fine apical d and ad which are curled apically in male [Lowland rain forest species] .... *H. violacea* (Macquart)

– Alar and thoracic squamae whitish, with pale brown tinge; hind tibia with stout strong apical d and ad in male [Lowland forest species] .... *H. infumata* (Bigot)

23. – Male: mid and hind femora remarkably stout, hind one curved and heavily fringed; mid and hind tibiae with strongly developed fringe; mid tibia with characteristic apical projection. Female: posterior margin of tergite 5 subequal to length of lateral margin of same tergite [Lowland rain forest species].. *H. fumipennis* (Walker)

– Male: mid and hind femora normal, hind one slightly curved; mid and hind tibia short and rather sparsely fringed, mid one without chitinous projection. female: posterior margin of tergite 5 about half length of lateral margin of same tergite [Lowland forest species] .... *H. tumrasvini* Kurahashi

24. – Large flies having very prominent hypopygium in male; mesopleuron with a few golden hairs among mesopleural bristles. Christmas Island. Indian Ocean .... *H. jucunda* (Kirby)

– Smaller flies; hypopygium sometimes prominent, but not markedly conspicuous; mesopleuron without golden hairs.. 25

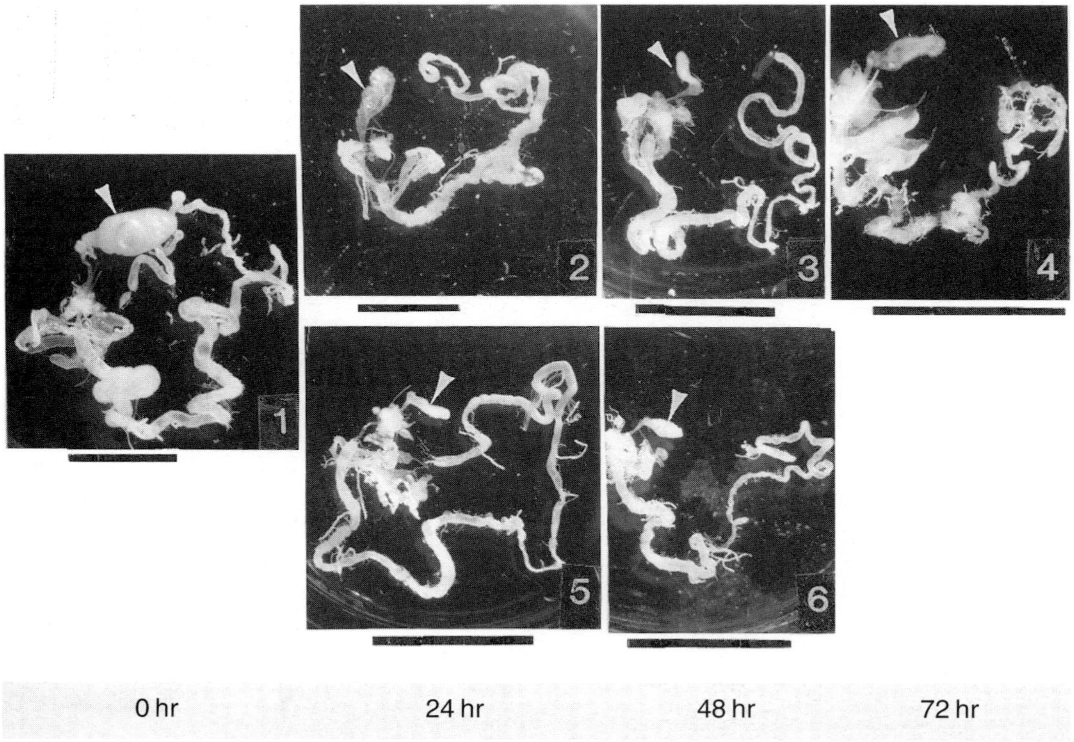

1 2 3 4 5 6

0 hr 24 hr 48 hr 72 hr

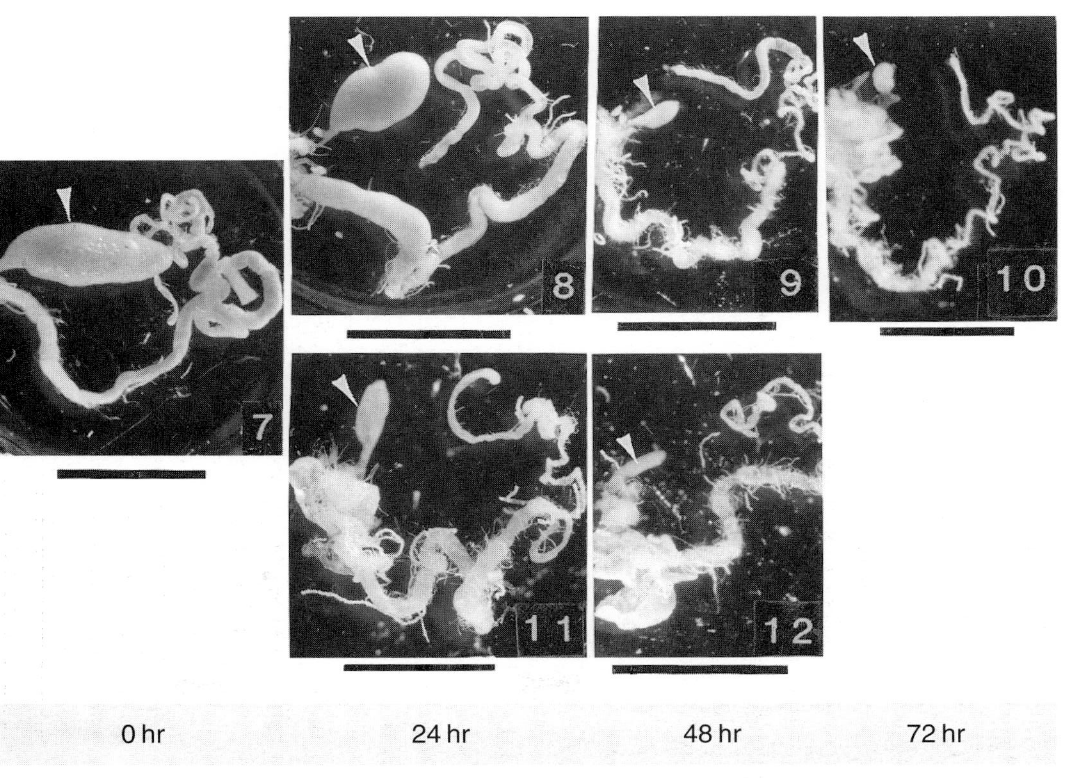

Fig. 3.42a. Reduction in size of the crop (arrows) in postfeeding larvae of *Phormia regina* (top 2 rows) and *Calliphora vicina* (bottom 2 rows) at 19 °C (2, 3, 4 and 8, 9, 10), and at 29 °C (5, 6 and 11, 12). Bars = 5 mm, except #4 = 1.8 mm. See Fig. 2.2 for details of the digestive tract.

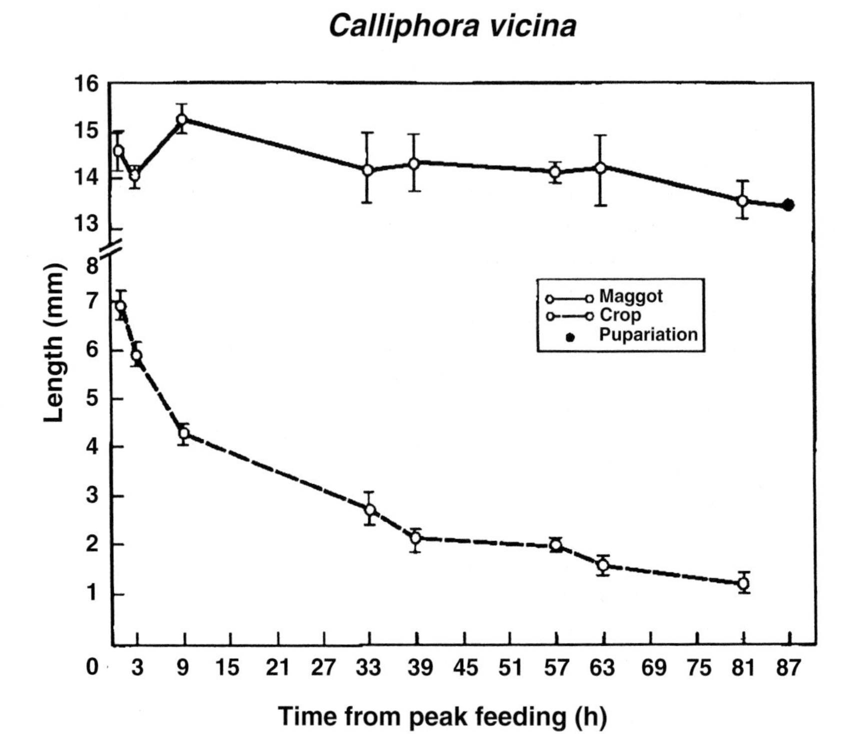

Fig. 3.42b. Maggot length versus crop length from peak of feeding until pupariation in *Calliphora vicina* at 22 °C.

25. – Third antennal segment entirely bright orange [Savanna or dry forest species] .... *H. pulchra* (Wiedemann)
  – Third antennal segment generally fuscous brown 26

26. – Thoracic squama pure white; male head holoptic; male abdomen densely convered on ventral surface with long hairs, the longest one nearly as long as length of arista; lateral lobes of epandrium oval with long fine hairs in male; facial tomentum silver white; female parafacialia silvery white; abdomen thinly dusted, usually metallic bluish green [Lowland forest species] .... *H. tagaliana* (Bigot)
  – Thoracic squama brownish white; male head subholoptic, with eyes separated by distance greater than width of third antennal segment; hairs on ventral surface of abdomen shorter than length of arista; lateral lobes of epandrium elongate, testaceous, very sparsely haired; facial tomentum greyish; female parafacialia grey; abdomen heavily dusted, usually metallic copper green [Synanthropic species] .... *H. ligurriens* (Wiedemann)

27. – Wings strongly infuscated along costal margin; parafacialia with row of fine hairs. Philippines .... *L. fumicosta* Malloch
  – Wings without demarcated costal infuscation; parafacialia bare 28

28. – Basicosta yellow 29
  – Basicosta black 30

29. – Male sternites with tuft of long hairs; male abdomen usually arched in profile; female body usually brassy or coppery on greenish background, with dense pruinosity; cerebrale in male bearing one occipital hair-like seta on each side [Synanathropic species] .... *L. cuprina* (Wiedemann)
 – Male sternites without tuft of long hairs; male abdomen not conspicuously arched in profile; female body usually metallic green, sometimes more or less with coppery tinge; cerebrale in male bearing five to eight hair-like setae on each side [Synanthropic species] .... *L. sericata* (Meigen)

30. – Abdomen without distinct black marginal bands; genal hairs long throughout 31
 – Abdomen with broad black marginal bands; genal hairs short along upper extremity 32

31. – Wing usually more strongly brownish-tinged; tergite 3 sometimes with lateral purple marginal band [Forest species] .... *L. salazarae* Kurahashi
 – Wing hyaline or more or less pale brownish; tergite 3 with no lateral marginal band [Forest species] .... *L. porphyrina* (Walker)

32. – Male: narrowest part of frons less than width of ocellar triangle; parafacialia narrower than width of third antennal segment along upper extremity 33
 – Male: narrowest part of frons as wide as width of ocellar triangle; parafacialia as broad as or broader than width of third antennal segment along upper extremity 35

33. – Alar and thoracic squamae both fuscous brown [Montane forest species] .... *L. hainanensis* Fan
 – Alar squama yellowish brown, thoracic one fuscous brown 34

34 – Thoracic squama largely infuscated; occiput with only one row of black postocular setae; frons index 0.19–0.20 in female [Montane forest species] .... *L. sinensis* Aubertin
 – Thoracic squama pale, brownish on disc; occiput with more than two irregular rows of black postocular setae [Montane forest species] .... *L. bazini* Séguy

35. – Female: alar and thoracic squamae entirely fuscous, the former with tuft of blackish hairs on inner lower margin; parafacialia in male at least as broad as width of third antennal segment; frons index in female 0.24–0.25 [usually found in montane forests more than 1500 m above sea level] .... *L. papuensis* Macquart
 – Female: alar squama fuscous to yellowish brown with tuft of dark brown to black hairs on inner lower margin, thoracic one fuscous to brown; parafacilia in male more or less broader than width of third antennal segment [Lowland forest species] .... *L. calviceps* Bezzi
 – Female: alar and thoracic squamae entirely or largely whitish, the former usually paler than the latter, with tuft of yellow hairs on inner lower margin; parafacialia in male as broad as width of third antennal segment [Lowland forest species] .... *L. bismarckensis* Kurahashi

36. – Presutural ial absent, rarely weakly developed 37
 – Presutural ial well developed 39

37. – Face, gena and postgena yellowish orange; epistome remarkably projecting forward, nearly by width of third antennal segment .... *Calliphora* (*Paracalliphora*), 38
 – Face, gena and postgena black; epistome very slightly projecting forward. Northern Vietnam [Montane and synanthropic species] .... *Aldrichina grahami* (Aldrich)

38. – Humeri reddish; thorax reddish in part; legs yellowish brown on coxae, apices of femora, and bases of tibiae; abdomen reddish on lateral sides. Indonesia (Sulawesi) [Montane forest species] .... *C. hasanuddini* Kurahashi and Selomo
   – Humeri bluish black, metallic, concolorous with thoracic dorsum; thorax entirely bluish black; legs largely blackish; abdomen entirely metallic blue. Malaysia, Indonesia (Sumatra, Java) and Philippines (Mindanao) [Montane forest species] .... *C. fulviceps* van der Wulp

39. – Posthumeral bristles two; alar and thoracic squamae both whitish at least in part 40
   – Posthumeral bristles three; squamae entirely blackish to fuscous brown except for pale margin 41

40. – Postsutural acr two; squamae whitish at base, largely fuscous brown. Himalayas [Alpine species] .... *C. himalayana* Kurahashi and Thapa
   – Postsutural acr three; squamae entirely or largely whitish. Himalayas [Alpine species] .... *C. chinghaiensis* Van and Ma

41. – Gena reddish on anterior two-thirds; parafacialia reddish; basicosta usually yellowish brown, but variable, sometimes blackish; mesothoracic spiracle orange; male head subholoptic, eyes separated by more than width of ocellar triangle [Synanthropic species] .... *C. vicina* Robineau-Desvoidy
   – Gena entirely black in ground colour; parafacialia fuscous; basicosta black; mesothoracic spiracle usually fuscous to blackish, rarely yellowish on lower part or entirely in one case; male head holoptic 42

42. – Posterior surface of postgena clothed with black hairs only. Himalayas [Alpine species] .... *C. loewi* Enderlein
   – Posterior surface of postgena yellowish-haired, intermixed with black hairs. Himalayas [Alpine species] .... *C. pattoni* Aubertin
   – Posterior surface of postgena clothed with yellowish hairs, so postgena entirely yellowish-haired. [Subalpine to alpine, synanthropic species] .... *C. vomitoria* (Linnaeus)

Key to the Sarcophagidae of the Oriental Region

(H. Kurahashi)

*This key is provided only for the male flies of forensic importance commonly found in the continental areas of the Oriental region such as Thailand and Malaysia; it does not cover all the species of Oriental flesh flies. (See also Roback, 1954; Sugiyama and Kano, 1984.)

1. – Hind coxa hairy on posterior surface; ntp 4, two strong primary bristles, two smaller subprimary bristles; sternites 3 to 4 fully exposed and overlapping ventral margins of correponding tergite .... Subfamily Sarcophaginae 2
   – Hind coxa bare on posterior surface; ntp 2; sternites 3 to 4 more or less concealed by ventral margin of corresponding tergite [No forensic importance] .... Subfamilies Miltogramminae and Paramacronychiinae

2. – Rows of frontal bristles parallel-sided. Himalayas [Montane and synanthropic species] .... *Ravinia pernix* (Harris)
   – Rows of frontal bristles divergent below level of antennal bases 3

3. – Postsutural dc 3, equally well developed, arranged at similar interval 4
   Postsutural dc 4, arranged at regular intervals, anterior 2 moderately developed in subequal length 6

– Postsutural dc 4–5; anterior two to three pairs vestigial or at most reduced to half length of posterior 2; if dc 4, then they are arranged at irregular intervals 20

4 – Presutural ac 2–4, well developed; cercus distinctly bent dorsally [Larvae parasitic, no forenstic importance] .... *Blaesoxipha*

– Presutural ac usually absent or vestigial, at most weakly developed as presuturals; cercus slender, not bent dorsally [Probably forensic importance] .... *Pierretia*...5

5. – First longitudinal vein r1 setulose; costa section 5 with short spines along anterior margin; sternite 4 covered with normal hairiness .... *P. calicifera* (Boettcher)

– First longitudinal vein r1 bare; costal section 5 with short spines only on basal half of anterior margin; sternite 4 almost entirely covered with mat of short hairs and some long ones on lateral margin .... *P. melania* Shinonaga and Tumrasvin

6. – Propleuron bare; anterior paramere usually slender, sometimes stout, but not bifid anteriorly 7

– Propleuron hairy; anterior paramere large and stout, bifid anteriorly; presutural ac weakly developed and arranged in row anteriorly .... *Sarcorohdendorfia*...11

7. – Presutural ac 2–4, well developed; 2–4 upper frontal bristles including preverticals rather strongly developed, reclinate 15

– Presutural ac usually absent or vestigial, at most weakly developed 8

8. – Sternite 5 with protuberance on middle; aedeagus with large and stout stylus .... *Phallosphaera*...9

– Sternite 5 without protuberance; aedeagus with slender stylus 10

9. – Three rows of black postocular setae regular; third antennal segment three times as long as second; cercus very large, widely broadened on dorsal surface and curved at apical half, without tuft of long hairs near distal end .... *P. kurahashii* Shinonaga and Tumrasvin

– Only first row of black postocular setae regular; third antennal segment four times as long as second; cercus not so broad and stretched, with tuft of long hairs in subapical portion .... *P. gravelyi* (Senior-White)

10. – Aedeagus with stylus as long as or slightly longer than juxta .... *Sinonipponia hainanensis* (Ho)

– Aedeagus with stylus large, covered with microscopic white hairs .... *Lioproctia beesoni* (Senior-white) (in part)

11. – Sternite 4 usually with patch of brushy hairs; gena covered with black hairs more than anterior half 12

– Sternite 4 without patch of brushy hairs; gena almost entirely clothed with whitish hairs except for only a few black ones on anterior extremity .... *S. montana* Shinonaga and Tumrasvin

12. – Wing yellowish tinged along anterior margin and veins posteriorly; cercus slightly curved and with groove basally; juxta with basal apophyses .... *S. flavinervis* (Senior-White)

– Wings hyaline; cercus strongly bent and without groove basally; juxta without apophyses 13

13. – Hairs on sternite 4 very long and equal in length to those on sternite 3; inner

ridge of posterior surface of cercus with some fine long hairs and outer ridge with spines; ventralia two-wave-like shaped and with hook-like apical portion .... *S. multivillosa* Shinonaga and Tumrasvin
- Hairs on sternite 4 much shorter than those of sternite 3; inner ridge of posterior surface of cercus bare, only with spines at outer ridge; ventralia well developed 14

14. – Occiput with many (12–14) black setulae below poc; ventralia serrated on anterior margin of apical half .... *S. inextricata* (Walker)
- Occiput with two to four black setulae below poc; ventralia lobe-like with no serration .... *S. antilope* (Boettcher)

15. – Hind tibia with one av, with fringe of long hairs sparsely arranged on anteroventral and posteroventral surfaces; aedeagus with well developed lobe-like vesica; stylus very long, thread-like, about twice as long as juxta .... *Fengia ostindicae* (Senior-White)
- Hind tibia with two av, sometimes one av, but without fringe; aedeagus without vesica .... *Sarcosolomonia*...16

16. – Tergite 3 always with median marginal bristle; cercus with subapical spines .... *S. shinonagai* Kano and Sooksri
- Tergite 3 without median marginal bristle; cercus without subapical spines 17

17. – First longitudinal vein r1 setulose 18
- First longitudinal vein r1 bare 19

18. – Anterior paramere with flange; ventralia bifid at apex .... *S. rohdendorfi* Nandi
- Anterior paramere without flange; ventralia pointed at apex .... *S. trifulcata* Shinonaga and Tumrasvin

19. – Middle of cercus with free part bent upward anteriorly; juxta small .... *S. aureomarginata* Shinonaga and Tumrasvin
- Middle of cercus normal, forked at apex; juxta large .... *S. crinita* (Parker)

20. – Arista pubescent, longest hairs as long as basal diameter; ntp 2; st 1 + 1 [Sea shore species] .... *Leucomyia alba* (Schiner)
- Arista long plumose; ntp 4; st 1 + 1 + 1 21

21. – Postsutural ac absent or fine 22
- Postsutural ac present as prescutellars 23

22. – Genital segments reddish orange; gena largely clothed with white hairs posteriorly; anterior part of alar squama with a tuft of white hairs at inner lower margin .... *Bercaea africa* (Wiedemann)
- Genital segments blackish; gena almost entirely clothed with black hairs; anterior part of alar squama with tuft of black hairs at inner lower margin .... *Parasarcophaga javana* (Macquart) (in part)

23. – Propleuron hairy 24
- Propleuron bare 31

24. – Sternite 5 with prominent conical protuberance in middle part .... *Rosellea khasiensis* (Senior-White)
- Sternite 5 without protuberance in middle part 25

25. – Hind tibia usually without fringe, at most weak, sparse, fringe present only on posterocentral surface, usually with two av; sternite 4 with rather long hairs on median part of posterior margin; ventralia large, globose, with numerous spines .... *Boettcherisca*...26

- Hind tibia with well developed fringe, if fringe poorly developed, then only one av present; sternite 4 without remarkable hairs on median part of posterior margin; ventralia not globose, without spine 28

26. – Black genal hairs not extending beyond anterior half; spines on apical part of cercus not extending to dorsal side ....*B. peregrina* (Robineau-Desvoidy)
- Black genal hairs extending beyond anterior half; spines on apical part of cercus extending to dorsal side 27

27. – Vesica reduced to small rounded shape.......................................*B. javanica* Lopes
- Vesica largely lobulated with small dorsal expansion...................*B. nathani* Lopes

28. – No membranous region between theca and corpus; lower half of ventralia highly sclerotized and serrated.....................................*Hosarcophaga serrata* (Ho)
- Between theca and corpus membranous; ventralia not serrated.........................29

29. – First longitudinal vein (r1) setulose; corpus poorly sclerotized; juxta largely rounded apically; ventralia reduced to small lobe; body length 8.0–8.5 mm [Sea shore species] ....*Alisarcophaga gressitti* (Hall and Bohart)
- Vein r1 bare; corpus highly sclerotized, slightly curved apically; ventralia well developed; body length 14.5–15.0 mm..30

30. – Hind tibia with well developed fringe on anteroventral and posteroventral surfaces; abdomen grey-dusted ....*Lioproctia pattoni* (Senior-White)
- Hind tibia with poor fringe on posteroventral surface; abdomen yellowish golden dusted on tergites 4 and 5.. ....*Lioproctia notasbilis* (Kano and Lopes)

31. – Sternites 2 and 3 densely clothed with long hairs; aedeagus with stylus large, covered with microscopic white hairs ....*Lioproctia beesoni* (Senior-White)
- Sternite 2 and 3 normally with sparse hairs; aedeagus with stylus slender, never covered with microscopic hairs..32

32. – Sternite 4 with mat of hairs...33
- Sternite 4 without mat of hairs..34

33. – Mid tibia with well developed fringe on antero- to postero ventrally, without v, rarely with slender av; hind tibia usually without av; adeagus very large; theca long; juxta rounded and wholly membranous; body length 13.0–17.0 mm ....*Seniorwhitea princeps* (Wiedemann)
- Mid tibia without or at most with short fringe posteroventrally; hind tibia with one stout av; aedeagus normal in size; theca short; juxta long and slender and slightly sclerotized; body length 7.0–10.5 mm ....*Harpagophalla kempi* (Senior-White)

34. – Sternite 4 sparsely haired; mid femur with comb-like pv in apical part; hind tibia usually fringed...*Parasarcophaga*.35
- Sternite 4 conspicuously dense-haired; mid femur without comb-like pv; hind tibia never fringed...47

35. – Antenna, palpus and genital segments bright orange; ventralia reduced to tubercles ....*P. ruficornis* (Fabricius)
- Antenna and palpus variable in colour, usually blackish; genital segments light to dark brown, orange-brown or black; ventralia developed 36

36 – Ventralia pedunculated...37
- Ventralia not pedunculated 39

37. – Palpus orange, if darkened, then hind tibia with two av on apical third .... *P. misera* (Walker)
– Palpus fuscous brown to black........ 38

38. – One row of black postocular setae; gena with only a few black hairs at anterior extremity; hind femur without fringe; hind tibia without av; ac 1 + 1 [Common synanthropic species] .... *P. taenionota* (Wiedemann)
– Three rows of black postocular setae, but only one row regular; gena with numerous black hairs anteriorly; hind femur with fringe on posteroventral surface; hind tibia with one av; ac 0–1 + 1 [Common synanathropic species) .... *P. albiceps* (Meigen)

39. – Juxta composed of small or slender median apophysis and pair of long and straight or curved arms; ventralia commonly sheet-like 40
– Juxta composed of large and broad median process and pair of long and curved arms; ventralia mostly filamentous .... *P. spinipenis* Shinonaga and Tumrasvin

40. – Sternites 2 & 3 with dense long hairs, longer than length of each sternite .... *P. javana* (Macquart) (in part)
– Sternites 2 & 3 shorter than length of each sternite 41

41. – Hind tibia without fringe........ 42
– Hind tibia with fringe 44

42. – Juxta small with lateral processes, bifid at apex.... *P. scopariiformis* (Senior-White)
– Juxta elongated, with median apophysis 43

43. – Membranous lobe of paraphallus rod-like, blunt at apex; mid tibia with two ad .... *P. iwuensis* (Ho)
– Membranous lobe of paraphallus divergent, curled at apex; mid tibia with one ad........ *P. yunnanensis* Fan

44. – One row of black postocular setae........ 45
– More than one row of black postocular setae 46

45. – Large fly; body length 13.0–14.0 mm; cercus enlarged, broadened laterally, with small spines along anterior margin of dorsal surface; vesica composed of three lobes; ventralia serrated; lateral arm of juxta pointed at apex .... *P. amplicercus* Shinonaga and Tumrasvin
– Medium-sized fly; body length 10.0–13.0 mm; cercus normal, not enlarged dorsally, without spines; vesica composed of one lobe; lateral arm of juxta bifid at apex .... *P. dux* (Thomson)

46. – Black genal hairs located only on anterior part of gena; cercus beak-shape at apex .... *P. brevicornis* (Ho)
– Black genal hairs extending beyond anterior half of gena; cercus pointed at apex........ *P. idmais* (Se'guy)

47. – Juxta united to corpus with suture, slightly sclerotized........ *Thyrsocnema*...48
– Juxta well differentiated from corpus, mostly membranous .... *Kanomyia bangkokensis* Shinonaga and Tumrasvin

48. – Tergites 4 and 5 covered with golden dusting; hind tibia with long posteroventral fringe; dc 4; body length 6.0–10.0 mm .... *T. bornensis* Shinonaga and Lopes
– Tergites 4 and 5 covered with usual grey dusting; hind tibia without fringe; body length 5.0–8.0 mm........ *T. caudagalli* (Boettcher)

Key to Third Instar Larvae of Carrion-Breeding Diptera of Forensic Importance in South-eastern Australia

(J. F. Wallman, Department of Biological Sciences, University of Wollongong, New South Wales 2522, Australia)

* South-eastern Australia herein constitutes the Australian Capital Territory (ACT), southern New South Wales (NSW), southern South Australia (SA), Tasmania (Tas) and Victoria (Vic). The key contains species commonly encountered in carrion in this region (although species may also occur in other Australian states). The key may misidentify larvae where intraspecific variation renders a character unreliable or where larvae belong to a species not included here. In any case, the morphological difference between some pairs of species is very subtle and such species may therefore be confused. Where there is doubt about the identity of larvae, they should be reared through to the adult stage, and/or a molecular approach to their identification should be adopted. The abbreviations 'AS' and 'TS', in reference to calliphorid larvae, are explained in Fig. 3.47a. Figures 3.43 to 3.49 illustrate the key features of the third instar larvae presented here (Holloway, 1985; Smith, 1986).

1. – Larvae slightly flattened dorsoventrally; with short processes laterally and dorsally; posterior spiracles each having four openings mounted on a brown, sclerotised tubercle (e.g.Fig.3.43a) (ACT, NSW, SA, Tas, Vic) .... Phoridae
   – Posterior spiracles with three openings, not mounted on a tubercle 2
2. – Last abdominal segment with three pairs of lateral tubercles and two pairs of posterior tubercles, one posterior pair dorsal to the spiracular disc and the other posterior pair ventral to the disc and projecting backwards conspicuously (Fig.3.6a(c)) (ACT, NSW, SA, Tas, Vic) .... *Piophila casei* (Linnaeus)
   – Last abdominal segment with tubercles absent laterally, but with at least four pairs of posterior tubercles (which may be flattened and inconspicuous) surrounding the spiracular disc (e.g.Fig. 3.43b) 3
   – Posterior spiracles in a deep cavity on the last abdominal segment (ACT, NSW, SA, Tas, Vic) .... *Sarcophaga* spp.
   – Posterior spiracles not in a deep cavity 4
4. – Posterior spiracles protrudent, with respiratory slits slightly sinuous; last abdominal segment with seven ventral anal tubercles (Fig.3.43c) (ACT, NSW, SA, Tas, Vic) .... *Hydrotaea rostrata* (Robineau-Desvoidy)
   – Posterior spiracles not protrudent, with straight respiratory slits; last abdominal segment with a pair of lateroventral anal lobes (Fig.3.43b) anal tubercles lacking (Calliphoridae) 5
5. – Posterior spiracular peritreme incomplete (Fig.3.44a) (*Chrysomya*) 6
   – Posterior spiracular peritreme complete (Fig.3.44b) 8
6. – Prominent processes present on all abdominal segments (e.g.Fig.3.44c) 7
   – Body segments without processes (ACT, NSW, SA) .... *Chrysomya megacephala* (Fabricius)
7. – Processes present dorsally, laterally and ventrally on AS1–8 as well as on TS2–3 (ACT, NSW, SA, Tas, Vic) .... *Chrysomya rufifacies* (Macquart)
   – Processes present dorsally and laterally on abdominal segments only (ACT, NSW, SA, Vic) .... *Chrysomya varipes* (Macquart)
8. – Oral sclerite totally unpigmented (e.g.Fig.3.45a) (*Lucilia*) 9
   – Oral sclerite at least partly pigmented (e.g. Fig.3.45b) (*Calliphora*) 10

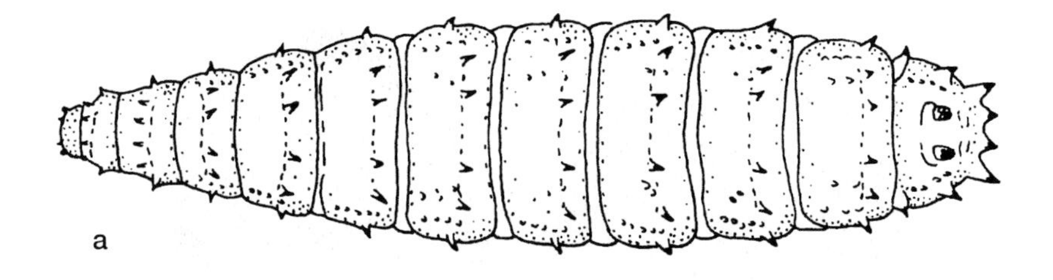

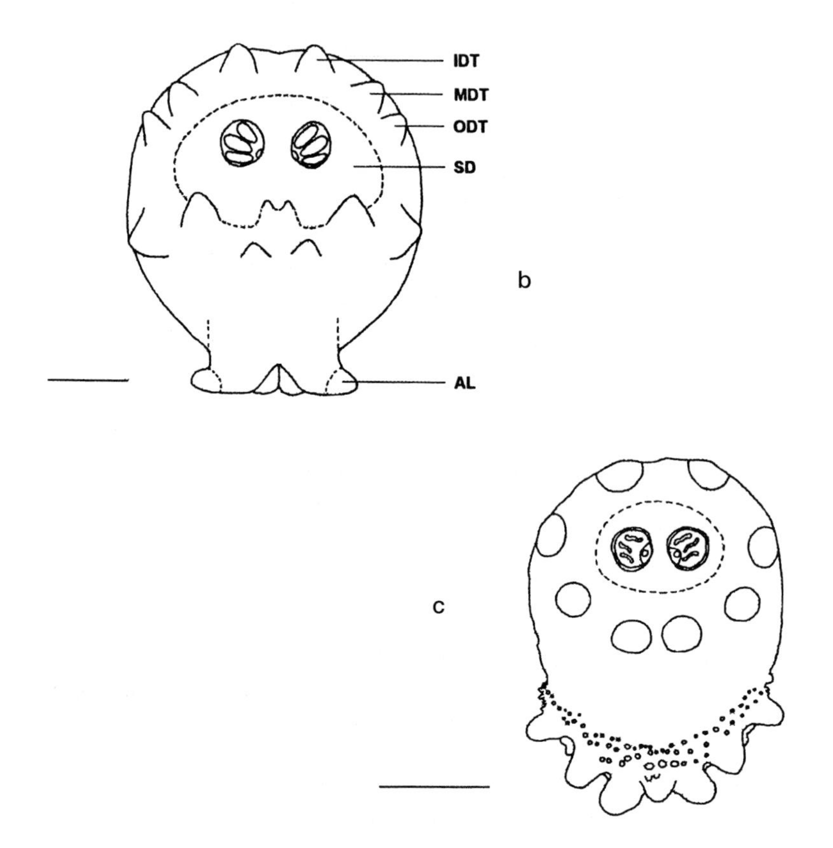

Fig. 3.43. a, Dorsal view of third-instar larva of *Megaselia* sp. (after Smith (1986)). b, Posterior view of third-instar larva of *Calliphora* sp. AL, anal lobe; IDT, inner dorsal tubercle; MDT, median dorsal tubercle; ODT, outer dorsal tubercle; SD, spiracular disc. Scale bar = 1.0 mm. c, Posterior view of third-instar larva of *Hydrotaea rostrata*. Scale bar = 0.5 mm

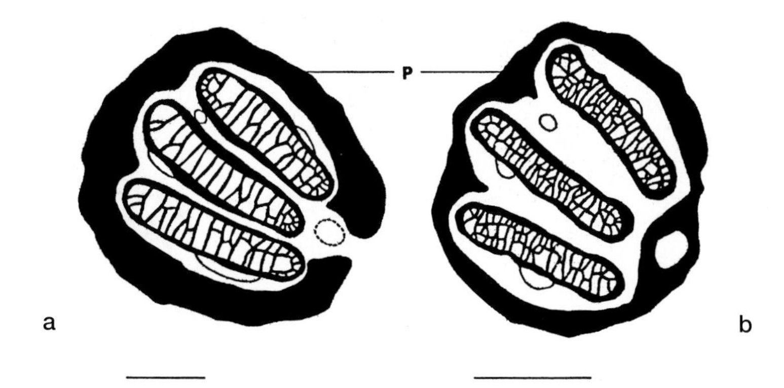

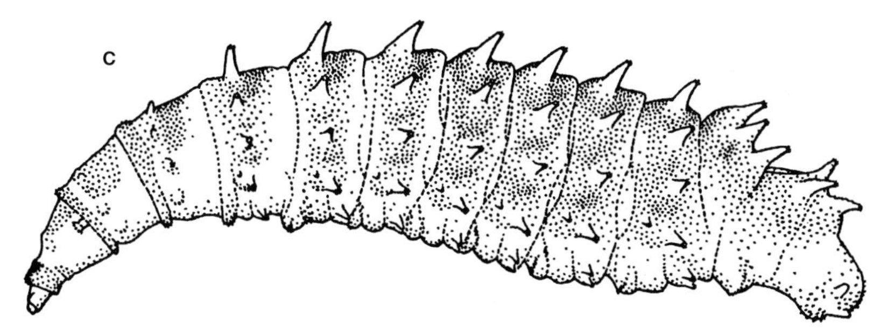

Fig. 3.44. a, *Chrysomya*-type posterior spiracle. Abbreviation: P, peritreme. Scale bar = 0.1 mm. b, *Calliphora*- and *Lucilia*-type posterior spiracle. P, peritreme. Scale bar = 0.1 mm. c, Lateral view of third-instar larva of *Chrysomya rufifacies.* (After Holloway 1985.)

9. – IDT, MDT and ODT on AS8 (Fig.3.43b) equidistant (ACT, NSW, SA, Tas, Vic) .... *Lucilia sericata* Meigen
   – MDT on AS8 closer to IDT than to ODT (ACT, NSW, SA, Tas, Vic) .... *Lucilia cuprina* (Wiedemann)

10. – Anterior spine band on AS5 always incomplete dorsally; microtubules inconspicuous dorsally on AS8 (e.g. Fig.3.46a);IDT, MDT and ODT on AS8 (Fig.3.43b) always equidistant 11
    – Anterior spine band on AS5 usually complete dorsally (sometimes incomplete in *Calliphora vicina*); microtubules conspicuous dorsally on AS8 (e.g. Fig. 3.46b); MDT on AS8 usually closer to ODT than to IDT 14

11. – Pleural spines (Fig.3.47a) present on AS1–7 (NSW, SA, Tas, Vic) .... *Calliphora maritima* Norris
    Pleural spines absent on AS1–7 12

12. – Spines arranged evenly (Fig.3.47b); anterior spine band on AS4 incomplete dorsally; anterior spines only sometimes present laterodorsally on AS5 (ACT, NSW, SA, Tas, Vic) .... *Calliphora hilli hilli* Patton
    – Spines arranged unevenly, mostly in sets of two or three (e.g. Fig. 3.47c);

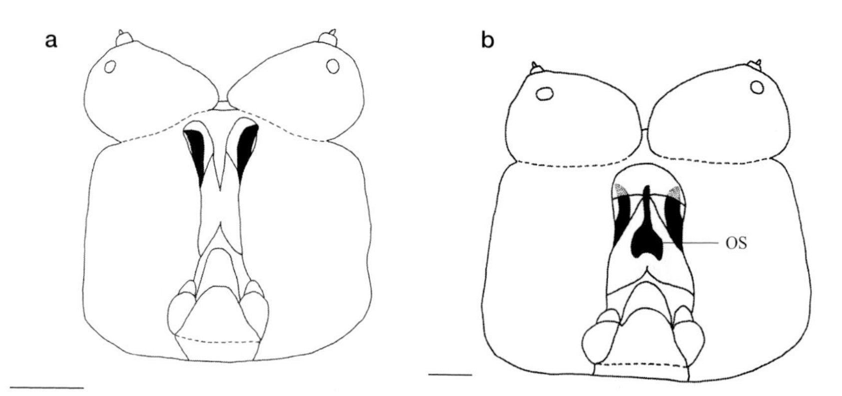

Fig. 3.45. a, Ventral view of head of third-instar larva of *Lucilia sericata*. Scale bar = 0.1 mm. b, Ventral view of head of third-instar larva of *Calliphora stygia*. OS, oral sclerite. Scale bar = 0.1 mm.

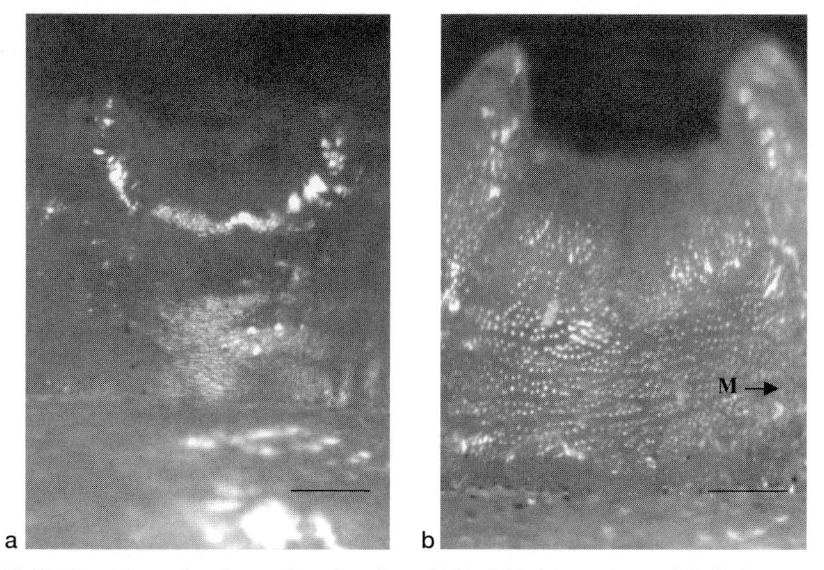

Fig. 3.46. a, Microtubercles on dorsal surface of AS8 of third-instar larva of *Calliphora dubia*. Scale bar = 0.15 mm. b, Microtubercles on dorsal surface of AS8 of third-instar larva of *Calliphora stygia*. M, microtubercle. Scale bar = 0.15 mm.

anterior spine band on AS4 complete dorsally; anterior spines always present laterodorsally on AS5 13

13. – Spines darker, with teeth smaller and only rarely in pairs (Fig.3.47c) (ACT, NSW, SA) .... *Calliphora dubia* (Macquart)
    – Spines paler, with teeth larger and more often in pairs (Fig.3.47d) (ACT, NSW, SA, Tas, Vic) .... *Calliphora augur* (Fabricius)

14. – Spines small with pointed teeth, sometimes double-pointed (Fig.3.48a);

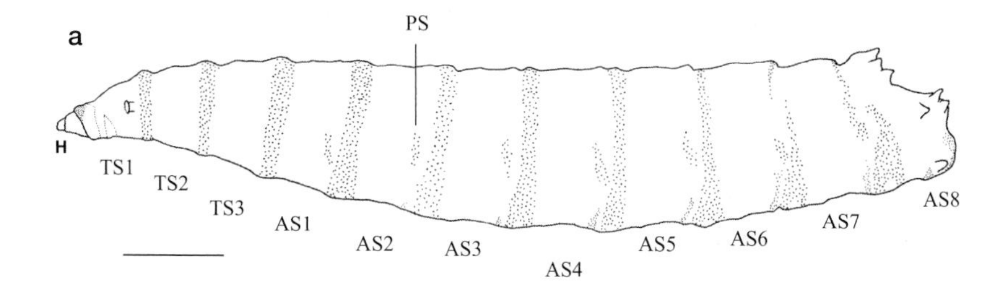

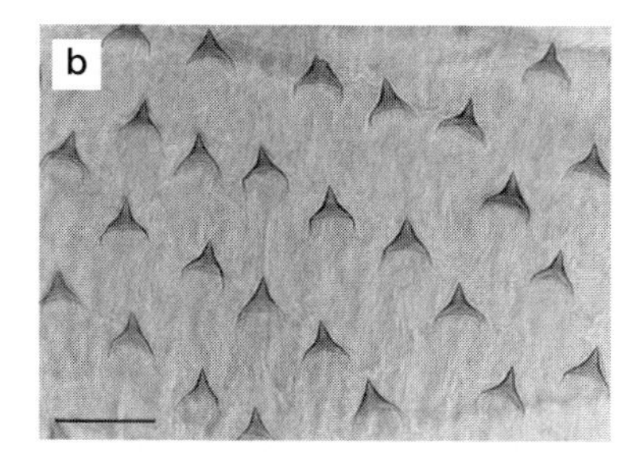

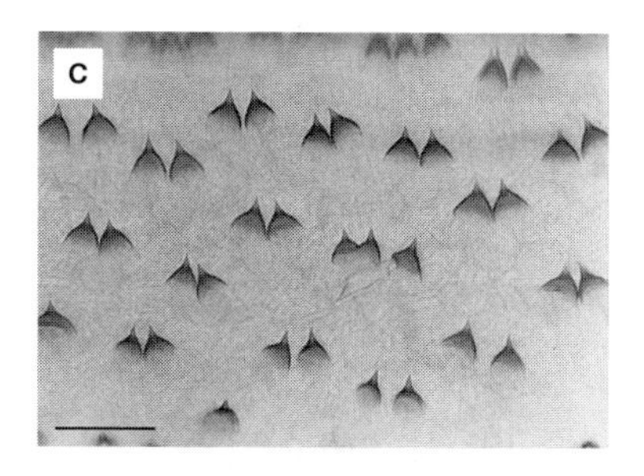

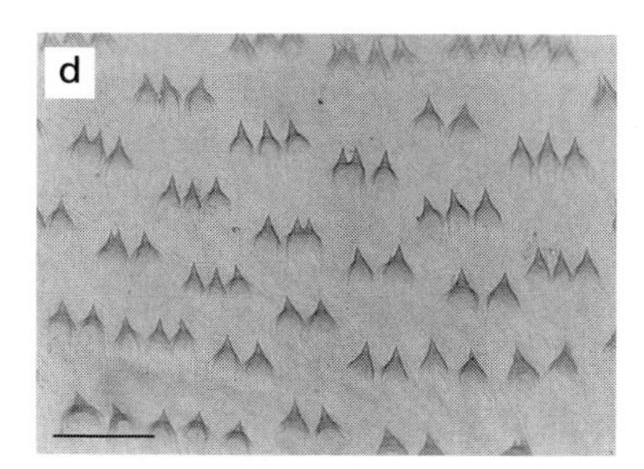

Fig. 3.47. a, Lateral view of third-instar larva of *Calliphora stygia*. AS, abdominal segment; H, head; PS, pleural spines; TS, thoracic segment. Scale bar = 2.0 mm. b, Spines in dorsal region of anterior spine band on AS1 of third-instar larva of *Calliphora hilli hilli*. Scale bar = 0.03 mm. c, Spines in dorsal region of anterior spine band on AS1 of third-instar larva of *Calliphora dubia*. Scale bar = 0.03 mm. d, Spines in dorsal region of anterior spine band on AS1 of third-instar larva of *Calliphora augur*. Scale bar = 0.03 mm.

spines dorsal to anus on AS8 not arranged in crude semicircular pattern (Fig.3.49a) (ACT, NSW, SA, Tas, Vic) .... *Calliphora vicina* Robineau-Desvoidy

– Spines large with rounded teeth, never double-pointed (e.g. Fig 3.48b); spines dorsal to anus on AS8 arranged in crude semicircular pattern (e.g. Fig.3.49b) 15

15. – Spines with raised base (Fig.3.48c) (SA) .... *Calliphora albifrontalis* Malloch

– Spines with base not raised (Fig. 3.48b) (ACT, NSW, SA, Tas, Vic) .... *Calliphora stygia* (Fabricius)

Key to Adults of Carrion-Breeding Diptera of Forensic Importance in South-eastern Australia*

*South-eastern Australia herein constitutes the Australian Capital Territory (ACT), southern New South Wales (NSW), southern South Australia (SA), Tasmania (Tas) and Victoria (Vic). The key contains species commonly encountered at carrion in this region

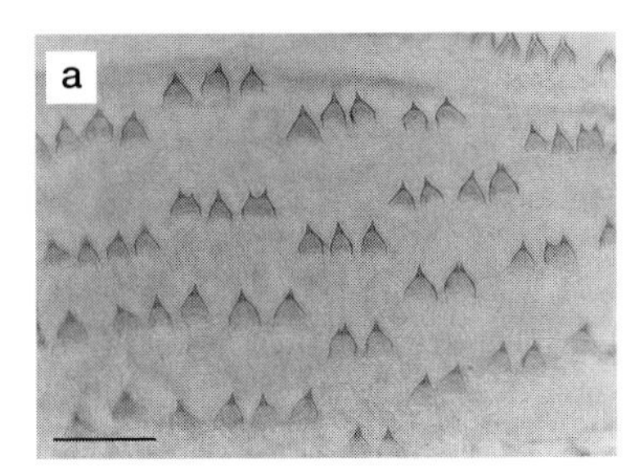

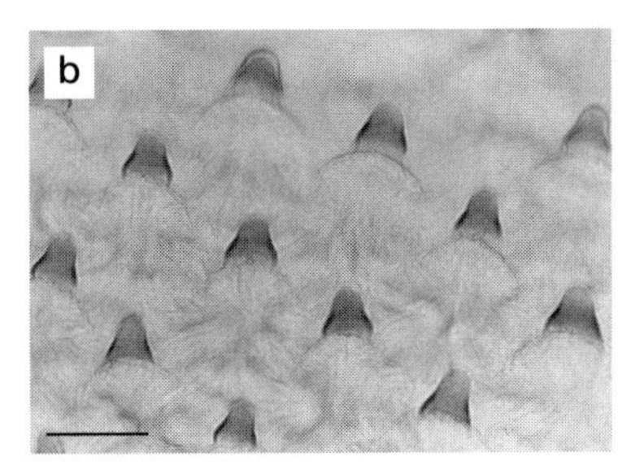

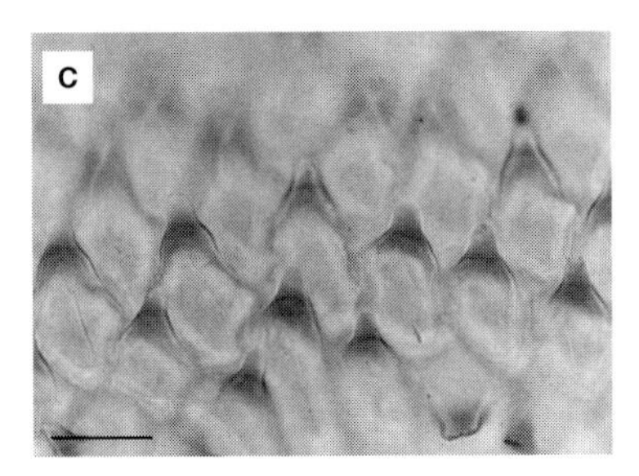

Fig. 3.48. a, Spines in dorsal region of anterior spine band on AS1 of third-instar larva of *Calliphora vicina*. Scale bar = 0.03 mm. b, Spines in dorsal region of anterior spine band on AS1 of third-instar larva of *Calliphora stygia*. Scale bar = 0.03 mm. c, Spines in dorsal region of anterior spine band on AS1 of third-instar larva of *Calliphora albifrontalis*. Scale bar = 0.03 mm.

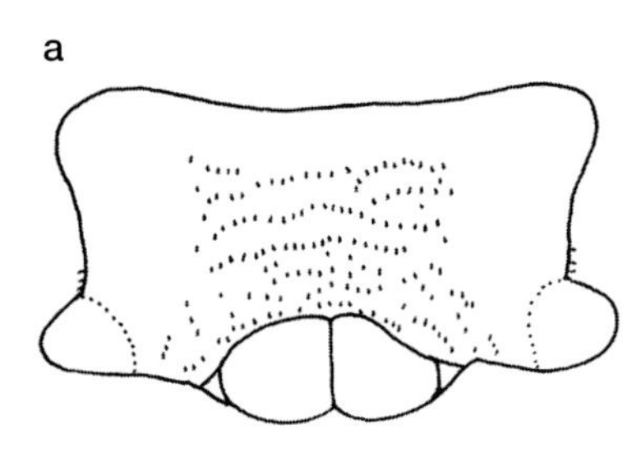

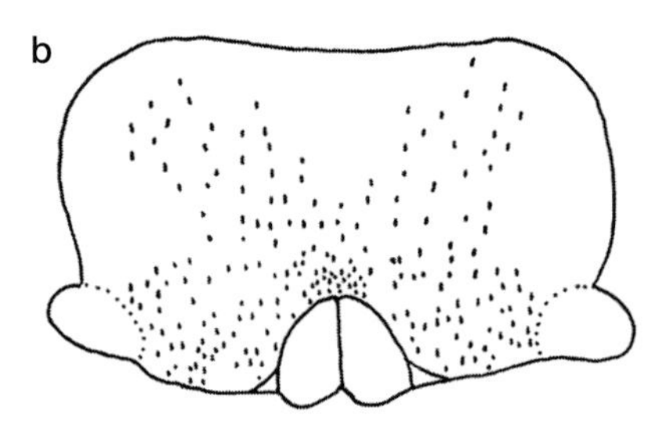

Fig. 3.49. a, Posterodorsal view of arrangement of posterior spine band on AS8 of third-instar larva of *Calliphora vicina*. Scale bar = 0.5 mm. b, Posterodorsal view of arrangement of posterior spine band on AS8 of third-instar larva of *Calliphora stygia*. Scale bar = 0.5 mm.

(although species may also occur in other Australian states) and is best applied to specimens reared from carrion, since some flies visit carrion to feed, rather than to reproduce. Even so, care should be taken with the identification of forms present in relatively small numbers, as well as in New South Wales, where tropical species may also be encountered. See Figs. 3.50 and 3.51.

1. – Wings with r veins (Fig.3.21) strongly thickened, terminating along with costa (Fig.3.21) at about middle of anterior margin, other veins much weaker, more or less parallel, running obliquely across wing; tiny humpbacked flies that run with a rapid, jerky motion (ACT, NSW, SA, Tas, Vic) ....Phoridae
   – Wings with venation not as above, more complex, with cross veins 2

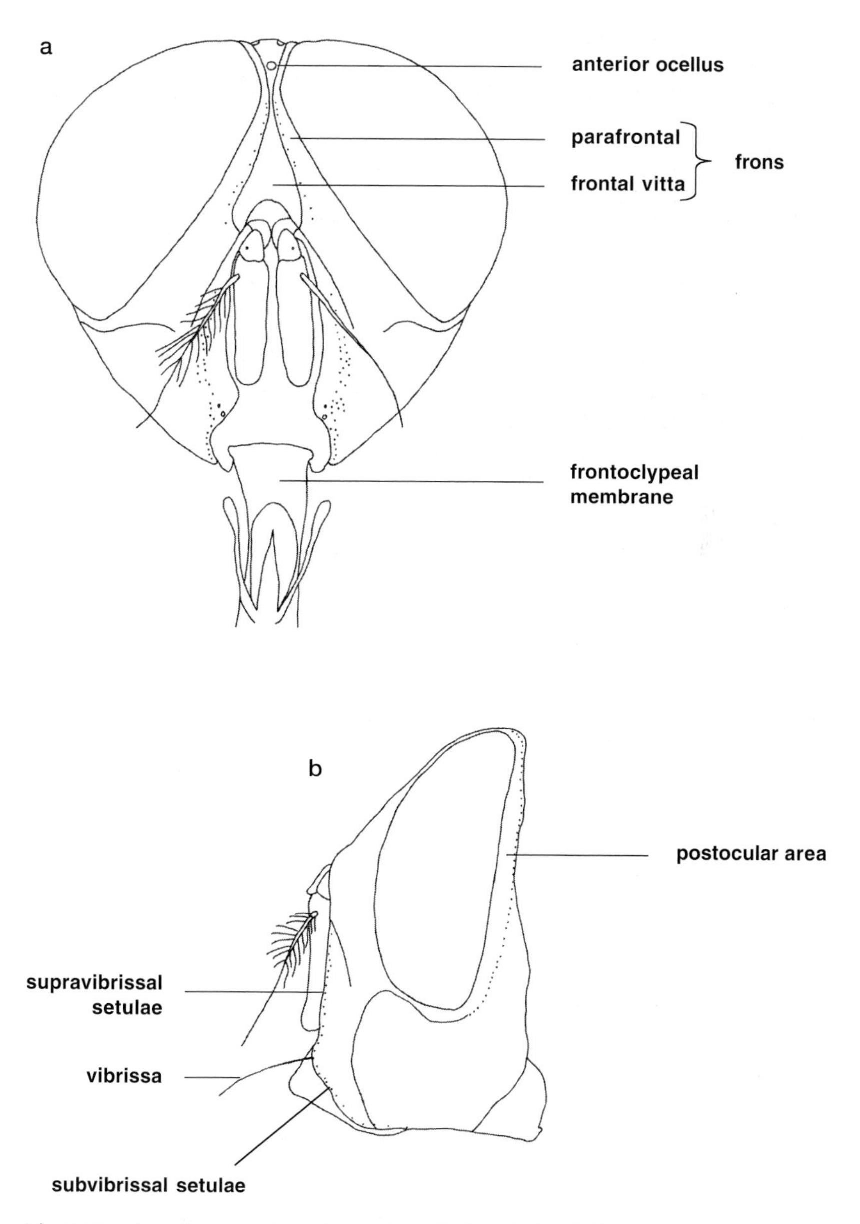

Fig. 3.50. a, Anterior view of head of male *Calliphora hilli hilli*. b, Lateral view of head of male *Calliphora hilli hilli*.

2. – Second antennal segment (Fig.3.18) without a linear seam dorsally; only one distinct pair of dorsocentral setae (Fig.3.20); small black and yellow species (ACT, NSW, SA, Tas, Vic)..*Piophila casei* (Linnaeus)
   – Second antennal segment with a distinct linear seam dorsally; at least 4 distinct pairs of dorsocentral setae..3

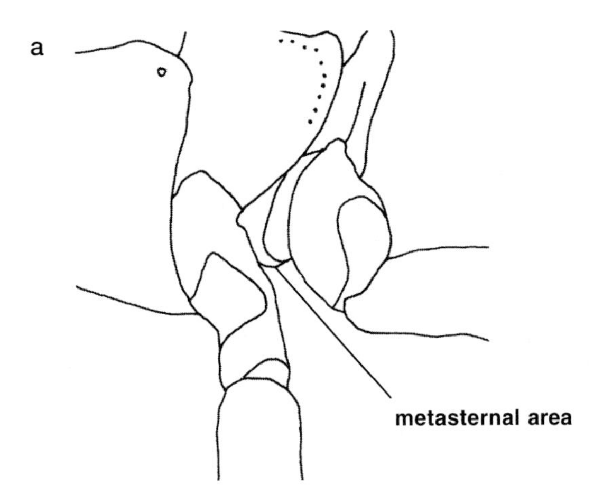

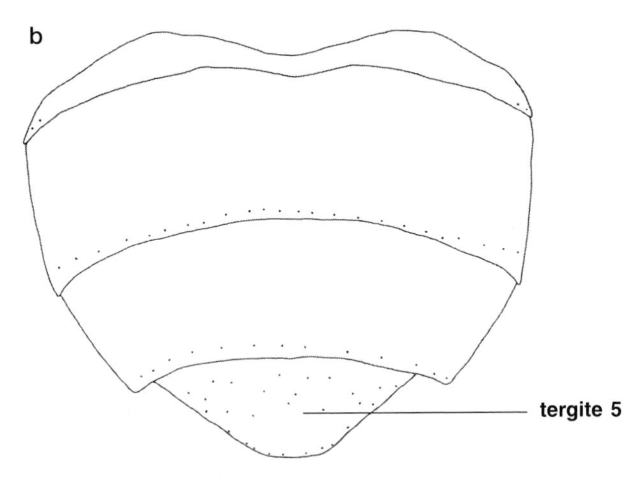

Fig. 3.51. a, Lateral view of the posteroventral region of thorax of *Calliphora hilli hilli*. b, Dorsal view of abdomen of *Calliphora hilli hilli*.

3. – Hypopleuron (Fig.3.19) with a patch of setulae posteriorly; small species with body black, except abdomen which is metallic blue to black with whitish pruinescence (ACT, NSW, SA, Tas, Vic) .... *Hydrotaea rostrata* (Robineau-Desvoidy)
   – Hypopleuron with a row of strong setae posteriorly 4

4. – Colour predominantly grey, never metallic; thorax with three dark longitudinal stripes; abdomen with a tesselated or checkerboard pattern (ACT, NSW, SA, Tas, Vic) .... *Sarcophaga* spp.
   – Colour often, at least in part, metallic blue or green, otherwise golden (Calliphoridae)...5

5. – Base of stem vein (Fig.3.21) setulose dorsally (*Chrysomya*) 6
   – Base of stem vein bare dorsally 8

6. – Mesothoracic spiracles (Fig. 3.19) dark brown to blackish; supravibrissal and subvibrissal setulae (Fig.3.50b) mostly black; male: eyes with ommatidia in upper two thirds enlarged and sharply demarcated from small ones in lower third (ACT, NSW, SA) .... *Chrysomya megacephala* (Fabricius)
   – Mesothoracic spiracles pale yellow, cream or white 8

7. – Smaller species, body length 7 mm or less; jowls (fig. 3.18) whitish (male) or yellow (female); femora (fig. 3.22) mostly dark brown and yellow to orange; male: fore femora whitish on inner surface, with prominent, white hairs dorsally (ACT, NSW, SA, Vic) .... *Chrysomya varipes* (Macquart)
   – Larger species, body length usually $>$7 mm; jowls orange-brown to black; femora metallic blue to black; male: fore femora entirely black, without white hairs dorsally (ACT, NSW, SA, Tas, Vic) .... *Chrysomya rufifacies* (Macquart)

8. – Lower calypter (Fig. 3.21) bare dorsally 9
   – Lower calypter hairy dorsally 10

9. – Frontoclypeal membrane (Fig.3.50a) light brown; metasternal area (Fig.3.51a) hairy; outer surface of fore femora and proximal half of under surface of mid femora metallic blue to black (ACT, NSW, SA, Tas, Vic) .... *Lucilia sericata* (Meigen)
   – Frontoclypeal membrane dark brown to blackish; metasternal area bare; outer surface of fore femora and proximal half of under surface of mid femora metallic green (ACT, NSW, SA, Tas, Vic) .... *Lucilia cuprina* (Wiedemann)

10. – Three pairs of anterior acrostichal setae (Fig.3.20) 11
    – Two pairs of anterior acrostichal setae 12

11. – Fore femora uniformly orange on outer surface; parafrontals and parafacials (Fig.3.18) with gold pruinescence; male: frons (Fig.3.50a) minimum width $<$ width of anterior ocellus (Fig.3.50a); ommatidia on anterior upper two thirds of eyes considerably enlarged to about twice width of others (ACT, NSW, SA, Tas, Vic) .... *Calliphora stygia* (Fabricius)
    – Fore femora brown to blackish on proximal quarter of upper-most half of outer surface; parafrontals and parafacials with silver pruinescence; male: frons minimum width $>$ width of anterior ocellus; ommatidia on anterior upper two thirds of eyes only slightly larger than others (SA) .... *Calliphora albifrontalis* Malloch

12. – Abdomen yellow-orange with contrasting metallic blue or green area dorsally 13
    – Abdomen of uniform colouration, lacking contrasting metallic area dorsally 14

13. – Dorsomedial area on abdomen with greenish-blue sheen; tergite 5 (Fig.3.51b) with yellowish pruinescence; male: frons (Fig.3.50a) minimum width $< 2\times$ width of anterior ocellus (ACT, NSW, SA, Tas, Vic) .... *Calliphora augur* (Fabricius)
    – Dorsomedial area on abdomen with rich blue or purplish sheen; tergite 5 with vivid white pruinescence; male: frons minimum width $>2\times$ width of anterior ocellus (ACT, NSW, SA) .... *Calliphora dubia* (Macquart)

14. – Abdomen metallic blue with grey pruinescence (ACT, NSW, SA, Tas, Vic) .... *Calliphora vicina* Robineau-Desvoidy
    – Abdomen not metallic blue 15

Table 3.16. *Duration of total larval feeding period of common Australian blowflies at constant temperatures*

Species	Temp. °C	Duration (days)	Species	Temp. °C	Duration (days)
Lucilia cuprina	15	9–15	*Chrysomya rufifacies*	20	9
	20	3.5–4		25	n.d.
	25	n.d.		28	3.3–5
	28	2–>4[a]		34	3–4
	34	2.5–4			
			Chrysomya varipes	20	12–13
Calliphora augur	9	~14		25	5–6
	15	6–7		34	>3
	20	4–5			
	25	3–4	*Chrysomya nigripes*	28	4–7
	28	2.5–4			
			Chrysomya saffranea[b]	28	3
Calliphora stygia	9	16–18			
	15	6	*Chrysomya megacephala*	28	3
	20	5–>10			
	25	3–6	*Hydrotaea rostrata*	9	70[c]
	28	3–5		15	10
		20		8–9	
		28		5–6	

Notes:

[a] Most larvae fed for 2–3 days.

[b] The data for *Ch. saffranea* include data for development at room temperature (mean 28 °C, range 20–35 °C).

[c] Most of the larvae pupated in the meat so the actual feeding period could have ended earlier than the time indicated. The period quoted includes the egg period.

Source: From O'Flynn, 1983.

15. – Postocular area (Fig.3.50b) with silver pruinescence; fore femora dark brown to blackish on proximal half of inner surface; abdomen with strong olive-green sheen (NSW, SA, Tas, Vic) .... *Calliphora maritima* Norris
    – Postocular area with gold pruinescence; fore femora orange on proximal half of inner surface; abdomen lacking olive-green sheen (ACT, NSW, SA, Tas, Vic) .... *Calliphora hilli hilli* Patton

The reader is urged to consult the following publications for additional information on timetables of development and other relevant aspects of the biology of flies from the Australasian region – Fuller (1932), Mackerras (1933), Norris (1965), Ishijima (1967), Monzu (1977), O'Flynn (1983), Nishida (1984),

Williams and Richardson (1984), Holloway (1985,1991), Spradbery (1991), Byrd and Butler (1997)and Wallman (1999). See also Table 3.16.

When the species and stage of the fly have been determined, and with relevant temperatures and other data in hand, the forensic entomologist is ready to calculate the age of the specimens, and estimate the time of death. From all that has been said it is worth emphasizing that the two are often not the same and are subject to contention, e.g. how much time elapses after death before eggs are laid, and what is the actual temperature experienced by the insects on the body.

Problems estimating time of death

Age Assessment of Maggots

Fly eggs often hatch within 24 hours, usually before the body is discovered, and are therefore of little value as evidence. A brief discussion of one case in which the age of fly eggs was crucial is given in Chapter 2. The chorion reveals nothing as the embryo develops and it is necessary to observe internal changes which are readily apparent only near the time of hatching.

LENGTH vs WEIGHT

A maggot's age is usually determined by its species, instar, length, and thermal history. Regarding length, the very initial process of collecting and preserving maggots can lead to error. For example, at autopsy the body is often routinely washed in a solution of San Veino and formalin. This does more than kill the maggots (some of which should have been rescued for rearing to the adult stage), it also shrinks them. When live third instar maggots of *Protophormia terraenovae* at peak of feeding are placed in San Veino they shrink from 15.6 mm in length to 11.2 mm. This reduces their apparent age by 11 hours; in formalin, the under-age error can be close to 17 hours. The spurious effect increases to 24 hours in young third instars of *P. terraenovae* and in *Calliphora vicina* larvae at peak of feeding (Tantawi and Greenberg, 1993a). Although shrinkage is less in 70% alcohol, here too the apparent age is reduced almost 10 hours in *P. terraenovae* and 19 hours in *C. vicina*. Telescoped segments are a sign of shrinkage and it may still be possible to physically stretch the cuticle to approximately the normal length of the specimen. To avoid the problem entirely, live maggots should be killed in hot water. They will then remain extended regardless of the preservative in which they are stored.

From hatching until pupariation, the larva's body weight and body length run parallel, increasing during the feeding stages and decreasing in the post-feeding larva (Nishida, 1984; Davies and Ratcliffe, 1994). Measuring a maggot's dry weight is suggested as an alternative to measuring its length. Wells and LaMotte (1995) recorded the dry weights of a series of *Cochliomyia macellaria* larvae of known ages reared at 28 °C. The objective of the study was to establish the age of a single larva by its weight. Instead, they found that it was difficult to distinguish among ages 0.75 to 1.25 days, between 1.5 and 1.75 days, between 2.0 and 2.5 days, and among ages 3.0 to 4.0 days. (For reference, larvae at 4.0 days were near but not at the postfeeding stage.) By choosing an entire age cohort, including runts, the weight distribution curve was broadened and the resolution between cohorts was diminished. In forensic practice, age determinations of larvae are based on the largest specimens and comparisons of weight versus length measurements should do the same. Nishida (1984) compared weight versus length in growth tables and graphs for five calliphorids and two sarcophagids at several temperatures. He selected a sample of larvae once a day and the length and weight of each larva (the largest?) were measured. The paper is in Japanese and I cannot say how the larvae were prepared for weighing. It is also not clear from a later account of Nishida *et al.* (1986) whether entire age cohorts were used or just the largest specimens, but their data for a number of species of carrion flies show a larger standard error for the weight measurements than for the length measurements of the larvae. Nevertheless, dry weight may be more useful than body length, especially in situations where larvae have not been properly prepared (Wells and Kurahashi, 1994; Davies and Ratcliffe 1994) and in the postfeeding stage when larval length frequently diminishes at a variable rate.

Under extreme conditions of weather (cold, rain), or limited fly access, a corpse might contain the larvae of a single fly. Usually, however, one finds maggots of mixed ages and species. Standard procedure is to select and identify samples of the largest larvae of each species to be used to estimate time of death. The entomologist then decides which species had deposited eggs (or larvae) first. This can be straightforward, given the usual mix of species. But consider the following possibility. A *Calliphora* has oviposited, or a *Sarcophaga* has larviposited, one day later than *Phormia* or *Phaenicia*, yet, by day 3 or 4, the larvae of the latecomers can be as large or larger than those of *Phormia* or *Phaenicia*. To resolve the problem, consult specific graphs of age versus length, and make dissections to evaluate the gut's fullness. A precocious larva, larger than the rest of its 'teammates', should not be a problem if an adequate sample is taken.

THERMAL HISTORY

Insect inventories of the human corpse, a la Mégnin (1894), or of animal models, provide a "general roadmap" for the entomologist to follow, but schedules of development for individual species of fly are the ultimate path leading to the postmortem interval. The thermal history of these cold-blooded creatures can be surprisingly complex. Flies lack a thermostat and their rate of development depends on the ambient temperature, but only partly. It is also influenced by the heat loading of a decomposing body in sun, the exothermic nature of a maggot aggregation, and possibly microbial action. At the crime scene, multiple temperature measurements of the body, the maggot mass, and the air should be recorded and compared with data from the nearest weather station. If the temperature data from the two sites are reasonably close, it is justified to use weather station data to obtain a retrospective of the temperatures at the crime scene. It does not necessarily follow, however, that these are the

Table 4.1. *Accumulated degree-hours (°C) in rabbits in sun or shade, based on hourly temperatures for 16 days in April*[a]

Position	Skin	Rectum	Underside	Air
Shade	4484	4830	4503	4486
	(+2.4%)	(+3%)	(0%)	–
Sun	6631	6191	6095	4985

Notes:

[a] Percentages are the differences in accumulated degree hours (rabbit in shade) when daily maximum–minimum temperatures are substituted for hourly temperatures over the 16-day period.

Source: (From Greenberg, 1991.)

same temperatures experienced by the insects on the body. Good temperature data are the foundation for reliable estimates of the postmortem interval. Single temperature readings of the body and the air, whether the body is indoors or outdoors, are likely to be inadequate. And too often, even minimal readings are not taken. The following case in Covington, Kentucky, hinged on the accuracy of the temperature data.

A young girl was last seen alive on April 21. Her body was discovered and removed from an abandoned house on May 1. No temperature data were recorded at that time. On May 2 and 3, the police returned to record a series of temperatures in the same room where the body had lain. The weather was cool and the temperature readings were in the 50s (°F; *c.* 10+°C). Based on these temperatures and the maximum size of the *Calliphora vicina* maggots collected from the body, the prosecution entomologist concluded that development would have taken a minimum of seven to nine days. We compared the temperatures taken by the police, hour-for-hour, with temperatures taken at the U.S. weather station not far away. The two sets of temperatures tracked very closely. Review of the data from the weather station from May 1 back to April 21 revealed that the weather was actually much warmer then, with temperatures in the 70s and 80s (°F; *c.* 21 and 27 °C). This significantly shortened the developmental period of the maggots and the postmortem estimate.

The sun effect To determine the sun's heat loading of a corpse, a group of my graduate students placed two rabbit carcasses of equal weight in sun or complete shade on the ground on campus. A Grass eight-channel polygraph recorded hourly temperatures for 16 days in April. The probes were placed in the skin, rectum, on the underside of the body in contact with the soil, and in air. Care was taken not to insert probes where maggots were present. A summary of the accumulated degree hours (ADH) for each probe is given in Table 4.1. (The forensic application of ADH is discussed later in this chapter.)

For the entire period average air temperature was 11% higher in sun than in shade, but the ADH of the rabbit in sun was 28 to 48% higher than that of the rabbit in shade (Greenberg, 1991). A number of additional factors would have to be considered in a real case, e.g. the insolating effect of dark clothes on a body, or of dark skin (the rabbits had white fur). Another possible factor is the nature of the ground – how much heat does it absorb, conduct, and radiate? In summer, the surface sand at a beach is warmer during the day and colder at night than soil nearby. The size of the carcass is an important consideration. Wallman (1999) has made an excellent study of thermogenesis of blowfly larvae that infest piglets in Australia. As he discusses, "Large and small carcasses exposed to solar radiation should gain heat at the same rate per unit area and thus might be expected to ultimately attain the same temperature. However, in reality, large carcasses should be warmer because larger objects lose less heat to the environment by radiation and convection." Large carcasses have a smaller surface-to-volume ratio and they do not heat up as quickly from absorption of solar radiation during the day, but at night they lose heat less rapidly from radiation and convection. By means of controlled experiments in summer, Wallman isolated the sun's effect and found that the mean/maximum increases in sun-exposed carcasses was 5 °C/11 °C higher than in shade. Even when carcass size and the time and place of exposure are the same, there may still be variability in the sequence of arrivals and composition of fly species, and the rate of decomposition (Tantawi *et al.*, 1998).

The bacterial effect Deonier (1940) concluded from outdoor studies of goats and sheep that: "Rapid decomposition of the carcasses from the action of bacteria was not in evidence when the temperatures were lower than those at which some of the species of flies are normally active. Although bacteria may have been responsible for some of the heat generated during larval development, heat from bacterial decomposition alone was not perceptible in any of the carcasses." Wallman's (1999) data on the core temperatures of uninfested piglets in shade do not differ significantly from ambient and support Deonier's observations. He indicates a need for further studies of uninfested carcasses to estimate the possible role of bacteria in heat generation because it is a fact that bacteria multiply prodigiously in the carrion soup (Greenberg and Miggiano, 1963). However, compared with the effect of the sun and the endogenous maggot heat, present evidence suggests that bacterial elevation of carcass temperature is probably negligible.

Maggot mass A maggot mass originates in the collective oviposition frenzy of females. After penetration of the body, the invading larvae become a feeding aggregation. The metabolic heat that is collectively generated was first reported in 1869 by M. Girard who measured a temperature excess of 32 °C in a box full of *Lucilia caesar* maggots (cited in Heinrich, 1993). Much depends on the size of the aggregate but excess heat is perceptible with as few as 25 larvae in 25 g of meat (Marchenko, 1988; Goodbrod and Goff, 1990). It increases with larval density, as clearly shown by Marchenko (1985) in third instar larvae of *Chrysomya albiceps*: 25 to 75 larvae produced an increase of 3 to 3.8 °C; 100 to 150 larvae, 6.4 to 10 °C; and 200 to 780 larvae, 13.6 to 19.5 °C. The reader should consult Goodbrod and Goff (1990) for excellent graphs of thermogenesis at various densities during larval development of *Chrysomya rufifacies* and *Chrysomya megacephala*. Even first instar larvae can produce substantial excess heat if their density is great enough (Wallman, 1999). Contrary observations are reported by Davies and Ratcliffe (1994): "Temperatures in two cases with 120–140 larvae reared at 20 °C were monitored with mercury thermometers and showed that temperatures within feeding larval groups were near ambient except for a short period near maximum growth rate when a rise of <1 °C was observed." Although larval density and thermogenesis depend on carcass size, a human corpse is usually not a food-limiting factor for the initial wave of forensically critical maggots. Williams (1984) cautions, however, that competition by overcrowding (in a baby, for example) may suppress the size of larvae while maintaining the length of time they will spend in the feeding stage. As long as a carcass is suitable, flies continue to lay eggs and the maggot mass quickly becomes a mix of sizes, ages, and species, augmenting the communal output of heat. Aggregation behavior not only facilitates the entry of larvae into a carcass but it accelerates their growth rate – another successful strategy in the ongoing carrion wars. Having said this, it is interesting to mention that *C. megacephala* and the predatory *C. rufifacies* infesting the same body sometimes segregate into separate maggot masses.

In midwinter field studies on large animals in Texas and Arizona, Deonier (1940) writes that "as larval development progressed, the temperature increased and remained high regardless of daily weather fluctuations." He recorded a maximum of 49 °C when the air temperature ranged between 9 °C and 22 °C. In one case, the maggot mass was active when the air temperature was −4 °C. This is important because it is well below the temperature at which bodies are kept in the morgue. An estimate of the postmortem period would certainly depend on whether the maggots continued to develop under these conditions. Marchenko (1973), studied 17 human cadavers and demonstrated a temperature excess of 18 °C to 45 °C in a maggot mass measured at night when air temperature was 4 °C to 15 °C. The fact that Marchenko (1985) later recorded no rise when the ambient temperature was 6 °C may reflect a lower larval density in his laboratory experiment.

Maggot mass temperatures averaged 16.7 °C above ambient in a pig exposed to the sun in summer, and 11.4 °C above ambient in a pig in shade (Shean *et al.*, 1993). Such a difference can have a dramatic outcome on the rate of corpse decomposition – a pig in the open was reduced to 17% of its body weight 18 days sooner than a pig in shade (Catts, 1992). Also in summer, Turner and Howard

Table 4.2. *Thermal death points (in °C) of third instar larvae of three species of blowfly under two different rates of heating. Given in brackets are the degree-minutes in excess of 25 °C during the heating period*

	0.2°C/min	0.6°C/min
Calliphora dubia	44 (1805)	45 (667)
Calliphora vicina	46 (2205	47 (807)
Calliphora rufifacies	47 (2420)	52 (1215)

Source: (From Wallman, 1999.)

(1992) recorded maggot mass temperatures in rabbit carcasses that reached 27 °C above ambient, with no appreciable difference between sun and shade. In a pasture in British Columbia, temperatures in infested pig carcasses remained over 50 °C for a number of days (Anderson and VanLaerhoven, 1996). The authors point out that the actual temperature at which a maggot develops is unknown, as maggots are constantly moving "down to the feeding site, then back out to the exterior of the mass." However, Turner and Howard (1992) did not observe any movement of maggots away from the maggot mass when the temperature exceeded 40 °C for 10 hours. And Marchenko (1973, 1975, 1988) observed ". . . an up to 50% shortening of the developmental period in a certain part of preimago stages . . ." compared to predictions based on ambient temperature data.

Temperatures of 39 °C and 45 °C have been considered lethal for larvae of *Calliphora* and *Phormia* respectively (Wigglesworth, 1967). Wallman (1999) reported much higher thermal death points that might enable maggots to develop at temperatures previously thought to be unsuitable (Table 4.2). A discussion of acclimation and other physiological mechanisms operant in the thermal death points of these insects is beyond the scope of this book, but it is not amiss to cite two examples – the first a laboratory experiment, the other a homicide – in which the heat tolerance of *Phormia* maggots was put to the test.

We packed an adult human skull with 1400 g of ground beef and retrieved it 24 hours later, with about 10000 eggs, from a cage of *Phormia regina*. With the calvarium replaced, we inserted a temperature probe through a small opening in the frontal bone and took readings in various areas within the cranium several times a day for one week, while we tracked larval growth (Figs. 4.1a,b). In this experiment, first instar larvae produced little excess heat, but second and third instar larvae produced significant heat that peaked at 18 °C above ambient, slightly before third instars reached maximum size. Despite the confined space and a temperature of about 41 °C, the larvae developed normally (Greenberg, 1991). The following homicide case generated additional evidence of this fly's temperature tolerance.

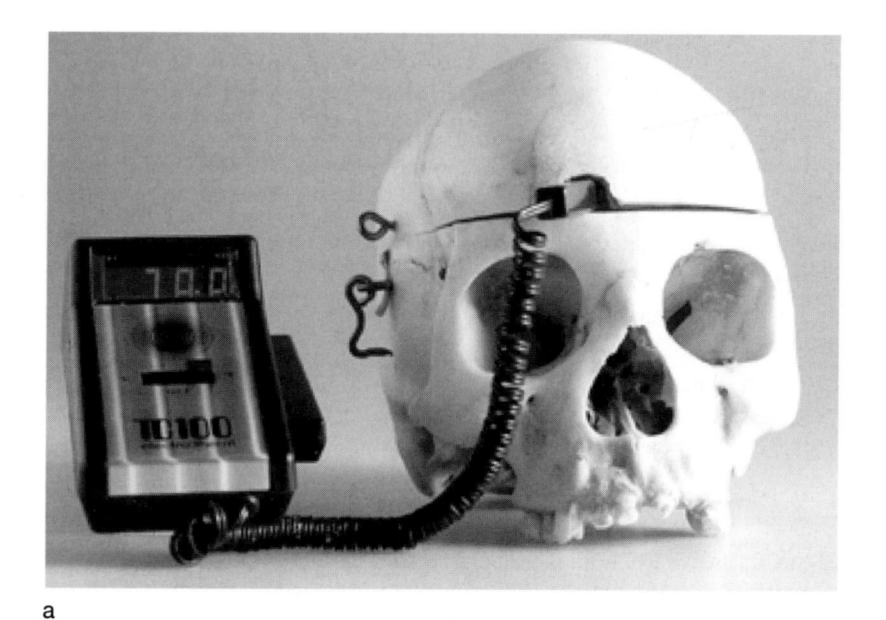

a

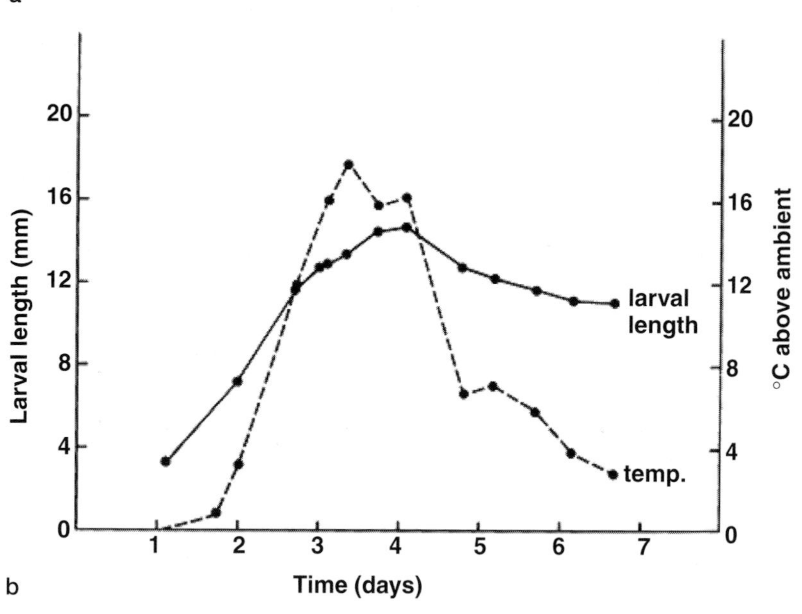

b

Fig. 4.1. *Phormia regina.* a, Monitoring temperature of a maggot mass in the cranium. b, Plot of excess heat generated in a seven-day period (Greenberg, 1991).

Victim in a van in August, 1988, a homicide victim was found inside a van parked in a shopping area in a Chicago suburb. The weather had been uniformly hot, and the temperature in the van was 49 °C when the body was found. The oldest third instar maggots on the body and clothes belonged to *Phormia regina.* The dissected gut and crop were still partially full, and the latter was 1.8 mm long, indicating that the maggots were well into the postfeeding stage. In the laboratory, I harvested several batches of eggs from our colony of this fly and inserted them into slits in a piece of meat and placed the meat on a bed of moist wood shavings in a quart jar. The mouth of the jar was covered with nylon mesh and the jar was placed in an incubator at 45 °C. The larvae required only 43 hours to hatch and reach the same stage as those taken from the victim. Three replicates gave confirmatory rearing times of 42, 45, and 43.5 hours. Given the maggots' confinement in a jar, it is possible that they actually experienced a temperature that was slightly higher than 45 °C. Under these conditions, the larvae developed extremely rapidly and were already in the postfeeding stage *less than two days after eggs were laid.*

Species such as *P. regina* and *Chrysomya rufifacies,* continue to aggregate as postfeeding larvae and pupae (Plate 11). Can these aggregations produce sufficient heat to accelerate development and derail the calculations of an unsuspecting entomologist? Spradbery (1992) has shown that pupae of the screwworm fly *Chrysomya bezziana,* stored in bulk at 25 °C, generated an excess temperature of 14 °C. What can we expect from a less dense cluster? What would be the temperature at the center and periphery of a pupal mass such as the one shown in Plate 10 and would any deviation from ambient prove to be significant? To my knowledge, this subject has not been investigated.

Accumulated Degree Hours

The concept of ADH has been used to optimize the timing of insecticide applications to control agricultural pests. It is based on the temperature-dependant rate of development of cold-blooded insects. The assumption is that the relation between the rate of development and temperature is linear in the midrange of the curve, with an upper lethal limit and a lower threshold below which development ceases. In the laboratory, to determine the total ADH of a fly, sum the number of hours from egg to adult and multiply by the temperature (°C), after subtracting the temperature of the developmental threshold. In the field, where temperatures vary, one multiplies the number of hours by the average of the hourly temperatures. ADH was first used in the following murder to determine time of death.

Late one Tuesday night in May, 1984, a young girl was allegedly seen with a man walking in the vicinity of an abandoned apartment building in downtown Waukegan, Illinois. Twenty-one days later, at 11 a.m., the badly decomposed body of a young girl was found under a blanket in a dark bathroom on the ground floor of the same building. The upper part of the body was skeletonized

and it was difficult for the coroner to assign a cause of death, except that a wire loop was drawn tightly around the cervical vertebrae. Miraculously, fingerprints could still be lifted from some of the fingers and they matched prints from classroom papers of a 9-year-old reported missing from Kenosha, Wisconsin.

The only insects at the scene were flies and all stages were collected and preserved. Most of the dead flies and empty puparia on floors and windowsills belonged to *Phormia regina*; a few were those of *Phaenicia sericata*. There were live teneral adults of *P. regina*, and its eggs and young larvae on the body indicated another cycle of the fly was under way. Given these facts alone, an estimate of the postmortem interval (PMI) could be made but it would have been less precise because the first life cycle had been completed.

Fortunately, a third fly, *Calliphora vicina*, was at the scene and its pupation period is the longest of the three species. Its pupae were scattered on the bathroom floor. We used Nuorteva's (1977) data on this fly to estimate the PMI, although his was an outdoor experiment with differences in carcass size and type. Still his average ambient temperature of 15 °C was closer to the temperature at the crime scene than anything we had. The use of ADH seemed appropriate in this case because the adults within the puparium were close to emergence. Starting with 11880 ADH for the first adult emergence we calculated back by subtracting each day's ADH (temperature × time) at the crime scene and arrived at 5 a.m. Thursday, 31 May, as the possible time of oviposition. At the time, nocturnal oviposition was not considered a realistic alternative and, given the cool weather and interior placement of the body, we concluded that oviposition probably had occurred the previous day.

Since then, ADH has been widely used, and some of its limitations have been revealed, as we describe below.

1. In our rabbit study, precise temperatures were recorded right from the start, directly at several sites on each rabbit and in air, without maggots to confound the temperature data. But a death scene investigation is not a controlled scientific experiment. What is the source of the temperature data? Is there a weather station some distance from the crime scene? If so, how do temperatures taken on and above the body compare, hour-by-hour, with temperatures recorded at the weather station? If hourly temperatures at the two places track closely, one is justified in using weather station data for the period in question. Even months or years later, it is still possible by this means to reconstruct a reasonable temperature history at the crime scene. This procedure was not followed in the Waukegan case. It was assumed that the bathroom on the ground floor of a boarded-up apartment building would be cave-like, with only small variations in temperature.
2. An accurate ADH depends on knowing when eggs were laid and when adults emerged.

3. ADHs based on rearings at a single, uniform temperature do not reflect the variable temperature insects might well experience on a body. Davies and Ratcliffe (1994) report an acceleration in larval growth at alternating temperatures, e.g.10 °C and 20 °C, 15 °C and 26 °C, and 5 °C and 15 °C (with 12-hour cycles), compared with constant temperatures of 15, 20, and 26 °C. The flies were *Calliphora alpina, C. vicina, C. vomitoria, Phormia* (=*Protophormia*) *terraenovae*, and *Lucilia* (=*Phaenicia*) *sericata.* In our experiments, the variable temperature regime was set at 16 °C and 29 °C, 12 hours each, and was compared with a constant temperature of 22.5 °C. With variable temperature, there was a significant (<0.05) retardation in the total time of development of *Phaenicia sericata,* and a statistically insignificant retardation of *Phormia regina, Cochliomyia macellaria,* and *Chrysomya rufifacies.* These results differ from those of Davies and Ratcliffe (1994) but they cannot be directly compared because we summed the entire life cycle, not just the larval stages. Ahmad (1936) exposed pupae of *Calliphora vicina* to 5 °C and then to higher temperatures. "In all these experiments the pupae subjected to a low temperature of 5 °C. show statistically significant acceleration in development at high temperatures of 14.8 °C., 18.4 °C., and 23 °C. The longer the period of exposure to low temperature the greater is the difference from the control individuals. . . . The acceleration produced when the pupae are subjected for one day to 5 °C. and for the remaining period to14.8 °C., 18.4 °C., or 23 °C., amounts to 0.21, 0.18 and 0.20 days, respectively. But when subjected for 8 days to 5°C. the acceleration is 2.12, 1.45 and 0.85 days, respectively." Concerning *Muscina stabulans,* Ahmad concludes "there is very little response of the pupal stage to alternating temperatures." According to Byrd and Butler (1996), developmental data from constant temperature rearings (25 °C) of *Cochliomyia macellaria* can be applied to cycling temperatures (amplitude 5.5 °C) as long as mean temperature values are comparable. Their mean values differed by 1.7 °C and may not be comparable. Under the same test conditions, third instars of *Chrysomya rufifacies* developed more slowly at a constant 25 °C (Marchenko and Vinogradova, 1984; Byrd and Butler,1997). Variable temperatures are a forensic reality and require more study.
4. The presence of a maggot mass is tantamount to a variable thermal history.
5. In the laboratory, higher temperatures (35 °C) and a dry substrate may induce postfeeding larvae of *Phaenicia sericata* to delay pupariation or enter diapause. In nature, the vagility of the larvae might enable them to find a microenvironment suitable for uninterrupted development (Greenberg, 1991). Diapause, or arrested development, has been discussed in Chapter 2. In blowflies it may occur in the postfeeding larva near pupariation, and in the adult; in fleshflies it occurs in the pupal

stage. It can happen to an entire cohort or just a portion, and it can last for months. Luckily for the entomologist, diapausing larvae and pupae are rarely the key forensic indicators.

6. Even a temporary submergence of eggs, larvae, or pupae can retard development.
7. Temperature optima may differ for the larva and pupa. For example, *Calliphora* larvae, like other larvae, develop rapidly at temperatures between 29 °C and 35 °C, but their pupae suffer or fail to develop at these temperatures. Ahmad (1936) found reduced viability of *Calliphora vicina* pupae at 27 °C and 30 °C, but pupae of *Muscina stabulans* were not affected. An ADH based on rearings of *Calliphora* at or close to the critical temperature may be inaccurate. For this reason Kamal's (1958) data for several species of *Calliphora* that were reared near the critical temperature are questionable.
8. An ADH based on the time span from egg to adult has a larger, built-in variability than the time span from egg to young larvae. The error is funnel-like, increasing with time. Therefore, the farther one extrapolates back – from adult to pupa to larva, etc. – the more the error increases. For example, if we use Kamal's (1958) "corrected" sums for the development of *Phormia regina*, the duration of egg to adult ranges from 7.6 to 21 days, whereas egg through second instar has a range of 1.2 to 3.2 days. One should use the best available data inclusive of the relevant stages only.
9. The ADH will be wrong if the assumed threshold temperature of development of a species is in error. In a homicide case in Covington, Kentucky, the prosecution entomologist assumed a threshold of 10–11 °C for *C. vicina* while in another case in Toronto, he used 6 °C. Both are incorrect. Published work indicates a threshold for the species at 3.5 °C (Davies and Ratcliffe (1994) and 2 °C (Vinogradova and Marchenko, 1984). In Switzerland, the fly entered a dark cave with a constant temperature of 5 °C and oviposited on a body (Faucherre *et al.*, 1999).
10. Laboratory-determined ADHs do not account for possible alterations in development rate when maggots feed on a corpse that contains drugs. (See next section.)

It hardly needs stating that most of the above factors are equally applicable to other models of maggot growth as a function of temperature (Nainys *et al.*, 1982). Catts (1990) evaluates the forensic use of ADH as follows: "Unless corpse temperatures are known precisely, ADH estimates are specious and should be avoided because they tend to give a false perception of scientific accuracy that can be misleading." When death scene investigators neglect to take temperatures of the corpse, the forensic entomologist may have to make PMI estimates based on imperfect temperature data. If ADH is to be used at all, it should be used with the caveats discussed above.

Flies and Drugs

In 1890, Webster noted live phorid flies, *Conicera* sp., on a cadaver that had been buried for two years. Stomach analysis of the cadaver revealed arsenic, and he commented "That the larvae of these flies might subsist upon the flesh of bodies killed by arsenic is by no means surprising, as they are doubtless tenacious of life." He saw no forensic use for this tenacity. In 1972, Nuorteva and Häsänen reported the transfer of detectable amounts of mercury from fish to sarcosaprophagous flies. This was followed by an assay of flies to determine whether a murder victim had lived in a mercury-polluted area (Nuorteva, 1977). It turned out that the levels of 0.12–0.15 ppm in the blowfly larvae were below the natural level of 0.2 ppm. In 1958, Utsumi *et al.* reported delayed skeletonization of rats poisoned with cyanide, arsenic, parathion, or bromdiethylacetyl urea, an hypnotic. They commented "that the beetles and the larvae of flies can scarcely feed or grow" on such carcasses.

Beyer *et al.* (1980) may have been the first to report the recovery of an abuse drug from flies. Maggots of *Cochliomyia macellaria,* taken from the badly decomposed body of a young woman who was suspected of a drug overdose, contained 100 μg/g of phenobarbital. "The complete absence of suitable tissue specimens for analysis led us to use the insect larvae as a source material for analysis."

The insecticide malathion is widely used commercially and in the home. Gunatilake and Goff (1989) analyzed pooled third instar larvae of *Chrysomya megacephala* and *Chrysomya rufifacies* that were taken from a body in an advanced stage of decomposition. Malathion poisoning was suspected. The insecticide was recovered from body fat and gastric contents of the corpse and from the larvae; the yield from the latter was 2050 μg of malathion per gram of larvae.

The opiate laudanum, was found in the hair of the poet Keats who used the drug to control pain (Baumgartner *et al.*, 1989). Introna *et al.* (1990) collected 40 livers from bodies known to be positive for opiates (morphine) and seeded them with eggs of *Calliphora vicina.* Third instar larvae were then analyzed by radioimmunoassay. Their opiate concentration ranged from 8 μg to 1208 μg/kg of larvae and correlated well with the opiate concentration of each liver in which the larvae developed (also, Kintz *et al.*, 1990c). Kintz *et al.* (1990b) recovered bromazepam and levomepromazine from the larvae of *Piophila casei,* the cheese skipper, a fly that typically occurs during the late stages of decomposition. Previously, the same group (Kintz *et al.*,1990a) recovered triazolam, oxazepam, phenobarbital, clomipramine, and alimemazine from unspecified calliphorid larvae. The drugs were isolated from the corpse of a man who had died about 67 days before.

Cocaine has been detected in pre-Columbian mummies from Peru (Springfield *et al.*, 1991). The first forensic isolation of cocaine and the cocaine derivative benzoylecgonine, was made by Nolte *et al.* (1992) from third instar

C. vicina. The larvae had been taken from a badly decomposed body of a person who was missing for five months in Connecticut. The authors point out that cocaine has a short *in vivo* half-life of 0.8–1.25 hours, depending on the route of administration. Therefore, its presence in both the larvae and skeletal muscle of the suicide indicates that the drug was used in the immediate few hours before death.

It is clear from these cases that carrion flies (and other insects) can serve as specimens for drug analysis, and are especially useful when there is advanced decomposition. Obviously, the mere presence of a drug in insects is not proof of the cause of death. However, in the suicide that Beyer *et al.* (1980) describe, the circumstantial evidence is compelling: "Two days before she was last seen she had a prescription filled for 100 tablets of phenobarbital. This was the empty bottle that was found in the purse adjacent to the body. The individual had a past history of five attempts at suicide. A handwritten note . . . was strongly suggestive of suicide."

The most important use of forensic entomology is the determination of time of death, and here, drugs play a surprising, if problematic, role. Goff *et al.* (1989) studied the effect of cocaine and its major metabolite benzoylecgonine, on the development of the fleshfly *Boettcherisca peregrina.* The larvae developed on rabbits injected with 35 mg (median sub-lethal dose), 69 mg (median lethal dose), or 137 mg cocaine (twice median lethal dose). From 0 to 30 hours, there was no effect compared with controls. From 30 to 70 hours, the actively feeding second and third instar larvae developed 12–18 hours more rapidly only at the two higher doses. Time of pupariation was correspondingly shortened, so that larval development was accelerated up to 24 hours. Duration of pupation was unaffected, as was viability. It is important to note that the maggots feeding on a corpse with a sub-lethal dose of cocaine did not have significantly altered development times. A pharmacological study of dose versus rate of development of maggots is therefore needed to determine a threshold. If properly developed, the method could be useful in establishing time of death in cocaine-related suicides, assuming the drug is above threshold in the insect. Otherwise, a sub-threshold level in the insect could be misinterpreted.

When rabbits were injected with 6, 12, 18, and 24 mg of heroin (morphine), which bracket the range of heroin-related human deaths, the development of *B. peregrina* maggots feeding upon the livers was accelerated up to 29 hours and pupariation occurred six hours earlier. But the pupation period was extended by an average of 18–36 hours. It is not clear whether there was a real difference between the minimum duration of pupation in the controls and in the 6-mg group (Goff *et al.*, 1991).

In a third study (Goff *et al.*, 1992), larvae of *Parasarcophaga ruficornis* were fed on tissues of rabbits receiving 71.4 and 142.9 mg of methamphetamine (median and twice median lethal dose). With both doses, the larvae developed more rapidly from hours 30 to 60, than larvae on 37.5 mg, or the controls.

According to the authors, larval development was shortened by up to 18 hours. Yet when one compares the minimum duration of the larval stage for both the controls and the maximum dose it is the same –138 hours. Pupal mortality is also lowest for these two groups. These results are not explained.

Evidently, abuse drugs, at certain concentrations, have physiological effects in flies that can alter their development rate and therefore, an estimate of the PMI. As research in this area expands it will be important to correlate with some precision, the drug threshold and the stages of a species when the drug's effect takes hold. Otherwise, the introduction of such evidence will tend to cloud rather than clarify the issue of time of death.

In Chapter 5 forensic entomology moves from the calm atmosphere of scientific inquiry to the open belligerence of litigation.

5

The Fly in Court

Medicolegal entomology, although born about a century ago, was dormant until relatively recently. The last quarter of the twentieth century witnessed a phenomenal growth of interest, as the courts have recognized the validity of medicolegal entomology and more and more entomologists have entered the field. This recognition is far-reaching and includes countries of the Far East, the Middle East, Africa, and Latin America as well as Europe and North America. Often, in the practice of a new science, its flaws and shortcomings are unknown, minimized or ignored by enthusiastic advocates. This is particularly true in the adversarial arena of a trial. In headline homicide cases, the expert entomologist usually does not testify unopposed. Nowadays the testimony of opposing entomologists and the conclusions they reach are likely to be minutely scrutinized and dissected in court. This benefits criminal justice, and, by exposing shortcomings overlooked in the laboratory, can promote good science.

Hospital/nosocomial cases

Our focus on the forensic biology of the fly now moves from the laboratory to the courtroom. The first examples are hospital or nosocomial cases. In my experience these cases are usually settled out of court, with minimum publicity.

A SERIOUS MISSTEP

A construction worker was injured on July 3 when he fell from a truck with broken flooring. He was taken to a Chicago area hospital where a number of surgical procedures were performed, including an open reduction of a tibial fracture of the right leg and a bone graft. After almost two weeks of fever the patient's temperature returned to normal and bleeding from the incision stopped. He was discharged July 23 and went home. On July 29, he was at the doctor's office, his sutures were removed and he was fitted with a long leg cast. On August 2, the patient was brought back to the hospital's emergency room complaining: "I woke up this morning and there were worms crawling from underneath my brace." The emergency room report describes the wound as malodorous. When the cast was removed, the wound had a yellowish drainage and was "grossly" infested with larvae. Subsequently, there were medical complications, including a procedure lasting 7.5 hours to graft abdominal muscle to the afflicted area. The patient's travails ultimately resulted in a malpractice suit. What concerns us here is the presence of maggots in the wound. Under other circumstances they might have been viewed as serendipitous agents of maggot therapy (see Chapter 1), but no one was inclined to view the infestation as anything but malignant. Lawyers on both sides wanted answers to a single question: where did the infestation originate? Was it during the patient's stay in the hospital from July 3 to the 23? In the doctor's office July 29? Or in the patient's home? The infestation could not have occurred in the hospital from July 3 until July 23, or in the patient's home from the July 23 until his visit to the doctor's office on July 29 because nothing was reported by the patient, or staff, or by the doctor during the period of hospitalization when he examined the wound, removed the sutures, and prescribed a cast. The period in question is July 29 until August 2. The other entomological evidence rests on the patient's report on when he saw maggots crawling out of the cast. Maggots were not collected and preserved in the emergency room, nor were photographs taken. At that time of year and place in Chicago the most abundant blowfly is *Phaenicia sericata*. Most likely, the maggots had crawled out of the cast because they had finished feeding and were dispersing. It is less likely that a high larval density had caused various instars to 'boil' out of the enclosed space. Assuming a temperature between 29 °C and 35 °C at the wound under the cast, it would take the larvae a fraction over three days to reach the postfeeding stage after eggs were laid (Fig. 3.23). Counting back three days from the morning of August 2, we arrive at July 30 as the likeliest time of oviposition, one day after the casting when the patient was home. The flies had ready access to an apartment that had no screens on the windows. The case progressed to a medical malpractice suit, with the entomology tangential to more serious medical issues, including the possibility of amputation.

When a maggot infestation occurs in a patient hospitalized for several weeks, the hospital is clearly suspect. Considering the huge patient populations in hospitals, nosocomial myiasis (fly infestation of a hospital patient) is relatively rare and when it does occur hospital staffs would rather not talk about it. Myiasis occurs in hospitals with or without screen doors and windows, with debilitated patients and the terminally ill as the usual targets. The gravid female fly is attracted most frequently to the nose because the insertion of a nasogastric tube can result in the accumulation of mucous, blood and/or odors of decomposition. Eggs are laid and the maggots develop unnoticed at least until they exit from the nose and disperse. The shock and outrage of a family member who witnesses this exodus is understandable.

In cases of myiasis, the entomologist establishes the species of the maggots and/or reared adults. Measurements are made of maggot length, and of crop length in dissected specimens. Evidence of blood in the crop should be recorded. An estimate of the age of the largest maggots is made from appropriate rearing schedules, assuming a nostril temperature of approximately 30 °C. An out-of-court settlement is usually reached after the entomologist is deposed and the opposing lawyer has had a chance to probe the entomological arguments.

Accidental Death or Homicide

We now move to court cases involving accidental death or homicide; here the entomology is frequently contested. We present the relevant facts and evidence on which the opposing entomological opinions are based, and, we leave it to the reader to assess the strengths and weaknesses of the arguments. In Part II the legal issues relevant to the practice of forensic entomology and science in general, are examined in detail.

Our case experience suggests the following advice to the novice forensic entomologist preparing for court. Carefully analyze the following: (1) Police reports – including the crime scene environment from a fly's "eye view", e.g. urban or rural, proximity of dwellings and food processing plants, mode and frequency of garbage disposal, and local weather; (2) the indoor death scene, e.g. fly access (tightness of windows, screens), presence of pets, household sanitation, and local weather; (3) pathology report for evidence of the size and site(s) of the maggot infestation of the body, and evidence of drugs; (4) weather station data to reconcile with on-site conditions; (5) photographs at the scene and at autopsy; and (6) insect specimens.

Turn a deaf ear, if humanly possible, to the opinions, often uninvited, of the police and lawyers. If you are working for the defense, write a draft report before examining the report of the prosecution entomologist. Then critique the other report for similarities and differences. Evaluate the scientific validity of your differences based primarily on the published literature (especially data that have been confirmed in other laboratories), and secondarily on unpublished data and professional experience. Point out the similarities as well as the differences in your final reports. Evidence permitting, the report should end with clear-cut conclusions. On the witness stand, speak with authority, but simply and clearly. Avoid scientific jargon. Jurors might disregard what they do not understand. It is imperative that you also simplify the entomology for your attorney. Most lawyers are eager to acquire courtroom competence and welcome a tutorial. Under cross-examination, give brief, non-defensive, objective responses. Verbose academics may be admonished and kept on a short leash by the judge and opposing lawyer.

With these preliminary suggestions in mind, it may prove useful to consider the following cases, placing yourself in the role of an expert asked to assist in the case. How would you, as a scientist and a would-be expert witness in forensic entomology, evaluate these facts? What conclusions would you draw, and with what degree of certainty? Are there problems with the information, such as gaps in coverage, unexplained facts, poor handling of the samples, etc? How would you prepare for deposition or in-court testimony as the proponent or opponent of this material?

A MISSING CHILD

In Covington, Kentucky, a six-year old girl was reported missing on April 21, 1989. The last person allegedly saw her alive at 17:00 on that date. The prosecution alleged that the child died in the early evening hours of April 21. Her body was found on May 1, by individuals in the neighborhood searching in an abandoned house for aluminum cans to recycle. The police report states, and photographs show, that the first floor doors and windows were boarded up and windows on the second floor were open. Access was by two doors along one side of the house and both doors had been forced open. Dogs were able to come and go, and there were dog feces and all kinds of trash in the room where the body lay. There were smear and splatter patterns of dried blood on one wall which, according to the investigator, "did not appear consistent with having been derived from the child's body". It was reported to him that the house had previously been used as a dog-fighting arena. Nevertheless, samples were taken for verification of origin and they proved negative for human blood.

The defendant, at the time, was 23 years of age and resided across the river from Covington in Ohio. He had no previous criminal record; however, on April 20 he allegedly molested a child near where he lived by putting his finger in the vagina of a 19-month-old baby. This incident occurred a day before the girl in Covington turned up missing. According to his attorney, the defendant has an IQ of 72, is an extremely slow learner, was in slow classes all his life, cannot see well and cannot drive a motor vehicle due to his lack of motor skills. The defendant says he has never been over in the Kentucky area. According to the defense, the foreign hairs found on the body did not match his, nor did the footprints or fingerprints at the scene. The finger of guilt originated among jailhouse informants who said that the defendant told them he sexually molested the girl and killed her.

The state's pathologist observed that "The left thigh has been virtually skeletonized from a point 2" below the femoral crease to the upper patella". He could not rule out sexual trauma because of the large number of maggots in the vaginal area. The pathologist for the defense countered that there was no possible way, under the circumstances, of discovering whether sexual trauma had occurred in the vaginal area.

The defendant was in jail for the sexual molestation of the 19-month-old baby on the morning of April 25, 1989. The deceased was not found until May 1, 1989. The prosecution contends that the child died during the early evening hours of April 21, 1989. The defense believes that the child could have died and was more likely to have died April 25 or later.

Two issues confront the opposing entomologists – time of death and sexual molestation.

Two key insect samples were collected and preserved in 70% alcohol at autopsy on May 2, 1989, at about 2 p.m. (Sample 1). About 60 larvae from the head area identified as third instar *Calliphora vicina, Phaenicia* sp., and *Phormia regina.* (Sample 2). About 75 larvae from the thigh area identified as in (Sample 1).

Following are the summary and conclusions of the prosecution entomologist, followed by that of the defense entomologist. Consider them both, and look for strengths and weaknesses.

Summary of conclusions

The insect evidence examined in this case indicates that the most probable time when female blow fly oviposition (egg deposition) would have taken place was after sunrise on the morning of April 24, 1989. Calliphorid larvae retrieved from the decedent and the scene where the body was recovered were identified from reared larval specimens as adult Calliphor[i]a vicina. *It is very probable that death would have occurred several days prior to sunrise on Monday April 24, 1989, due to the cool weather (highs only reaching the 50s and 60s*[°F]*) and the seclusion of the body in a boarded up building. These two factors could limit the female blow flies from immediately accessing the body for oviposition, thus causing a delay from the time death occurred to the time when insect activity actually began.*

National Weather Service Station Data (NWS Cincinnati, OH)

Hourly climatological data sheets from the National Weather Service station at Cincinnati, Ohio for the time period 18 April, 1989 through 3 May, 1989 are included at the end of this report. *Temperature data taken at the scene by officers of the Covington Police Department are as follows:*

May 2, 1989

Time	*Room temperature*
1445	*57 °F*
1555	*57 °F*
1715	*56 °F*

May 3, 1989

Time	*Room Temperature*
1300	*55 °F*
1405	*56 °F*
1430	*60 °F*
1500	*60 °F*
1600	*60 °F*

Discussion

The insect evidence examined in this case indicates that the most probable time when female blow fly oviposition (egg deposition) would have occurred was after sunrise on the morning of April 24, 1989. Calliphorid larvae retrieved from the decedent and the scene where the body was recovered were identified from reared larval specimens as adult Calliphor[i]a vicina. *The developmental time interval for this species to reach the late third instar larvae based on the measured temperatures (upper 50s to low 60s* [°F]*) for the habitat where the body was found* *was approximately seven to nine days minimum time (based on rearing data from Reiter, 1984, and Kamal, 1958). It is very probable that death would have occurred several days prior to sunrise on Monday April 24, 1989, due to the cool weather (highs only reaching the 50s and 60s* [°F]*) and the seclusion of the body in a boarded up building. The temperatures in the building would be more constant than temperatures in the open, with higher minimum temperatures and lower maximum temperatures, due to the darkened shaded effect of the boarded windows and the ability of the house to create an insulating effect. These two factors could limit the female blow flies from immediately accessing the body for oviposition, thus causing a delay from the time death occurred to the time when insect activity actually began.*

Bibliographic references cited in the above report are Kamal (1958) and Reiter (1984).

On August 2, 1993, the entomologist for the defense received a package that contained reports by a forensic anthropologist, the medical examiner, the police, the prosecution entomologist, the National Weather Service data for Cincinnati, Ohio, trial transcripts of the above individuals, photographic slides taken at the discovery scene and a crime scene video. A package followed which contained the insect specimens detailed above. Following is the report of the defense entomologist

1. Fly habits and accessibility to the building – Calliphora vicina, *the blue bottle fly, is common in North America from Mexico City to Alaska. It is a cold-loving species and one of the first blow flies to appear in March and April and one of the last species to disappear in the fall. It can't stand the heat, so summer finds it active at only the higher elevations and latitudes and at night. During frost-free winters on sunny days (in Russia), adults will fly at shade temperatures of 5° to 6 °C (41 °F to 43 °F). I have found them active and flying on sunny days in mid-winter in Italy and in the Midwest when temperatures were 5° to 8 °C (41 °F to 46 °F). Adults have been observed emerging from pupation at 0° to 2 °C (32° to 35.6 °F). Under the prevailing temperatures during the period in question, this species of fly would have been on the wing and reproductively active.*

Calliphora vicina *is an eusynanthropic fly meaning that it is closely tied to the human environment. It is the fly often found on the windows of one's home trying to get out. It frequents dumpsters, feeds on table scraps, and breeds in carrion, e.g., animal carcasses (including that of the decedent), and in garbage dumps. The flies would have been in the general area and would not have had to go far to discover the body. In fact, the flies could have been in the building because of the presence of dog feces, to which they are attracted for moisture, salts, and protein. Dogs had access to the house via door panels and one of the windows which was not boarded up.*

Flies bent on feeding and laying eggs on a corpse will enter far more unlikely places than in this case. I have consulted in numerous homicide cases, including the following, to support this observation:

1. C. vicina *laid eggs on a body wrapped in a quilt and covered with plastic garbage bags, tied with rope, and placed in the unheated basement of a 2-story apartment building in Chicago, beginning of February. There was no obvious access except a small break in a window (Beverly Brown homicide, 1990).*
2. C. vicina *laid eggs on the corpse of an unknown female in a closet of a boarded-up 2–story building in Chicago in April.*
3. C. vicina *and two other species of blow flies laid eggs on the body of a 9-year-old girl, the day after the night she was last seen alive. The body was in a pitch black bathroom, behind a closed door, of a boarded-up apartment (Lake County, IL; Alton Coleman, the convicted killer, has death sentences in Indiana and Ohio, as well).*

To summarize from some of my cases, C. vicina *specifically, and blow flies generally, have entered cars with windows shut, trunks of cars, and houses with doors and windows shut. They are active during the day, both in the light and in the dark, and at night.*

2. Weather data relevant to conditions at the crime scene – I quote [the prosecution entomologists] *report on page 6: "It is very probable that death would have occurred several days prior to sunrise on Monday, April 24, 1989, due to the cool weather (highs only reaching the 50s and 60s) and the seclusion of the body in a boarded up building." Having dealt with fly access, let's now deal with temperature at the crime scene.*

The body was transported from the crime scene at approximately 0015 hours, May 2, 1989, but "Temperature data taken at the scene by officers of the Covington Police Department . . ."([prosecution entomologist's] *report, page 3) were taken on May 2 and 3, 1989. In the following table we have added the temperature taken at the same time at the National Weather Service station in Cincinnati for comparison.*

Time	*Room temperature (°F)*	*Weather station (°F)*
May 2		
1445	*57*	*51*
1555	*57*	*52*
1715	*56*	*50*
May 3		
1300	*55*	*58*
1405	*56*	*62*
1430	*60*	*60*
1500	*60*	*60*
1600	*60*	*62*

The temperatures recorded by the National Weather Service track fairly closely the temperatures recorded at the crime scene. We are therefore justified in using weather data from the National Weather Service in our analyses. According to these data, maximum/minimum temperatures on May 2 were 53/39 °F and on May 3, 63/35 °F. But the weather from April 21 to May 1 was much warmer. Here are the temperatures (°F) for this period:

Date	*Maximum*	*Minimum*
April 21	*73*	*50*
22	*71*	*47*
23	*63*	*46*
24	*73*	*48*
25	*85*	*58*
26	*84*	*64*
27	*82*	*61*
28	*78*	*57*
29	*76*	*54*
30	*66*	*49*
May 1	*62*	*47*

The days were quite warm, especially from April 25 through April 29, and conditions were ideal for blow fly activity. Even on the 27th, when it rained, it was sunny 64% of the day. ". . . cool weather (highs only reaching the 50s and 60s)" applied only to May 2 and 3 after the body had been removed. Therefore, reliance on temperatures recorded on May 2 and 3 is clearly misleading, and entomological analyses based on such data are invalid. Furthermore, the tree in front of the house had only partially leafed out and would not have blocked sunlight. This, and the dark roof, would have further increased the interior temperature.

3. Entomological analyses and conclusions *– In this section we deal with methods used to assess the age of the oldest larvae sent to me, which were in vials # 1 and 2.*

A1. *Maggot length is not mentioned in the* [prosecution entomologist's] *report. Yet the age of actively feeding maggots is routinely determined by their length, species, and thermal history. Thermal history simply means that, within physiological limits, the rate of insect development is accelerated at warm temperatures and retarded at cool temperatures.*

One needs to compare the temperature at which the maggots developed with known rates of development to calculate the age of the actively feeding maggots taken from the body. I have graphs for about a dozen species of flies at 10°, 12.5°, 19°, 22°, 25°, 29°, and 35 °C.

A2. *The largest maggots of C. vicina in vial # 1 (taken from head area) had the following lengths (mm): 16, 17.3, 17.0, 16.5, 17.1, 18, 16.2, 16, 17.4, and 17.2. The largest maggots of Phormia regina from vial # 1 were: 11.0, 11.0, 11.0, 11.3, 10.8, 10.0, and 8.6mm.*

A3. *The largest maggots of* C. vicina *in vial # 2 (taken from the thigh area) were: 16.0, 15.0, 16.0, 16.0, 14.5, 15.0, 15.5, and 12.5 mm. The largest maggots of* P. regina *from vial # 2 were: 10.5, 10.0, 11.0, 10.8, 9.0, 9.0, 10.0, 9.3, 8.7, and 10.2 mm.*

A4. *The beetles and stray flies, which do not appear teneral (newly emerged), in vials # 3, 4, and 5 contribute nothing to the analysis.*

A5. *Actively feeding maggots have an intestinal tract and crop which are engorged with food, and they do not have a layer of white fat deposited on the inside of their cuticle (skin). When third instar maggots cease feeding, they enter a postfeeding stage which lasts two or more days, during which they gradually empty their crop and digestive tract and deposit a layer of fat next to the inner surface of the cuticle. We have published studies measuring the rate of emptying of the crop, as a function of time, in several species. In* C. vicina *the crop is nearly 7 mm long at peak of feeding, shrinking to ca. 1 mm just before pupariation. Length of the postfeeding maggot is not a reliable criterion of age: that is why dissection of the gut is essential. There is no mention in* [the prosecution entomologist's] *report that he dissected maggots to assess whether they were still actively feeding or had entered the postfeeding stage; if the latter, how old were they and by what means did he assess their age? I dissected maggots randomly selected from among those whose lengths I had measured, and these are the results:*

Vial # 1 : C. vicina – *crop length: 3.5, 4.3, 4.0, and 3.5–4.0 mm; maggots had complete deposits of fat; hindguts not completely empty. A color photograph of the body (presumably at autopsy) shows similar diminished crops of postfeeding maggots.*

P. regina – *all specimens were still in the actively feeding stage. These were 10 to 11 mm long with full guts. At peak of feeding they would be 15–16 mm.*

Vial # 2 : C. vicina – *crop length ranged from 5 to 6 mm; salivary glands and guts fully engorged. At the beginning of the postfeeding stage.*

P. regina – *all specimens were still in the actively feeding stage.*

A6. *Conclusions from the above observations:*

1. *Larvae of* C. vicina *and* P. regina *infested both head and thigh areas.*
2. *Female* C. vicina *first laid eggs in the head area, and later in the thigh area.*
3. *Female* P. regina *laid eggs later than* C. vicina, *but at about the same time in both head and thigh areas. Being a secondary invader makes this species of secondary importance in estimating time of death.*
4. *The evidence for the presence of younger larvae of* C. vicina *in the thigh area rather than in the head area strongly suggests that oviposition may have occurred here following dog attack of this area. Generally, natural orifices of the head are the first to receive eggs because they offer accessibility, moisture, and delicate tissues for initial invasion of larvae. The huge maggot mass in the thigh area is easily explained*

when one considers the oviposition frenzy which would follow the exposure of blood, fluids, and tissue from dog attack of the thigh. The conjecture of vaginal trauma as the reason for this maggot mass is far fetched.

In the Jerri-Ann Richards homicide in Pawtucket, Rhode Island, the 4-month-old infant was raped before she was murdered, yet C. vicina *did not lay eggs in the vaginal area, but around the head, even though she was unclothed. Incidentally, the murder occurred in November and outdoor temperatures were colder than in the Iles case.*

B. *Another crucial aspect of the forensic entomology overlooked by* [the prosecution entomologist] *is maggot mass. As we mentioned earlier, warm temperatures speed up development, and cool temperatures slow it down. A maggot mass produces considerable heat and raises the temperature experienced by the maggots at least 17 °C (ca. 31 °F) above ambient (8. Greenberg, "Flies as Forensic Indicators," J. Medical Entomology, 28:565–577, 1991, figure 1). [In the] BBC TV documentary, "The Natural History of Crime (Murder)." the thermometer reached 44 °C (ca. 112 °F) while the air temperature was 33 °C (91 °F). The effect is the same in the cold. As long ago as 1940, Deonier observed (outdoors in Texas and Arizona in January and February) that a maggot mass was 18 °C (65 °F) and quite active when the air temperature was −4 °C (25 °F). The effect of a maggot mass on temperature has thus been known and accepted for years. Photographs of the body taken at the scene show a large maggot mass in the head area and another in the thigh area. This is factored into our calculation of the age of the oldest maggots, which were from the head area.*

C. [The prosecution entomologist] *relies on rearing data of C. vicina from Reiter (1984) and Kamal (1958). Apparently, he did not have data of his own. Kamal reared his flies at 80 °F. How can* [he] *base his conclusions on work done at 80 °F when he says highs in the room only reached the 50s and 60s? Nowhere does he mention by what means (graphs, extrapolations, etc.) he got from hither to yon. Kamal's work has been discredited by a Professor in the Entomology Department at Washington State University, where Kamal took his degree. The Professor examined Kamal's reared specimens and found them to be stunted, possibly as a result of inadequate larval food which could have altered the time it took the flies to develop. Kamal's published results (Table 1 copy of paper enclosed) have numerous arithmetic errors. He gives 14 days as the total minimum time for* C. vicina *to develop from egg through the pupa stage, but his figures add up to 15.5 days. Kamal gives his average time as 18 days, but the correct addition is 21 days; his maximum time is 25 days, but his numbers add up to 33.7 days!*

Christian Reiter (1984), working in Austria, is a good scientist, and his work on C. vicina *is carefully done (copy enclosed). His Table 1 and Figures*

2 and 3 summarize the relevant data on egg and larval development up to the stage of the oldest larvae in this case, which were 17.6 mm long with crops partially full when taken from the body and preserved:

Temperature (°C)	14–16*	18–19	22–23	25[1]	30
Total time (days)	8.8	6.8	5.6	4.1	4.2

* "upper 50s low 60s" [of prosecution entomologist] = approx. 15 °C
[1] B. Greenberg, J. Med. Entomology, 28:565–577, 1991, Table 3.

Please keep in mind, the above data do not include the effect of maggot mass and that during the period when the maggots were developing on and in the body, weather data show average temperatures in the 60s and 70s, with maxima in the 80s on 3 days. Given these facts, temperatures above 20 °C (68 °F) would have been experienced by the maggots and are the only realistic ones in this case. With Reiter's data, it took 4.2 to 5.6 days from the time of oviposition to reach the stage of the oldest preserved maggots.

I have made a careful study of the rate of shrinking of the maggot's crop in the postfeeding period. In C. vicina *the fully distended crop is nearly 7 mm long. At room temperature, it takes about 21 hours to reach the minimum-size crop we measured in our dissections. At higher temperatures, it would take fewer hours to reach the same stage.*

I conclude that it took a minimum of 4.1 days from the time the first eggs were laid for the larvae to reach the oldest stage we found. The body was placed in the cooler at 40 °F at approximately midnight, May 1 1989. We will give the prosecution the benefit here by assuming maggot development stopped at midnight, although maggot development would have continued for many hours for the following reasons:

1. *Maggot mass. Last week I had occasion to check the temperature of a maggot mass I had put into a cold room at 4.2 °C (39 °F); the temperature was 14.4 °C (58 °F) and the maggots were active.*
2. *The body, on a section of rug, was placed on a piece of plastic, then put into a body bag which was sealed into an aluminum transfer case. The multi-layers of insulation would have retained heat and moisture, proper conditions for continued maggot development.*

Counting back 4.1 days from midnight of May 1 we arrive at approximately 10 PM April 27 when the first eggs were laid. Given the presence of dog feces which is a fly attractant, it is reasonable to assume that flies were in close proximity and even in the house. Given the unsanitary conditions and prevailing temperatures which reached into the 80s, the body was probably not there more than 24 hours when egg-laying began. The murder would have occurred on April 27.

If we assume the less probable alternative of 5.6 days, we arrive at about 10 AM April 26 when the first eggs were laid. Given a 24-hour period as the maximum time for the flies to find the body and lay eggs, the murder would have occurred on April 25. Please keep in mind that carrion flies have so keen a sense of smell, it would make a bloodhound blush. They can zero in on carrion from a mile away in just a few hours. Egg-laying usually begins within a few hours after the flies' arrival.

Now that you have read both reports what questions occur to you? Is there any additional factual information that might be helpful? To what extent do you agree with the conclusions of these entomologists? How would you respond to or amplify upon their findings if you were asked to provide a deposition or to testify at trial?

In the real-life case, there were three trials. In the first trial, the prosecution entomologist presented his report and testified unopposed. The jury found the defendant guilty. The case was appealed all the way to the State Supreme Court and, because of irregularities, was sent back for re-trial. In the second trial the defense presented the findings of an entomologist whose report is given above. The prosecution entomologist expanded his testimony to include a discussion of the application of accumulated degree-hours (ADH) to arrive at his estimate of the time of death. *Calliphora vicina* was the primary insect indicator in this case and he argued that 10 or 11 °C would have to be subtracted from each temperature reading because that was the fly's threshold of development. This would extend the estimated time of death and bring it closer to the time the child was last seen alive. The defense entomologist countered that the actual threshold for this species was not clearly established by science but that the literature indicated it was probably closer to 3 °C.

Not indicated in his report was the defense entomologist's testimony regarding the charge of sexual molestation. This was obviously brought because of the defendant's alleged previous sexual molestation of the 19-month-old baby and the presence of a maggot mass in the groin area of the murder victim. The entomologist described two of his previous cases to the court. The first one involved the rape and murder of a 4-month-old baby. Her naked body was found in an alley in an abandoned factory district in a town in Rhode Island. It was in November when the weather was decidedly cool. There were eggs and first instar larvae of *Calliphora vicina* in the neck folds and on the eyelids of the baby, but there were none in the vaginal area (or anywhere else). In the other case, a woman was reported missing by her husband on August 28, 1991, in a city in Oklahoma. Her body was found a few days later in an abandoned storm-damaged house about 30 miles away. One end of a nylon strap was bound around her neck and the other end was tied to the clothes bar of a closet. The medical examiner ruled asphyxiation after being unable to pinpoint any other cause of death. There was a large maggot mass in the groin area despite the fact that she had not been sexually molested. Neither of these cases proved that the young girl in Kentucky had not been sexually molested but taken together they showed that the presence of maggots in the vaginal area was not proof of sexual molestation.

In fact, the maggot mass in the groin area could be simply explained. The dog attack of the child's thigh exposed the area to a second frenzy of oviposition after the first round in the head area. This is shown by the fact that the maggots that developed in the groin area were smaller and therefore younger than those in the head area.

The outcome of the second trial was a hung jury – all but one juror voted for acquittal – and the case went back for a third time. The entomological arguments were essentially unchanged and this time the jury returned a verdict to acquit. It would be presumptuous to suggest that the entomology evidence was outcome determinative in this case. Certainly there were other arguments for acquittal, e.g., the foreign hairs found on the body did not match the defendant's; nor did footprints and fingerprints at the scene. Taken all together, it convinced the jury of his innocence. But only the jurors will ever know the extent to which forensic entomology helped to decide the case. And that is typical of life as an expert witness. Did you "win" or "lose" the case? Were you at least a factor? What would have been the result without your input? You will rarely, if ever, know the answers to these questions.

THE COVERUP

In the middle of the night in a small town in eastern Canada, everyone in the house was asleep except a 19-year-old woman. She went into the bathroom and filled the tub with about 12 inches (*c.* 30.5 cm) of water. When she disrobed it was obvious that she was pregnant, and she was in labor.

What had gone on before, what happened that night, and the events that followed made this no ordinary case for an entomologist. For months, leading up to that night in early October, to everyone who asked her whether she was pregnant – people at school, co-workers at harvest time, and friends at parties where she indulged in alcohol and marijuana – her answer was always "no".

Now she was alone with reality. She sat in the tub and occasionally got out and walked around the house. Between the hours of 1 and 5 in the morning, while sitting in the tub, she gave birth to a boy. For a half-hour she sat in the tub, immobile. Then she reached for a pair of scissors, cut the umbilical cord and got out of the tub and cleaned herself off. The infant lay under the bloody water. Next, she lifted the baby out of the water, wrapped him in newspaper and put him into a plastic grocery bag. Then she drained the tub, placenta and bloody water went down the drain. She placed the plastic bag under a tree outside her house and went back to bed. No one woke and no one knew that she had given birth. She kept her secret as she hemorrhaged for two days.

When she returned to school everyone noticed that she no longer appeared to be pregnant.

About two months later, on a morning in early December, a man noticed a white plastic bag in the grass at the side of his house. That afternoon he picked up the bag and dumped its contents near the barn, believing it contained animal remains suitable for his cats. Out fell the baby. Only later would he learn that this was his grandson.

The pathologist reported that the 7lb (*c.* 3kg) baby was full term. There was no congenital anomaly or trauma that could have resulted in death. Sections of both lungs revealed uniform alveolar expansion with expansion of the bronchi and bronchioles. The baby had been born alive and asphyxia was considered a possible cause of death. There was little decomposition and the pathologist estimated that death had occurred about three days before.

Given that the public had been witness to the rise and fall of her abdomen, the young woman was an obvious suspect, but under police interrogation she persisted in her denials. When finally she agreed to a DNA comparison with that of the baby, and the two DNAs matched, she confessed and was charged with second degree murder.

What had happened during the two months between the birth of the baby and its discovery? No insects were reported to be on the body, in the newspaper wrapping, or on the inside or outside of the plastic bag. Was it possible that after two months out of doors in the fall there would have been no fly activity and very little decomposition? This is highly unlikely because day-by-day weather analysis indicated many days in October and November when flies would have been active. The entomologist's analysis was more a series of conjectures, and his involvement ended there. She was convicted but questions remained that have not been answered. What do you think happened?

A TRIPLE MURDER

The family – father, mother, daughter, and son – lived in a town not far from Toronto, Canada. On June 10, 1992, the dead bodies of all but the son were discovered by a neighbor in the upper story of their two-story residence. The middle-aged female's body was on the upper floor hall landing. The body of a middle-aged man was on the bed in the master bedroom, covered, except for the head, with a blanket. The naked body of the teen-age female lay on a bed in another bedroom. The pathologist reported that all three had died in the residence at the same time, the parents by blows to the head and the daughter by unknown causes. The victims had not been seen alive since about 17:00 June 5, 1992, except for one witness who claimed to have seen them on June 6 and 7. All windows and doors were closed at the time of discovery, except for a sliding door on the first floor that was open 1 or 2ft (c. 31 to 61 cm). According to the presiding judge's review of the circumstances, the son was in the residence "... from a few hours prior to 1:47 a.m. on June 6, until approximately 8:15 on June 6. He was in

Ottawa on the day the bodies were discovered. He had been in Ottawa since approximately 2:00 p.m. on June 6. He was located in Ottawa on June 11, and charged as a young offender with three counts of "first degree murder" on June 12, 1992. The trial began on April 18, 1995 and continued at intervals until December 17, 1996. On August 2, 1996 the crown abandoned the charges of first degree murder, and the trial continued on second degree murder".

The judge summarizes the entomological facts and evidence subsequently made available to the forensic entomologists (names deleted) for the crown and for the defense. "The bodies of the three victims were found decomposing and infested with fly larvae (maggots), as well as fly eggs . . . an entomologist at the Royal Ontario Museum attended at the . . . residence on June 10, 1992, [and] collected several live and dead adult flies. The live specimens were killed . . . in a cyanide bottle. Some larvae and eggs were also collected from the dead bodies. Some of these larvae were killed in Kahle's solution. All specimens were delivered to the Royal Ontario Museum . . . at the morgue . . . more larvae and eggs were collected from the dead bodies prior to performance of autopsies. All the specimens were taken to the Royal Ontario Museum where certain of the live larvae were killed as before in Kahle's solution and the remaining larvae plus the eggs were placed in special baggies where they were reared . . . through to the emergence of the adults from the pupae. Rearing was done at controlled temperature and humidity, and the baggies were monitored each twenty-four hours. The emerged adult flies were then killed and pinned. . . . A quick calculation . . . indicates that a total of two hundred and sixty three larvae, more than one hundred and fifty eggs and one hundred and forty two adults were processed."

The following entomological observations and procedures described by the entomologist from the Royal Ontario Museum supplement the judge's account. Three eggs were seen scattered on the face of the adult female, and fly larvae were seen in the mouth between the lower lips and gums. Some of these specimens were collected at 20:00 and placed on beef liver to be reared to the adult stage for identification of the species; this included the three eggs. The liver was put into an aluminum foil pouch inside a zip-lock plastic bag and kept at 24 °C ± 2. A sample of the larvae was preserved in Kahle's solution. Live adults were also captured.

Eggs were not seen on the man or girl and further sampling was postponed until just prior to autopsy at noon the following day. The bodies were held at 4 °C from approximately 02:00 June 11, 1992 until the autopsies. At that time eggs and larvae were recovered from the man, additional samples of eggs and larvae were taken from the woman, and eggs were recovered from the girl. "Large maggot masses were not seen at either the scene or the morgue"

The salient data derived from the above collections are:

- *Mother* – Adult flies reared from specimens taken at the scene were *Phaenicia sericata* and *Sarcophaga falculata*, both eclosed June 29, 1992.

(In all the rearings, eclosions occurred at an undetermined time prior to 09:00 on the date indicated.) Of two larvae killed at autopsy, one was *Phormia regina* (6.6 mm), the other was an unspecified sarcophagid (6.3 mm). Eggs and larvae planted on beef liver at the morgue yielded adults of *P. regina* from June 22 to 26; *P. sericata* from June 24 to 29; and *Calliphora livida* and *Sarcophaga falculata* on 29 June. Live adults of these species were collected at the scene.

- *Father* – As indicated, all sampling was at the morgue. Several larvae of *P. regina* were killed and measured 1.1 to 2.6 mm; others were reared and eclosed as *P. regina* on 24 and 26 June. An egg mass yielded adults of the same species on 26 and 29 June.
- *Daughter* – Eggs were collected from the hair at the morgue and reared to yield *P. regina* on 24 and 25 June.

The temperature in the residence during the time the bodies lay there became a sticking point. The crown's expert produced a computer model of what the temperature must have been, based on October readings. The model predicted that the temperature in the residence at 15:00 on June 10, 1992, should have been 75 °F (*c.* 24 °C). When a detective entered the house at 15:00 on that date the thermometer on the thermostat in the lower hallway registered 85 °F (*c.* 30 °C). The detective later testified that the temperature was noticeably higher as he ascended to the second floor where the bodies lay. At 19:15 that evening the temperature was 77 °F (25 °C) in the hallway a few feet from where the body of the woman lay.

Following is the report of the crown's entomologist.

Summary of Conclusions [mother]

The insect evidence recovered from the scene, the remains at the scene and the remains during autopsy indicate that the first adult fly activity and deposition of life stages (eggs and larvae) most likely occurred after 1700 hrs. June 5,1992 for some of the Diptera (fly) species and no later than 0500 hrs. June 7,1992 for another species. Four Diptera species used for the analysis were collected and identified from the second floor of the house, as eggs and larvae from the body at the scene, and as eggs and larvae from the body at autopsy. These species included: 3 Calliphorids; Calliphora livida, Phaenicia sericata, *and* Phormia regina*; 1 Sarcophagidae;* Sarcophaga falculata. Calliphora livida *and* S. falculata *indicate a time consistent with the longer portion of the time period for their respective developmental rates and the likelihood of their presence in the house if death occurred shortly after the decedent was last seen. The other two species,* P. sericata *and* P. regina, *point to colonization at the shorter end of the range, June 6, 1992 AM and June 7, 1992 at 0500 hrs.*

Factors taken into consideration for these estimates were based upon known developmental times for these or similar species, climatological occurrences for the time period in question, temperature regimes and energy requirements (accumulated degree hours) within the house where the bodies were located, and temperature regimes and energy requirements while the immature insects were developing in the laboratory.

The conclusion of when death occurred for this case, when based upon the four species of carrion flies studied and the temperatures present to drive immature development, in this forensic entomologist's professional opinion, was between the evening of June 5, 1992 and the early morning of June 6, 1992.

DISCUSSION

The insect evidence recovered from the scene and during autopsy (four fly species identified and used in the analysis), indicate the oldest oviposition (egg laying) by adult female flies on the adult female victim would have occurred sometime during the time period from 1700 June 5, 1992 for some of the species to 0500 June 7, 1992 for P. regina. *Two species of Diptera,* Calliphora livida *and* Sarcophaga falculata, *intimate that their respective developments began a short time after the decedent was last seen at 17:00 on June 5, 1992, sometime after darkness. While it is not common for calliphorids or sarcophagids to oviposit or larviposit (Calliphoridae lay eggs while Sarcophagidae deposit larva) during darkness in rural environments, it is possible, under conditions where artificial lighting is present in urban environments, for this oviposition (larviposition) to occur (Greenberg 1991). The discovery of three dead adult female flies at the scene (Sample # 5), where the specimen of the species* C. livida *was found at the hall window and the specimen of* S. falculata *was found in a spider web, suggest that these specimens could have been present and most likely alive in the house at the time of death of the decedent and would, therefore, have been "ready and waiting" to deposit eggs and larvae at the earliest available time after the person was killed. Also, the species* Phaenicia sericata *(common in urban environments) indicates that oviposition of that species occurred during the morning of June 7, 1992.* Phormia regina *is known to arrive at decomposing tissue somewhat later than the other three species found. This species, most likely, would not be deposting eggs for up to 18 to 24 hours or more following death. In all four cases of these identified species depositing eggs or larvae, death would have occurred sometime prior to fly activity and oviposition. Given the known climatological conditions (rainy cool weather for the initial portion of the time in question) and the limited immediate accessability (2nd floor of a house with only a patio door opened 12 to 24 inches* [c. 31 to 61 cm]*) of the remains to carrion flies, it would not be unlikely for there to be a considerable delay from when death occurred to the actual colonization of the remains by carrion flies. It is possible for this delay to be from 24 to 48 hours in duration.*

An extremely important aspect of the entomological analysis is for the accurate estimation of temperatures where the remains were found. This is directly linked to when the fly species started their next generation and that generation's continued development. Therefore, extensive temperature data were collected, analyzed, and calculations made with the help of computer modeling.

A hygrothermograph was used to assist in calibration of the temperature relationship between the outside environment and the specific location where the bodies were found on the second floor of the house. By comparing hourly temperatures from within the house and the corresponding hourly temperatures at the Environment Canada weather station at Buttonville (ca. 3 miles distance) this temperature relationship could be evaluated. These data were sent to "Enermodal Engineering Limited" in Canada. Prior to measuring temperatures in the house, physical conditions with respect to the house (door openings, lights, blinds and shades) were set identically to the day the bodies were found. Both rainy, cloudy, overcast days and days of full sunshine occurred from the time the decedents were last seen, and therefore, both types of days were used with the hygrothermograph temperature data gathering. Hourly temperatures on the 23rd and 24th of September 1992 were calibrated with the corresponding temperatures from the Buttonville weather station. These two days were days of good visibility and sunshine while the period of hygrothermographic measurement in the house on October 9 and 10, 1992 were days of rain, fog and cloud cover eventually becoming broken.

In addition, Enermodal Engineering Limited, a company of thermodynamic, climatological, and environmental engineers experienced in assessing heat transfers and fluxes occurring in enclosed and other structures, were contacted to develop a computer model for estimation of the temperatures that would have been present in the house during the time period in question (June 5, 1992 to June 10, 1992). Upon completion of the model, a report was sent [to the prosecution entomologist] with hourly temperature evaluations present in the house for the interval in question. These hourly temperatures were used by [the prosecution entomologist] to assess the energy units (degree hours) available for insect development.

By knowing the total life cycle requirements of the different species (from Kamal 1958) and calculating the total energy units available, the minimum time interval for development of the species can be determined. It has been seen in certain developmental data that when dealing with northern strains of certain calliphorid species, developmental energy requirements may be somewhat less than is needed by southern counterparts of the same species (E.P. Catts pers. comm.). It should be noted that Sarcophaga falculata *is not shown in Kamal's data, but a closely related sister species,* Sarcophaga bullata, *is shown. These species developmental times would be similar. Also,* C. livida *is not shown in these data, but Calliphora vicina, a closely related sister species of C. livida, is shown. These two species will have nearly identical developmental time intervals and energy requirements.*

In conclusion, in this forensic entomologist's professional opinion, the most probable time when death could have occurred in this case, when based upon the four species of carrion flies studied and the temperatures present to drive immature development, was between the evening of June 5,1992 and the early morning of June 6, 1992.

Summary of Conclusions [daughter]
The insect evidence recovered from the remains during autopsy indicate that the first adult fly activity and deposition of life stages (eggs) most likely occurred sometime during the early morning hours of June 7,1992. One Diptera (fly) species used for the analysis was collected as eggs from the body at autopsy. This species was: Calliphoridae; Phormia regina, *and points to colonization at a time period around 0500 hrs. on June 7,1992.*

Factors taken into consideration for this estimate were based upon known developmental times for this species, climatological occurrences for the time period in question, temperature regimes and energy requirements (accumulated degree hours) within the house where the bodies were located, and temperature regimes and energy requirements while the immature insects were developing in the laboratory

The conclusion of when death occurred for this case, when based upon this species of carrion fly studied and the temperatures present to drive immature development, in this forensic entomologist's professional opinion, was approximately 24 hours prior to early morning (0500) on Sunday, June 7, 1992.

DISCUSSION
The insect evidence recovered during autopsy (one fly species identified and used in the analysis), indicate the oldest oviposition (egg laying) by adult female flies on the teenage female victim would have occurred at approximately 0500 on June 7, 1992. Phormia regina oviposition most likely took place sometime in the early morning hours of June 7, 1992, possibly during darkness. While it is not common for calliphorids to oviposit during darkness in rural environments, it is possible, under conditions where artificial lighting is present in urban environments, for this oviposition/larviposition to occur (Greenberg 1991). Phormia regina *is known to arrive at decomposing tissue somewhat later than the other three species found in connection with the adult female remains (Case Report # 92–62922A) and it would be consistent not to see this species deposting eggs for up to 24 hours or more following death. In this case, death would have occurred sometime prior to fly activity and oviposition. Given the location of this body in the room to the west of the hallway, the known climatological conditions (rainy cool weather for the initial portion of the time in question) and the limited immediate accessability (2nd floor of a house with only a patio door opened 12 to 24 inches) of the remains to carrion flies, it would be possible for there to be a considerable delay from when death occurred to the actual colonization of the remains by carrion flies. It is possible for this delay to be from 24 to 48 hours in duration.*

(The discussion of temperature is the same as before and we move to the final paragraph).

In conclusion, the most probable time when death could have occurred in this case, when based upon the species studied and the temperatures present to drive immature development, in this forensic entomologist's professional opinion, was approximately 24 hours prior to 0500 hrs on Sunday, June 7,1992, therefore, at or prior to 0500 hrs on Saturday, June 6,1992.

Summary of Conclusions [father]

The insect evidence recovered from the remains during autopsy indicate that the first adult fly activity and deposition of life stages (eggs) most likely occurred sometime during the early morning hours of June 7,1992. One Diptera (fly) species used for the analysis was collected as eggs and larvae from the body at autopsy. This species was: Calliphoridae; Phormia regina. *Developmental time for* P. regina *points to colonization at some time period around 0500 hrs.on June 7,1992.*

Factors taken into consideration for this estimate were based upon known developmental times for this species, climatological occurrences for the time period in question, temperature regimes and energy requirements (accumulated degree hours) within the house where the bodies were located, and temperature regimes and energy requirements while the immature insects were developing in the laboratory.

The conclusion of when death occurred for this case, when based upon this species of carrion fly studied and the temperatures present to drive immature development, in this forensic entomologist's professional opinion, was approximately 24 hours prior to early morning (0500) on Sunday, June 7,1992, therefore, at or prior to 0500 hrs. Saturday, June 6, 1992.

DISCUSSION

The insect evidence recovered during autopsy (one fly species identified and used in the analysis), indicate the oldest oviposition (egg laying) by adult female flies on the adult male victim would have occurred at approximately 0500 on June 7, 1992. Phormia regina *oviposition most likely took place sometime in the early morning hours of June 6, 1992, possibly during darkness. While it is not common for calliphorids to oviposit during darkness in rural environments, it is possible, under conditions where artificial lighting is present in urban environments, for this oviposition/larviposition to occur (Greenberg 1991).* Phormia regina *is known to arrive at decomposing tissue somewhat later than the other three species found in connection with the adult female remains (Case Report # 92–62922A) and it would be consistent not to see this species depositing eggs for up to 24 hours or more following death. In this case, death would have occurred sometime prior to fly activity and oviposition. Given the location of this body in the room at the end of the hall, the known climatological conditions (rainy cool weather for the initial portion of the time in question) and the limited immediate accessability (2nd floor of a house with only a patio door opened 12 to 24 inches) of the remains to carrion flies, it would be possible for there to be a considerable delay from when death occurred to the actual colonization of the remains by carrion flies. It is possible for this delay to be from 24 to 48 hours in duration.*

[The discussion of temperature is as previously given].

In conclusion, the most probable time when death could have occurred in this case, when based upon the species of carrion fly studied and the temperatures present to drive immature development, in this forensic entomologist's professional opinion, was approximately 24 hours prior to 0500 hrs. on Sunday, June 7, 1992, therefore, at or prior to 0500 hrs. on Saturday, June 6, 1992.

The report of the entomologist for the defense follows.
First, I would like to establish the entomological basis for my conclusions drawn from the published literature and my own work on the forensic applications of blow fly biology.

1. *The 6.6 mm long 2nd instar larva of* Phormia regina *is the insect indicator in this case on which a valid estimate of the time of death can be based.*
2. *There is no scientific evidence that* Calliphora vicina *and* Calliphora livida *have similar developmental rates. Nor have they been considered the same species since Robineau-Desvoidy recognized them as distinct species in an 1830 publication. Besides, only eggs of* C. livida *were recovered from the bodies, indicating a later arrival for this fly.*
3. *There is no scientific evidence that* Sarcophaga argyrostoma *(=*falculata*) and* Sarcophaga bullata *have the same developmental rates. True, they belong in the same group of the genus* Sarcophaga, *but this is not based on physiology, and there are 34 members of the group. Therefore, it is unlikely that they would all have similar rates of development.* Sarcophaga cooleyi *is also in the same group and there is no obvious explanation as to why the prosecution entomologist (PE) chose to rely on* S. bullata *rather than* S. cooleyi, *as both are in Kamal's (1958) table. Kamal (p.265) states that* S. bullata *developed more slowly than* S. cooleyi *or* S. shermani, *another member of the same group. The largest* Sarcophaga *larvae (taken from [the mother's] body) were up to 6.8 mm long, slightly longer than* P. regina. *But larvae of* Sarcophaga *grow to ca. 20 mm compared to 15–16 mm for* P. regina, *and* Sarcophaga *larvae get a headstart of ca. one day by being deposited as larvae, not eggs. Therefore, the* Sarcophaga *larvae, like the* livida *eggs, are later arrivals and should not figure in a PMI estimate.*

Kamal's (1958) publication, on which the PE relies, has been discredited by leading forensic entomologists such as Prof. E. Paul Catts and Dr. Z. Erzinçlioglu. Here are some of the obvious problems with Kamal's work.

1. *Duration of development is incorrectly summed for the 11 species of flies in Table 1.*
2. *Kamal, page 265: "A high mortality rate of prepupae (post-feeding larvae) was recorded by all* Calliphora *species. No success was achieved in rearing any of the* Calliphora *species beyond the third generation. Data for the three species of* Calliphora *were obtained from two or more separate colonies." 'Prepupae' of* Calliphora *spp. fail to pupariate or pupariate and die at 29 °C or above (Greenberg 1991, Reiter 1984). Kamal's rearing temperature was probably not 26.7 °C (80 °F) but 29 °C or higher, causing "A high mortality rate..."*
3. *Kamal's flies were examined by Prof. Catts, a dipterist and forensic entomologist in the same department. They were found to be 1/4 to 1/2 below normal size, evidence of inadequate larval nutrition and/or crowding. Kamal admits this (p.269).*
4. *This was Kamal's dissertation and nothing was ever again published on flies.*
5. *The range of 10 to 22 hours as the incubation period for the eggs of* P. regina, *for example, raises a red flag. Ten hours is too brief for 26.7 °C. Either the temperature was higher or the eggs were laid and not collected for 6 to 8 hours and were then treated as if they were just laid.*
6. *There is an unusually wide range of time for the beginning stages of development (egg through 2nd instar) of the four relevant species.*

   S. bullata *(?*S. falculata*): 38–60 hours;*
   P. regina*: 29–76 hours;*
   P. sericata*: 33–92 hours;*
   C. vicina / *C. livida): 54–90 hours.*

Why does the PE continue to rely on Kamal's data when excellent studies by Reiter, Marchenko, and Kano and Nishida are published? He ignores the publication, "Flies as Forensic Indicators" (1991), and, oddly enough, he ignores a good study of P. regina's *developmental rate by Tracy Cyr (Master's Thesis, 1993) which includes flies he provided from Indiana.*

Thus far, there are three serious flaws in [the PE's] analysis. First is the double assumption that certain species are equivalent and therefore must have the same developmental rates. There is no scientific evidence to support this double assumption. Second, is his total reliance on Kamal's questionable data while testifying that he has done hundreds of rearings of P. regina. *Are his own data even less reliable than those of Kamal? Since leaving Purdue he has not had access to a lab for such studies. The third flaw is his reliance on ADH (accumulated degree hours) to estimate the PMI.*

Using ADH in [this] case, where the oldest specimen has completed less than 20% of its scheduled development, opens a Pandora's box. Science follows the principal of parsimony or Ockam's Razor. Seek the simple explanations, the most complicated explanations are usually wrong. ADH should not have been used. But having been used, it was used poorly. Following are arguments against using ADH in this case.

1. *The time span from egg to adult has a wider range of variability, therefore error, than the time span from egg through the 2nd instar. For example, using Kamal's "corrected" sums for* P. regina, *total development time (egg to adult) has a range of 7.6 to 21 days, whereas egg through 2nd instar has a range of 1.2 to 3.2 days. Why would anyone choose to use an ADH for the total period when better, more precise data are available for the specific stages in question?*
2. *If* C. livida *behaves like other* Calliphora *species which have been studied, different stages respond very differently to the same temperatures. For example, while other kinds of blow flies develop well at and above 29 °C,* Calliphora *postfeeding larvae pupariate and die, or simply fail to pupariate. This was not considered by [both PEs] as a possible factor in the rearings, and there is a question about the actual temperatures in the rearing containers.*
3. *There is no scientific evidence that* Sarcophaga argyrostoma *is the physiological equivalent of* Sarcophaga bullata. *Therefore, ADH's based on the latter are not applicable.*
4. *There is no scientific evidence that* Calliphora livida *is the physiological equivalent of* Calliphora vicina. *Therefore ADH's based on the latter are not applicable.*
5. *In fact, Kamal says (p. 265) "Within the genus* Calliphora *we find that* C. vomitoria *developed more slowly than* C. vicina *and* C. terrae-novae. *Similarly,* Sarcophaga bullata *developed more slowly than* S. cooleyi *and* S. shermani.*"*
6. *An accurate ADH depends on starting with eggs of known age. There is uncertainty about the age of the eggs used for the rearings and the ADH's in this case. How old were the eggs when they were collected and placed on beef liver for rearing? They could have been laid within the hour or they could have been 15 or 20 hours old (some were taken at the morgue). Carefully controlled rearing data depend on knowing when eggs were laid, within one or two hours, at most. Without this knowledge, and given the sporadic, not continual, rearing observations by Morris et al., there can easily be an error of more than half-a-day, and possibly close to a full day, just in egg hatching time. The same applies to adult emergence time.*
7. *[The PE] contradicts himself concerning the developmental threshold as a basis for figuring the ADH of* C. vicina.
   a) *In this trial and in testimony given in two re-trials in Covington, KY, he claims 10° to 11 °C as the threshold.*
   b) *In the Yashnev trial in Toronto (p.7 of his testimony), he says "Below this*

temperature (6 °C), development ceases for these species of blow flies," *referring to* Calliphora *species. Which is correct? We believe neither.*

c) *I have been unable to discover any controlled experiments, published or unpublished, by [the PE] on developmental thresholds of* C. vicina *or any other blow fly.*
d) *Competent published work by Davies and Ratcliffe (1994) indicates a threshold close to 0 °C.*

8. [The PE's] *PMI's are based on rearings of the eggs and larvae in aluminum foil pouches inside baggies. This is a confined space with little or no ventilation, except pinholes for air. Based on the number of adults eclosing, there were probably more than 60 larvae in some of these rearings. (Samples of larvae were removed 3 separate times. How many total larvae removed?) Such a number of larvae in this tight situation could conceivably raise the temperature as a result of their metabolic heat, and the rearing temperature would no longer be the nominal 24 °C but something else. Were thermometers inserted into the midst of the maggots to address this question? In effect the rearings were not under a known and controlled temperature. This invalidates any ADH's derived from the rearings. See Fig. 1, "Flies as Forensic Indicators."*

To estimate the PMI it is not necessary – indeed, it is confusing – to present data for flies other than P. regina *because this fly was the first to lay eggs (before* Sarcophaga *larviposited) and the eggs were laid on [the mother's] body. The simplest, most direct scientific method for estimating the PMI in this case is to determine how long it would take for freshly laid eggs of* P. regina *to reach a larval length of 6.6 mm at the probable temperature range in the upstairs hallway. The relevant data for* P. regina *have been published by Nishida (1984), Greenberg (1991), and Cyr (1993), summarized in Table 1.*

Table 1. Phormia regina*: Minimum duration (hr) from known time of oviposition to 6.6 mm larvae at several constant temperatures.*

	20 °C	22 °C	25 °C	26.6 °C	29 °C	30 °C	35 °C
A+		70					31
B°	72		72			43	36
C*				66			

+ Greenberg (1991), table 2, and graph at 35 °C.
°Nishida (1984).
**Cyr, M.S. thesis (1993). Oviposition through 2nd instar: 66 hr is the average minimum (range 62–70 hr) of separate rearings of flies from Washington State, Indiana, Texas and Louisiana.*

The bodies were at 4 °C at the morgue from 0200 hr until autopsy shortly after noon June 11, 1992. Larvae were collected and killed, just prior to autopsy, at

1200 hr. It is unlikely that significant growth of the maggots occurred during refrigeration (4 °C) because the maggots were on the surface (neck area) and no maggot masses were observed.

Two crucial factors in an estimate of time of death are: when did P. regina *lay eggs on [the mother's] body; and at what temperature did the eggs develop? Because there was no maggot mass, ambient temperature in the upstairs hallway would be operative. The reasonable temperature estimate in the upstairs hallway is a range between 25 °C (77 °F) and 29 °C (84 °F) and 30 °C (86 °F), based on police observations (85 °F recorded at 14:57 hours, June 10, 1992, downstairs thermostat; hotter upstairs than downstairs). If the flies were already present in the house (suggested by the presence of dead flies on the window sills), oviposition would probably have occurred within 2 hours after death of [the mother], certainly within 24 hours. Weather was not a deterrent if flies came into the house from outdoors. The presence of three decomposing bodies would have increased the attraction odors, and the conservative 24-hour delay in oviposition obtains. Given these assumptions, we derive the following estimates of time of death by counting back from 0200 hours June 11, 1992 (when [the mother's] body was placed at 4 °C), as shown in Table 2.*

Table 2.

Time of death when oviposition is:

Temperature	*2 hr postmortem*	*24 hr postmortem*
25° C	*0001 hr, June 8, 1992*	*0200 hr, June 7, 1992*
30° C	*0500 hr, June 9, 1992*	*0700 hr, June 8, 1992*

First instar larvae of P. regina *(1.1 to 2.6mm) were the oldest fly stage found on the body of [the father]. Eggs of* P. regina *were the oldest fly stage found on the body of [the daughter]. One cannot assume a sequence of death from these data. One can only state that the flies oviposited first on [the mother], then on [the father], and finally on [the daughter].*

The judge found the son guilty of the triple murders, explaining that the entomological arguments on both sides were not conclusive, but that the PE's estimate of the time of death fitted better with other circumstances in the case. He followed this with a statement that seems to defy logic. "The [PE's] method, of course, employed a much larger sample, which one might suppose would enhance the chance of success, but it produced results over a wide range as to time of death, both ends of which were totally incompatible with my findings of time of death based on the other evidence. [The PE] was fully aware of this fact. He carefully explained during the course of his testimony that he had adjusted the results of his calculations by the A.D.H. method in the light of the other data with which he had been supplied, for example the fact that witnesses had seen the deceased persons alive on June 5. He asserted that such adjustment, in applying the A.D.H. method, constituted normal and proper procedure."

Based on our previous crtique of ADH, nothing could be farther from the truth. And nothing could be more antithetical to the spirit and the practice of science, and indeed, any area of human knowledge! To pick and choose data in order to win the case was a blatant misuse of forensic entomology. And where was the judge, presumably seasoned and wise, who abandoned his own background in biology (which he discussed openly during the trial), to embrace instead a "science" of convenience?

By way of a word of solace to those not as yet disillusioned from exposure to the legal system, let us make clear that the next three chapters can be read on more than one level. For those who care, or at least are paid to care, we supply extensive citations to the applicable legal authority in the form of statutes, code provisions, and case law. These chapters contain the type of detailed legal information necessary to evaluate, to plan, and to pursue a case from start to finish. But if you are not a lawyer, rest easy. You can skip the legal technicalities and citations, which have mostly been exiled to the footnotes for your convenience. Leave the footnotes for the lawyers. Instead, you can focus on the main body of the text for an overview of the law of scientific evidence in general and forensic entomology in particular. You will gain an understanding of the legal framework within which scientific experts must operate. And you will learn what it takes to survive in the litigation game.

Entomology References

Ahmad, T. 1936. The influence of constant and alternating temperatures on the development of certain species of insects. *Proc. Nat. Inst. Sci. India*, **2**: 67–91.

Anderson, G. S. 1995. The use of insects in death investigations: analysis of cases in British Columbia over a five year period. *Can. Soc. Forensic Sci. J.*, **28**: 277–292.

Anderson, G. S. and VanLaerhoven, S. L. 1996. Initial studies on insect succession on carrion in southwestern British Columbia. *J. Forensic Sci.*, **41**: 617–625.

Ash, N., and Greenberg, B. 1975. Developmental temperature responses of the sibling species *Phaenicia sericata* and *Phaenicia pallescens*. *Ann. Entomol. Soc. Am.* **68**: 197–200.

Babin, R. W., Kahane, J. C., and Freed, R. E. 1990. Exercise in paleo-otolaryngology: head and neck examination of two Egyptian mummies. *Ann. Otol. Rhinol. Laryngol.* **99**: 742–748.

Baumgartner, D. L. and Greenberg, B. 1983. The primary screwworm [m] fly, *Cochliomyia hominivorax* (Coquerel) (Diptera, Calliphoridae), in Peru. *Rev. Brasileira Biol.*, **43**: 215–221.

Baumgartner, D. L. and Greenberg, B. 1984. The genus *Chrysomya* (Diptera: Calliphoridae) in the New World. *J. Med. Entomol.*, **21**: 105–113.

Baumgartner, D. L. and Greenberg, B. 1985. Distribution and medical ecology of the blow flies (Diptera: Calliphoridae) of Peru. *Ann. Entomol. Soc. Am.*, **78**: 565–587.

Baumgartner, W. A., Hill, V. A., and Blahd, W. O. 1989. Hair analysis for drugs of abuse. *J. Forensic Sci.*, **34**: 1433–1453.

Bergeret, M. 1855. Infanticide. Momification naturelle du cadavre. *Ann. Hyg. Méd. Leg.* **4**: 442–52.

Beyer, J. C., Enos, W. F., and Stajic, M. 1980. Drug identification through analysis of maggots. *J. Forensic Sci.*, **25**: 411–412.

Blackith, R. E. and Blackith, R. M. 1990. Insect infestations of small corpses. *J. Nat. Hist.*, **24**: 699–709.

Bohart, G. E. and Gressitt, J. L. 1951. *Filth-inhabiting flies of Guam*. Honolulu: Bernice P. Bishop Museum Bulletin 204.

Braack, L. E. O. and Retief, P. F. 1986. Dispersal, density and habitat preference of the blow-flies *Chrysomyia albiceps* (Wd.) and *Chrysomyia marginalis* (Wd.) (Diptera: Calliphoridae). *Onderstepoort J. Veter. Res.* **53**: 13–18.

Brier, B. 2001. A thoroughly modern mummy. *Archeology*, (January/February): 44–50.

Browne, L. B. 1958. The choice of communal oviposition sites by the Australian sheep blowfly (*Lucilia cuprina*). *Austral. J. Zool.*, **6**: 241–247.

Byrd, J. H. and Butler, J. F. 1996. Effects of temperature on *Cochliomyia macellaria* (Diptera: Calliphoridae) development. *J. Med. Entomol.*, **33**: 901–905.

Byrd, J. H. and Butler, J. F. 1997. Effects of temperature on *Chrysomya rufifacies* (Diptera: Calliphoridae) development. *J. Med. Entomol.*, **34**: 353–358.

Byrd, J. H. and Butler, J. F. 1998. Effects of temperature on *Sarcophaga haemorrhoidalis* (Diptera: Calliphoridae) development. *J. Med. Entomol.*, **35**: 694–698.

Byrd, J. H. and Castner, J. L. (Eds) 2001. *Forensic entomology. The utility of arthropods in legal investigations.* Boca Raton: CRC Press.

Catts, E. P. 1990. Analyzing entomological data. In *Entomology and death – a procedural guide.* ed. E.P. Catts and N. Haskell, chapter 8. Clemson, SC: Joyce's Print Shop.

Catts, E. P. 1992. Problems in estimating the postmortem interval in death investigations. *J. Agric. Entomol.*, **9**: 245–255.

Chastel, A. 1984. *Musca depicta.* Milan: Franco Maria Ricci.

Cheng, Ko. 1890. *Zhe yu gui jian* [Cases in the history of Chinese trials] (English transl.) China Lu shih, publisher, no page numbers.

Cheng, Ko. 189(?) *Zhe yu gui jian bu* [Additional cases in the history of Chinese trials] (English transl.). Reprinted 1985, Beijing: Chung-hua shu chu.

Coates, R. 1842. Contributions to surgery. On the ravages of the fly in wounds and ulcers and the means of prevention and remedy. *Med. Examiner*, **1**(9): 129–132.

Cockburn, A., Barraco, R. A., Reyman, T. A. and Peck, W. H. 1975. Autopsy of an Egyptian mummy. *Science*, **187**: 1155–1158.

Coe, M. 1978. The decomposition of elephant carcasses in the Tsavo (East) National Park, Kenya. *J. Arid. Env.*, **1**: 71–86.

Cragg, J. B. 1955. The olfactory behaviour of *Lucilia* species (Diptera) under natural conditions. *Ann. Appl. Biol.*, **44**: 467–477.

Cragg, J. B. and Cole, P. 1952. Diapause in *Lucilia sericata* (Mg.) Diptera. *J. Exp. Biol.*, **29**: 600–604.

Cunha e Silva, S. L. and Milward-de-Azevedo, E. M. V. 1992. Estudo comparado do desenvolvimento de dois morfotipos larvais de *Cochliomyia macellaria* (Fabricius) (Diptera, Calliphoridae). *Rev. Bras. Zool.*, **9**: 181–186.

Curry, A. 1979. The insects associated with the Manchester mummies. In *Manchester mummy project. Multidisciplinary research on ancient Egyptian mummified remains.* ed. A. R. David, pp. 113–117. Manchester: Manchester University Press.

Cyr, T. 1993. *Forensic differences among geographic races of* Phormia regina

(Meigen) (Diptera: Calliphoridae). M.S. thesis, Washington State University, Dept. Entomology.

DasGupta, B. and Roy, P. 1969. Studies on the behaviour of *Lucilia illustris* Meigen as a parasite of vertebrates under experimental conditions. *Parasitol.*, **59**: 299–304.

David, R. and Tapp, E. (Eds) 1984. *Evidence embalmed. Modern medicine and the mummies of ancient Egypt.* Manchester: Manchester University Press.

Davies, L. and G. G. Ratcliffe. 1994. Development rates of some pre-adult stages in blowflies with reference to low temperatures. *Med. Veter. Entomol.*, **8**: 245–254.

Dear, J. P. 1979. A revision of the Toxotarsinae (Diptera, Calliphoridae). *Papeis Avulsos Zool.*, (São Paulo), **32**: 145–182.

Dear, J. P. 1985. A revision of the New World Chrysomyini (Diptera, Calliphoridae). *Revista Brasil. Zool.*, **3**: 109–169.

De Jong, G. D., 1994. An annotated checklist of the Calliphoridae (Diptera) of Colorado, with notes on carrion associations and forensic importance. Jour. Kansas Entomol. Soc., **67**: 378–385.

Denlinger, D. L. 1978. The developmental response of flesh flies (Diptera: Sarcophagidae) to tropical seasons: variation in generation time and diapause in East Africa. *Oecologia*, **35**: 105–107.

Denlinger, D. L. and Tanaka, S. 1988. The impact of diapause on the evolution of other life history traits in flesh flies *Oecologia*, **77**: 350–356.

Deonier, C. C. 1940. Carcass temperatures and their relation to winter blowfly populations and activity in the Southwest. *J. Econ. Entomol.*, **33**: 166–70.

Dewaele, P., Leclercq, M. and Disney, R. H. L. 2000. Entomologie et médecine légale: les Phorides (Diptères) sur cadavres humains. Observation inédite. *J. Méd. Légale Droit Médical*, **43**: 569–572.

Dicke, M. 2000. Insects in Western art. *Am. Entomol.*, **46**: 228–236.

Dirrigl, F. J. and Greenberg, B., 1995. The utility of insect remains to assessing human burials: a Connecticut case study. *Archeol. Eastern N. Am.*, **23**: 1–7.

Dobzhansky, T. 1950. Evolution in the tropics. *Am. Scient.*, **38**: 209–221.

Early, M. and Goff, M. L., 1986. Arthropod succession patterns in exposed carrion on the island of O'ahu, Hawaiian Islands, USA. *J. Med. Entomol.*, **23**: 520–531.

Ellison, J. R. and Hampton, E. N. 1982. Age determination using the apodeme structure in adult screwworm flies (*Cochliomyia hominivorax*). *J. Insect Physiol.*, **28**: 731–736.

Erzinçlioglu, Y. Z. 1985 Immature stages of British *Calliphora* and *Cynomya*, with a re-evaluation of the taxonomic characters of larval Calliphoridae (Diptera). *J. Nat. Hist.*, **19**: 69–96.

Erzinçlioglu, Y. Z. 1987a. The larvae of some blowflies of medical and veterinary importance. *Med. Veterin. Entomol.*, **1**: 121–125.

Erzinçlioglu, Y. Z. 1987b. Recognition of the early instar larvae of the genera

Calliphora and *Lucilia* (Dipt., Calliphoridae). *Entomol. Monthly Mag.*, **123**: 197–198.

Erzinçlioglu, Y. Z. 1988. The larvae of the species of *Phormia* and *Boreellus*: northern, cold-adapted blowflies (Diptera: Calliphoridae). *J. Nat. Hist.*, **22**: 11–16.

Erzinçlioglu, Y. Z. 1989a. The value of chorionic structure and size in the diagnosis of blowfly eggs. *Med. Veter. Entomol.* **3**: 281–285.

Erzinçlioglu, Y. Z. 1989b. The early larval instars of *Lucilia sericata* and *Lucilia cuprina* (Diptera, Calliphoridae): myiasis blowflies of Africa and Australia. *J. Nat. Hist.*, **23**: 1133–1136.

Erzinçlioglu, Y. Z. 1990a. On the interpretation of maggot evidence in forensic cases. *Med. Sci. Law*, **30**: 65–66.

Erzinçlioglu, Y. Z. 1990b. The larvae of two closely-related blowfly species of the genus *Chrysomya* (Diptera, Calliphoridae). *Entomol. Fennica*, **1**: 151–153.

Ettershank, G., Macdonnell, I., Croft, R. 1983. The accumulation of age pigment by the fleshfly *Sarcophaga bullata* Parker (Diptera: Sarcophagidae). *Aust. J. Zool.*, **31**: 131–138.

Evans, A. C. 1935. Some notes on the biology and physiology of the sheep blowfly, *Lucilia sericata* Meig. *Bull. Entomol. Res.*, **26**: 115–122.

Evans, A. C. 1936. Studies on the influence of the environment on the sheep blow-fly *Lucilia sericata* Meig. IV. The indirect effect of temperature and humidity acting through certain competing species of blow-flies. *Parasitol.*, **28**: 431–439.

Fan, X. (Ed.) 1992. *Key to the common flies of China. 2nd* edn, Shanghai : Shanghai Institute of Entomology, Academia Sinica. (In Chinese.)

Faucherre, J., Cherix, D. and Wyss, C. 1999. Behavior of *Calliphora vicina* (Diptera, Calliphoridae) under extreme conditions. *J. Insect Behav.*, **12**: 687–690.

Ferrar, P. 1987. *A guide to the breeding habits and immature stages of Diptera Cyclorrhapha.* Part 1, text. Part 2, figures. *Entomonograph*, vol. 8, ed. L. Lyneborg, Leiden: E.J. Brill/Scandinavian Science Press.

Ferreira, M. J. M. 1978. Sinantropia de dípteros muscoideos de Curitiba, Paraná. *Rev. Bras.Biol.*, **38**: 445–454.

Finell, N. and Jarvilehto, M. 1983. Development of the compound eyes of the blowfly *Calliphora erythrocephala*: changes in morphology and function during metamorphosis. *Ann. Zool. Fennici*, **20**: 223–234.

Fraenkel, G. and Bhaskaran, G. 1973. Pupariation and pupation in cyclorrhaphous flies (Diptera) : terminology and interpretation. *Ann. Entomol. Soc. Am.*, **66**: 418–422.

Fraenkel, G. and Hsiao, C. 1968. Manifestations of pupal diapause in two species of flies, *Sarcophaga argyrostoma* and *S. bullata*. *J. Insect Physiol.*, **14**: 689–705.

Fraser, A., Ring, R. A. and Stewart. R. K. (1961). Intestinal proteases in an insect, *Calliphora vomitoria* L. *Nature*, **192**: 999–1000.

Fuller, M. E. 1932. The larvae of the Australian sheep blowflies. *Proc. Linn. Soc.* NSW, *57*: 77–91.

Fuller, M. E. 1934. *The insect inhabitants of carrion: a study in animal ecology. Commonwealth of Australia.* Council for Scientific and Industrial Research Bulletin No. 82.

Gagné R. J. and Miller, S. E. 1981. *Protochrysomyia howardae* from Rancho Labrea, California, Pleistocene, new junior synonym of *Cochliomyia macellaria* (Diptera: Calliphoridae). *Bull. Southern Calif. Acad. Sci.*, **80**: 95–96.

Germonpré, M. and LeClercq, M. 1984. Des pupes de *Protophormia terraenovae* associées à des mammifères pléistocènes de la Vallée flamande (Belgique). *Bull. Inst. Roy. Sci. Natur. Belgique*, **64**: 265–268.

Gerwin, A. C. M. van, Browne, L. B., Vogt, W. G., Williams, K. L., Gerwin Van, A. C. M., and Barton-Browne, L. 1987. Capacity of autogenous and anautogenous females of the Australian sheep blowfly, *Lucilia cuprina*, to survive water and sugar deprivation following emergence. *Entomol. Exp. Appl.*, **43**: 209–214.

Giebultowicz, J. M. and Saunders, D. S. 1983. Evidence for the neurohormonal basis of commitment to pupal diapause in larvae of *Sarcophaga argyrostoma. Experientia*, **39**: 196–194.

Goddard, J. and Lago, P. K. 1985. Notes on blow fly (Diptera: Calliphoridae) succession on carrion in northern Mississippi. *J. Entomol. Sci.*, **20**: 312–317.

Goff, M. L., Brown, W. A., Hewadikaram, K. A., and Omori, A. I. 1991. Effect of heroin in decomposing tissues on the development rate of *Boettcherisca peregrina* (Diptera, Sarcophagidae) and implications of this effect on estimation of postmortem intervals using arthropod development patterns. *J. Forensic Sci.*, **36**: 537–542.

Goff, M. L., Brown, W. A., and Omori, A. I. 1992. Preliminary observations of the effect of methamphetamine in decomposing tissues on the development rate of *Parasarcophaga ruficornis* (Diptera: Sarcophagidae) and implications of this effect on the estimation of postmortem intervals. *J. Forensic Sci.*, **37**: 867–872.

Goff, M. L., Omori, A I. and Goodbrod, J. R. 1989. Effect of cocaine in tissues on the development rate of *Boettcherisca peregrina* (Diptera: Sarcophagidae). *J. Med. Entomol.*, **26**: 91–93.

Goodbrod, J. B. and Goff, M. L. 1990. Effects of larval population density on development and interactions between two species of *Chrysomya* (Diptera: Calliphoridae) in laboratory culture. *J. Med. Entomol.*, **27**: 338–343.

Graham-Smith, G. S. 1914. *Flies in relation to man. Non-bloodsucking flies.* Cambridge University Press. [See p. 84.]

Green, A. A. 1951. The control of blowflies infesting slaughter-houses. I. Field observations of the habits of blowflies. *Ann. Appl. Biol.*, **38**: 475–494

Greenberg, B. 1971. *Flies and disease.* vol. 1, *Ecology, classification and biotic associations.* Princeton : Princeton University Press.

Greenberg, B. 1973. *Flies and disease.* vol. 2. *Biology and disease transmission.* Princeton : Princeton University Press.

Greenberg, B. 1984. Two cases of human myiasis caused by *Phaenicia sericata* (Diptera: Calliphoridae) in Chicago area hospitals. *J. Med. Entomol.*, **21**: 615.

Greenberg, B. 1988. *Chrysomya megacephala* (F.) (Diptera: Calliphoridae) collected in North America and notes on *Chrysomya* species present in the New World. *J. Med. Entomol.*, **25**: 199–200.

Greenberg, B. 1990. Nocturnal oviposition behavior of blow flies (Diptera: Calliphoridae). *J. Med. Entomol.*, **27**: 807–810.

Greenberg, B. 1991. Flies as forensic indicators. *J. Med. Entomol.* **28**: 565–577.

Greenberg, B. 1998. Reproductive status of some overwintering domestic flies (Diptera: Muscidae and Calliphoridae) with forensic implications. *Ann. Entomol. Soc. Am.*, **91**: 818–820.

Greenberg, B. and Miggiano, V. 1963. Host-contaminant biology of muscoid flies. IV. Microbial competition in a blowfly. *J. Infect. Dis.*, **112**: 37–46.

Greenberg, B. and Paretsky, B. 1955. Proteolytic enzymes in the house fly, *Musca domestica* (L.). *Ann. Entomol. Soc. Am.*, **48**: 46–50.

Greenberg, B. and Singh, D. 1995. Species identification of calliphorid (Diptera) eggs. *J. Med. Entomol.*, **32**: 21–26.

Greenberg, B. and Szyska, M. L. 1984. Immature stages and biology of fifteen species of Peruvian Calliphoridae (Diptera). *Ann. Entomol. Soc. Am.* **77**: 488–517.

Greenberg, B. and Tantawi, T. I. 1993. Different developmental strategies in two boreal blow flies (Diptera: Calliphoridae). *J. Med. Entomol.*, **30**: 481–484

Gregor, F. 1971. Key and figures to adult flies (Chapter 3); Figures of fly larvae (Chapter 4). In *Flies and Disease,* vol. I, *Ecology, classification , and biotic associations.* ed. B. Greenberg, Princeton : Princeton University Press.

Guimarães, J. H. and Papavero, N. 1999. *Myiasis in man and animals in the Neotropical Region. Bibliographic database.* São Paulo: Fundação de Amparo à Pesquisa do Estado de São Paulo.

Guimarães, J. H., Prado, A. P., and Linhares, A. X. 1978. Three newly introduced blowfly species in southern Brazil (Diptera, Calliphoridae). *Rev. Bras. Entomol.*, **22**: 53–60.

Gunatilake, K. and Goff, M. L. 1989. Detection of organophosphate poisoning in a putrefying body by analyzing arthropod larvae. *J. Forensic Sci.*, **34**: 714–716.

Hall, D. G. 1948. *The blowflies of North America..* Baltimore, MD: The Thomas Say Foundation, Monumental Printing Co.

Hall, M. J. R. 1995. Review. Trapping the flies that cause myiasis: their responses to host stimuli. *Ann. Trop. Med. Parasit.*, **89**: 333–357.

Hall, R. D. 2001. Introduction: perceptions and status of forensic entomology. In *Forensic entomology. The utility of arthropods in legal investigations.* New York: CRC Press.

Hall, R. D., and Doisy, K. E. 1993. Length of time after death: effect on attraction and oviposition or larviposition of midsummer blow flies (Diptera: Calliphoridae) and flesh flies (Diptera: Sarcophagidae) of medicolegal importance in Missouri. *Ann. Entomol. Soc. Am.* **86**: 589–93.

Hamilton-Patterson, J. and Andrews, C. 1979. *Mummies: death and life in ancient Egypt*, pp. 46–7. New York, Viking Press.

Heinrich, B. 1993. *The hot-blooded insects: strategies and mechanisms of thermoregulation.* Cambridge, MA: Harvard University Press.

Hightower, B. G. and Alley, D. A. 1963. Local distribution of released laboratory-reared screw-worm flies in relation to water sources. *J. Econ. Entomol.*, **56**: 798–802.

Hinton, H. E. 1960. Plastron respiration in the eggs of blowflies. *J. Ins. Physiol.*, **4**: 176–183.

Holloway, B. A. 1985. Immature stages of New Zealand Calliphoridae. In Calliphoridae (Insecta: Diptera) ed. J. P. Dear, *Fauna of New Zealand* 8: 12–14, 80–83.

Holloway, B. A. 1991. Identification of third-instar larvae of flystrike and carrion-associated blowflies in New Zealand (Diptera: Calliphoridae). *New Zealand Entomol.*, **14**: 24–28.

Homer. *The Iliad.* Translated by R. Lattimore, 1951. Chicago: University of Chicago Press.

Hough, G. de N. 1897. The fauna of dead bodies with special reference to Diptera. *Brit. Med. J.* (Dec. 25), 1853–1854.

Huchet, J.-B. 1995. Insectes et momies égyptiennes. *Bull. Soc. Linn. Bordeaux*, **23**: 29–39.

Introna, F. Jr., Lo Dico, C., Caplan, Y. H. and Smialek, J. E. 1990. Opiate analysis in cadaveric blowfly larvae as an indicator of narcotic intoxication. *J. Forensic Sci.*, **35**: 118–122.

Isherwood, I., Jarvis, H. and Fawcitt, R. A. 1979. Radiology of the Manchester mummies. *In Manchester Museum mummy project. Multidisciplinary research on ancient Egyptian remains*, ed. A. R. David, pp.25–64. Manchester: Manchester University Press.

Ishijima, H. 1967. Revision of the third stage larvae of synanthropic flies of Japan. (Dipt.: Anthomyiidae, Muscidae, Calliphoridae and Sarcophagidae). *Jap. J. Sanit. Zool.*, **18**: 47–100.

Ives, A. R. 1991. Aggregation and coexistence in a carrion fly community. *Ecol. Monogr.*, **6**: 75–94.

James, M. T. 1947. *The flies that cause myiasis in man.* U.S. Dept. Agric., Misc. Publ. No. 631.

James, M. T 1970. Family Calliphoridae. in *A catalogue of the Diptera of the Americas south of the United States.* No. 102: 1–28. ed. N. Papavero. São Paulo: Museu de Zoologia, Universidade de São Paulo.

Jirón, L. F. 1979. Sobre moscas califóridas de Costa Rica. *Brenesia,* **16**: 221–222.

Jirón, L. F. and Cartín, V. M. 1981. Insect succession in the decomposition of a mammal in Costa Rica. *NY Entomol.* Soc. **89**: 158–165.

Johnson, M. D. 1975. Seasonal and microseral variations in the insect populations on carrion. *Am. Mid. Nat.,* **93**: 79–90.

Johnston, W. and Villeneuve, G. 1897. On the medico-legal application of entomology. *Montreal Med. J.,* **26**: 81–90.

Kamal, A. S. 1958. Comparative study of thirteen species of sarcosaprophagous Calliphoridae and Sarcophagidae (Diptera). I. Bionomics. *Ann. Entomol. Soc. Amer.,* **51**: 261–271.

Kano, R., Field, G. and Shinonaga, S. 1967. Sarcophagidae (Insecta: Diptera). *Fauna Japonica.* Tokyo: Biogeograph. Soc. Japan.

Kano, R. and Shinonaga, S. 1968. Calliphoridae (Insecta: Diptera). *Fauna Japonica.* Tokyo: Biogeograph. Soc. Japan.

Keister, M. L. 1953. Some observations on pupal respiration in *Phormia regina. J. Morph,* **93**: 573–587.

Kilmer, A. D. 1987. The symbolism of the flies in the Mesopotamian flood myth and some further implications. In *Language, literature, and history. Philological and historical studies.* Presented to Erica Reiner. ed. F. Rochberg-Halton, pp. 175–180. New Haven, CT: American Oriental Society.

Kintz, P., Godelar, B., Tracqui, A., Mangin, P., Lugnier, A. A. and Chaumont, A. J. 1990a. Fly larvae: a new toxicological method of investigation in forensic medicine. *J. Forensic Sci.,* **35**: 204–207.

Kintz, P., Tracqui, A., Ludes, B., Waller, J., Boukhabza, A., Mangin, P., Lugnier, A. A. and Chaumont, A. J 1990b. Fly larvae and their relevance in forensic toxicology. *Am. J. Forensic Med. Pathol.,* **11**: 63–65.

Kintz, P., Tracqui, A., and Mangin, P. 1990c Toxicology and fly larvae on a putrefied cadaver. *J. Forensic Sci. Soc.,* **30**:243–246.

Kneidel, K. A. 1984. Influence of carcass taxon and size on species composition of carrion-breeding Diptera. *Am. Midl. Nat.,* **lll**: 57–63.

Kobayashi, H. 1922. On the further notes of the overwintering of house flies. *Jap. Med. World,* **2**: 193–196.

Kovacs, M. G. 1985. *The epic of Gilgamesh. Tablet XI,* p.102. Stanford: Stanford University Press.

Kritsky, G. 1985. Tombs, mummies and flies. *Bull. Entomol. Soc. Am.* (Summer): 18–19.

Kueppers, H. 1982. *Color atlas: a practical guide for color mixing.* Woodbury, NY: Barron's. Original German edn, *Dumont's Farben Atlas,* 1978, DuMont, Cologne, Germany.

Kurahashi, H. and Ohtaki, T. 1989. Geographic variations in the incidence of pupal diapause in Asian and Oceanian species of the fleshfly *Boettcherisca* (Diptera: Sarcophagidae). *Physiol. Entomol.,* **14**: 291–298.

Lambert, B. L. 1997. *Traces of the past. Unraveling the secrets of archeology through chemistry,* pp.249,250. Helix books. Reading, MA: Addison-Wesley.

Lambremont, E. N., Fisk, F. W. and Ashrafi, S. 1959. Pepsin-like enzyme in larvae of stable flies. *Science,* **129**: 1484–1485.

Lane, R. P. 1975. An investigation into blowfly (Diptera: Calliphoridae) succession in corpses. *J. Nat. Hist.,* **9**: 581–588.

Leek, F. F., 1969. The problem of brain removal during embalming by the ancient Egyptians. *J. Egyptian Archeol.,* **55**: 112–116.

Levot, G. W., Brown, K. R., Shipp, E. 1979. Larval growth of some calliphorid and sarcophagid Diptera. *Bull. Entomol. Res.,* **69**: 469–475.

Lister, A. M. 1993. The Condover mammoth site: excavation and research 1986–93. *Cranium,* **10**: 61–67.

Liu, D. and Greenberg, B. 1989. Immature stages of some flies of forensic importance. *Ann. Entomol. Soc. Am.,* **82**: 80–93.

Lord, W. D. July 18, 1986. Testimony in the trial. *State of Connecticut v. Erich Seebeck/Adam John,* New London. Conn. Pp. 53, 62, 69, 70, 73.

Lord, W. D. and Burger, J. F. 1983. Collection and preservation of forensically important entomological materials. *J. Forensic Sci.,* **28**: 936–944.

Lord, W. D. and Burger, J. F. 1984. Arthropods associated with harbor seal (*Phoca vitulina*) carcasses stranded on islands along the New England coast. *Intl J. Entomol.,* **26**: 1282–1285.

Lothe, F. 1964. The use of larval infestation in determining time of death. *Med. Sci. Law,* **4**: 113–115.

Lundt, H. 1964. Ökologishe untersuchungen über tierische beseidlung von Aas im Boden. *Pedobiol.,* **4**: 158–180.

Mackerras, M. J. 1933. Observations on the life-histories, nutritional requirements and fecundity of blowflies. *Bull. Entomol. Res.,* **24**: 353–362.

MacLeod, J. 1947. The climatology of blowfly myiasis. I. Weather and oviposition. *Bull. Entomol. Res.,* **38**: 285–303.

Madeira, N. M. 1985. Habito de pupacão de Calliphoridae (Diptera) na natureza e o encontro do parasitoide *Spalangia endius* (Hymenoptera: Pteromalidae). *Rev. Brasil Biol.,* **45**: 481–484.

Marchenko, M. I. 1973. On the determination of the time of death based on cadaver entomofauna. In *Scientific conference of medicolegal experts,* pp. 78–79. Leningrad.

Marchenko, M. I. 1975. Methods of investigating cadaver entomofauna complex in medicolegal aspect. In *Current problems of forensic medicine and morbid anatomy,* pp.100–103. Tallinn.

Marchenko, M. I. 1985. Development of *Chrysomya albiceps* WD. (Diptera, Calliphoridae). *Entomol. Rev.*, **64**: 107–112. (Originally published in *Entomol. Obozreniye*, no.l, pp. 79–84.)

Marchenko, M. I. 1988. The use of temperature parameters of fly growth in medicolegal practice. General trends. In *Proc. Int. Conf. Med. Veter. Dipterol*, ed. J. Olejnicek, pp. 254–257. Proc. Int. Ceske Budejovice.

Marchenko, M. I. 1989. Method of retrospective determination of insect development onset in a cadaver. *Sud.-Med. Exp.*, **32**(1): 17–20.

Marchenko, M. I. and Vinogradova, E. B. 1984. The influences of seasonal temperature changes on the rate of cadaver destruction by fly larvae. *Sud. Med. Exp.*, **27**(4): 11–14.

Mariluis, J. C. and Peris, S. V. 1984. Datos para una sinopsis de los Calliphoridae neotropicales. *Eos*, **60**: 67–86.

Mariluis, J. C. and Schnack, J. A. 1985/86. Ecologia de una taxocenosis de Calliphoridae del area Platense (Provincia da Buenos Aires) (Insecta, Diptera). *Ecosur*, **12/13**: 81–89.

Mariluis, J. C., and Schnack, J. A. 1989. Ecology of the blow flies of an eusynanthropic habitat near Buenos Aires (Diptera, Calliphoridae). *Eos*, **65**: 93–101.

Maspero, G. 1904 *History of Egypt, Chaldea, Syria, Babylonia, and Assyria*, vol. 6, p.6. London: Grolier Press.

McKnight, B. E. 1981. *The washing away of wrongs: forensic medicine in thirteenth-century China.* Dissertation. University of Michigan, Ann Arbor.

Mégnin, J.P. 1894. *La faune des cadavres. Application de l'entomologie à la médecine légale. Encyclopedie Scientifique des aide-memoire.* Paris: G. Masson and Gauthier-Villars.

Miller, M. L., Lord, W. D., Goff, M. L.,Donnelly, B., McDonough, E. T., and Alexis, J. C. 1994. Isolation of amitryptiline and nortryptiline from fly puparia (Phoridae) and beetle exuviae (Dermestidae) associated with mummified remains. *J. Forensic Sci.*,**39**: 1305–1313.

Monzu, N. 1977. *Coexistence of carrion breeding Calliphoridae (Diptera) in Western Australia.* Doctoral thesis, University of Western Australia.

Motter, M G. 1898. A contribution to the study of the fauna of the grave. A study of on[e] hundred and fifty disinterments, with some additional experimental observations. Jour. *NY Entomol. Soc.*, **6**: 201–231.

Nainys, J.-V. J., Marchenko, M. I., and Kazak, A. N. 1982. A calculation method for estimating by entomofauna the period during which the body had remained in the place where it was found. *Sud. Med. Exp.*, **25**(4): 21–23. (In Russian).

Niezabitowski, E. K. 1902. Experimentelle Beiträge zur Lehre von der Leichen Fauna. *Viertebjahrschrift Gericht. Med. Offentl. Sanitat.* **23**: 44–50.

Nishida, K. 1984. Experimental studies on the estimation of postmortem intervals by means of fly larvae infesting human cadavers. *Jap. J. Legal Med.*, **38**: 24–41.

Nishida, K., Shinonaga, S. and Kano, R. 1986. Growth tables of fly larvae for the estimation of postmortem intervals. *Ochanomizu Med. J.*, **34**: 9–24.

Nolte, K. B., Pinder, R. D. and Lord, W. D. 1992. Insect larvae used to detect cocaine poisoning in a decomposed body. *J. Forensic Sci.*, **37**: 1179–1185.

Norris, K. R. 1959. The ecology of sheep blowflies in Australia. *Monogr. Biol.*, **8**: 514–544.

Norris, K. R. 1965. The bionomics of blowflies. *Ann. Rev. Entomol.*, **10**: 47–68.

Noury, P. 1932. Les dieux chasse-mouches protecteur de la santé. *Chron. Méd.*, **39**: 113–116.

Nuorteva, P. 1963. Synanthropy of blowflies (Dipt., Calliphoridae) in Finland. *Ann. Entomol. Fenn.*, **29**: 1–49.

Nuorteva, P. 1977. Sarcosaprophagous insects as forensic indicators. In *Forensic medicine: a study in trauma and environmental hazards*, ed. C.G.Tedeschi, W.G.Eckert, and L.G.Tedeschi, vol. II, chapter 47. Philadelphia: W.B. Saunders Co.

Nuorteva, P. 1987. Empty puparia of *Phormia terraenovae* R.-D. (Diptera, Calliphoridae) as forensic indicators. *Ann. Entomol. Fenn.*, **53**: 53–56.

Nuorteva, P. and Häsänen, E. 1972. Transfer of mercury from fishes to sarcosaprophagous flies. *Ann. Zool. Fenn.*, **9**: 23–27.

Nuorteva, P. and Nuorteva, S-L. 1982. The fate of mercury in sarcosaprophagous flies and in insects eating them. *Ambio*, **11**: 34–37.

Nuorteva, P., Schumann, H., Isokoski, M., and Laiho, K. 1974. Studies on the possibilities of using blowflies (Diptera: Calliphoridae) as medicolegal indicators in Finland. *Ann. Entomol. Fenn.*, **40**: 70–74.

O'Flynn, M. A. 1983. The succession and rate of development of blowflies in carrion in southern Queensland and the application of these data to forensic entomology. *J. Aust. Entomol. Soc.*, **22**: 137–148.

O'Flynn, M. A. and Moorhouse, D. E. 1979. Species of *Chrysomya* as primary flies in carrion. *J. Aust. Entomol. Soc.*, **18**: 31–32.

Oldroyd, H. 1964. *The natural history of flies.* London: Weidenfeld and Nicolson.

Omar, B., Marwi, M. A., Oothuman, P. and Othman, H. F. 1994a. Observations on the behaviour of immatures and adults of some Malaysian sarcosaprophagous flies. *Trop. Biomed.*, **11**: 149–153.

Omar, B., Marwi, M.A.,, Sulaiman, S. and Oothuman, P. 1994b. Dipteran succession in monkey carrion at a rubber tree plantation in Malaysia. *Trop. Biomed.*, **11**: 77–82.

Orfila, M. 1848. *Memoire sur les exhumations juridiques. Traité de médecine légale*, pp. 80–165. Paris: LabÈ.

Payne, J. A. 1965. A summer carrion study of the baby pig, *Sus scrofa* Linnaeus. *Ecology*, **46**: 592–602.

Payne, J. A. and King, E. W., 1968. Arthropod succession and decomposition of buried pigs. *Nature*, **219**: 1180–1181.

Pendola, S. and Greenberg, B. 1975. Substrate-specific analysis of proteolytic enzymes in the larval midgut of *Calliphora vicina*. *Ann. Soc. Entomol. Am.*, **68**: 341–345.

Perez, C. 1910. Recherches histologiques sur la métamorphose des muscides. *Calliphora erythrocephala* Mg. *Arch. Zool. Exp. Gen.*, **4**: 1–274.

Pettigrew, T. J. 1834. *A history of Egyptian mummies.* London: Longman, Rees, Orme, Brown, and Longman.

Porada, E. 1948. *Corpus of ancient Near Eastern seals in North American Collections. Vol. I. The collection of the Pierpont Morgan Library.* The Bollingen Series XIV. New York: Pantheon Books.

Povolný, D.1971. Synanthropy. definition, evolution and classification. In *Flies and disease.* vol. 1, *Ecology, classification and biotic associations,* ed. B. Greenberg. Princeton: Princeton University Press.

Putman, R.J. 1977. Dyamics of the blowfly, *Calliphora erythrocephala,* within carrion. *J. Anim. Ecol.*, **46**: 853–866.

Ranade, D. R. 1977. Studies on the external metamorphosis of *Musca domestica nebulo* Fabr. (Diptera-Cyclorrhapha-Muscidae). Part III. Imaginal discs and development of the thoracic appendages, the legs and wings. *J. Anim. Morphol. Physiol.*, **24**: 277–284.

Réaumur, R.A. F. de. 1738. *Memoires pour servir à l'histoire des insectes,* vol. 4. Amsterdam: P. Mortier.

Riegel, G. T. 1979. The fly whisk. *Entomol. Soc. Am. Bull.*, **25**:196–199.

Reiter, C. 1984. Zum Wachstumsverhalten der Maden der blauen Schmeisfliege *Calliphora vicina. Z. Rechtsmed.*, **91**: 295–308.

Ring, R. A. 1967. Maternal induction of diapause in the larva of *Lucilia caesar* L. (Diptera: Calliphoridae). *J. Exp. Biol.*, **46**: 123–136.

Roback, S. S. 1954. The evolution and taxonomy of the Sarcophaginae (Diptera, Sarcophagidae). *Biol. Monogr.*, **23**: 1–181.

Rodriguez, W. C. and Bass, W. M. 1983. Insect activity and its relationship to decay rates of human cadavers in east Tennessee. *J. Forensic Sci.*, **28**: 423–432.

Rodriguez, W. C. and Bass, W. M. 1985. Decomposition of buried bodies and methods that may aid in their location. *J. Forensic Sci.*, **30**: 836–852.

Rognes, K. 1991. Blowflies (Diptera: Calliphoridae) of Fennoscandia and Denmark. *Fauna Entomol. Scand.*, **24**: 1–272.

Roubaud, E. 1922. Sommeil d'hiver cédant à l'hiver chez les larves et nymphes de Muscides. *C.R. Acad. Sci. (Paris)*, **174**: 964–966.

Saunders, D. S. 1997. Under-sized larvae from short-day adults of the blow fly, *Calliphora vicina,* side-step the diapause programme. *Physiol. Entomol.*, **22**: 249–255.

Schlein, Y. 1979. Age grouping of anopheline malaria vectors (Diptera: Culicidae) by the cuticular growth lines. *J. Med. Entomol.*, **16**: 502–506.

Schlein, J., and Gratz, N. G. 1972. Age determination of some flies and mosquitos by daily growth layers of skeletal apodemes. Bull. Wld. Hlth Org., **47**: 71–76.

Shannon, R. C. 1926. Synopsis of the American Calliphoridae (Diptera). Proc. Entomol. Soc. Wash., **26**: 115–139.

Shean, B. S., Messinger, L. and Papworth, M. 1993. Observations of differential decomposition on sun exposed v. shaded pig carrion in coastal Washington State. J. Forensic Sci., **38**: 938–949.

Singh, D. and Greenberg, B. 1994. Survival after submergence in the pupae of five species of blow flies (Diptera: Calliphoridae). *J. Med. Entomol.*, **31**: 757–759.

Singh, D. and Bharti, M. 2001. Further observations on the nocturnal oviposition behaviour of blow flies (Diptera: Calliphoridae). *Forensic Sci. Int'l*, **120**: 124–6.

Sivasubramanian, P. and Biagi, M. 1983. Morphology of the pupal stages of the fleshfly, *Sarcophaga bullata* (Parker) (Diptera: Sarcophagidae). *Int. J. Insect Morphol. Embryol.*, **12**: 355–359.

Skelly, P. J. and Howells, A. J. 1987. Larval cuticle proteins of *Lucilia cuprina*. Electrophoretic separation, quantification and developmental changes. *Insect Biochem.*, **17**: 625–633.

Smeeton, W. M. I., Koelmeyer, T. D., Holloway, B. A., and Singh, P. 1984. Insects associated with exposed human corpses in Auckland. *New Zealand Med. Sci. Law*, **24**: 167–174.

Smirnov, E. S. 1940. La problème des mouches à Tadjikistane. *Med. Parasitol.*, **9**: 515–517.

Smith, K. G. V. 1986. *A manual of forensic entomology*. London: British Museum and Cornell University Press.

Sohal, R. S. 1981. Metabolic rate, aging, and lipofuscin accumulation. In *Age pigments* ed. R.S.Sohal. Amsterdam: Elsevier/north Holland Biomedical Press.

Spradbery, J. P. 1991. *A Manual for the diagnosis of screw-worm fly*. Goanna Print Pty Ltd.

Spradbery, J. P. 1992. *Studies on the prepupal and puparial stages of the old world screw-worm fly,* Chrysomya bezziana *Villeneuve (Diptera: Calliphoridae)*. CSIRO Division of Entomology Technical Paper 29. Canberra: CSIRO.

Springfield, A. C., Cartmell, L. W., Aufderheide, A., and Weems, C. 1991. Benzoylecgonine in pre-Columbian mummy hair (abstract). *Am. Acad. Forensic Sci., 43rd annual meeting, Anaheim, California.*

Stoffolano Jr., J. G., Greenberg, S. and Calabrese, E. 1974. A facultative imaginal diapause in the black blowfly, *Phormia regina*. *Ann. Entomol. Soc. Am.*, **67**: 518–519.

Subramanian, H. and Mohan, K. R. 1980. Biology of the blowflies *Chrysomyia megacephala, Chrysomyia rufifacies* and *Lucilia cuprina*. *Kerala J. Veterin. Sci.*, **11**: 252–261.

Sugiyama, W. and Kano, R. 1984. Systematics of the Sarcophagidae of the Oriental region based on the comparative morphology of the male genitalia (Diptera, Sarcophagidae). *Jap. J. Sanit. Zool.*, **35**: 343–356.

Sychevskaîa, V. I. 1965. Biology and ecology of *Calliphora vicina* R.-D. in central Asia. *Zool. Zhurn.* **44**: 552–560. (In Russian.)

Tantawi, T. I., El-Kady, E. M., Greenberg, B., and El-Ghaffar, H. A. 1996. Arthropod succession on exposed carrion in Alexandria, Egypt. *J. Med. Entomol.*, **33**: 566–580.

Tantawi, T. I. and Greenberg, B. 1993a. The effect of killing and preservative solutions on estimates of maggot age in forensic cases. *J. Forensic Sci.*, **38**: 702–707.

Tantawi, T. I. and Greenberg, B. 1993b. *Chrysomya albiceps* and *C. rufifacies* (Diptera: Calliphoridae): contribution to an ongoing taxonomic problem. *J. Med. Entomol.*, **30**: 646–648.

Tantawi, T. I., Wells, J. D., Greenberg, B., El-Kady, E. M. 1998. Fly larvae (Diptera: Calliphoridae, Sarcophagidae, Muscidae) succession in rabbit carrion: variation observed in carcasses exposed at the same time and in the same place. *J. Egypt. Ger. Soc. Zool.*, **25**: 195–208.

Tapp, E. 1984. Disease and the Manchester mummies – the pathologist's role. In *Evidence embalmed. Modern medicine and the mummies of ancient Egypt*. ed. R. David and E. Tapp, p. 89. Manchester: Manchester University Press.

Tessmer, J. W., Meek,C. L., Wright, V. L. 1995. Circadian patterns of oviposition by necrophilous flies (Diptera: Calliphoridae). *Southwest. Entomol.*, **20**: 439–445.

Tessmer, J. W. and Meek, C. L. 1996. Dispersal and distribution of Calliphoridae (Diptera) immatures from animal carcasses in southern Louisiana. *J. Med. Entomol.*, **33**: 665–669.

Tullis, K. and Goff, M. L. 1987. Arthropod succession patterns in exposed carrion in a tropical rainforest on O'ahu Island, Hawai'i. *J. Med. Entomol.*, **24**: 332–339.

Turnbull, I. F. and Howells, A. J. 1980. Larvicidal activity of inhibitors of DOPA decarboxylase on the Australian sheep blowfly, *Lucilia curprina*. *Aust. J. Biol. Sci.*, **33**: 169–181.

Turner, B. and Howard, T. 1992. Metabolic heat generation in dipteran aggregations: a consideration for forensic entomology. *Med. Vet. Entomol.*, **6**: 179–181.

Tyndale-Biscoe, M. and Kitching, R. L. 1974. Cuticular bands as age criteria in the sheep blowfly *Lucilia cuprina* (Wied.) (Diptera, Calliphoridae). *Bull. Entomol. Res.*, **64**: 161–174.

Ullyet, G. C. 1950a. Competition for food and allied phenomena in sheep blowfly populations. *Phil. Trans.*, **234**: 77–175.

Ullyet, G. C. 1950b. Pupation habits of sheep blowflies in relation to parasitism by *Mormoniella vitripennis*, WLK. (Hym. Pteromalid.). *Bull. Entomol. Res.*, **40**: 533–537.

Utsumi, K. 1959. Studies on arthropods congregated to animal carcasses, with regard to the estimation of postmortem interval. *Ochanomizu Med. J.*, **7**: 2780–2801.

Utsumi, K., Nakajima, M., Mitsuya, T. and Kaneko, K. 1958. Studies on the insects congregated to the albino rats died of different causes. *Ochanomizu Med. J.*, **7**: 2697–2707.

Van Buren, E. D. 1936–7. Mesopotamian fauna in the light of the monuments. Archeological remarks upon Landsberger's "Fauna des alten Mesopotamien". *Arch. Orientforschung*, **11**: 1–37.

Van Buren, E. D. 1939. *The fauna of ancient Mesopotamia as represented in art*, pp. 108, 109. Rome: Pontificium Institutum Biblicum.

VanLoerhoven. S. L., and Anderson, G. S. 1999. Insect succession on buried carrion in two biogeoclimatic zones in British Columbia. *J. Forensic Sci.*, **44**: 32–43.

Vinogradova, E. B. 1986. Geographical variation and ecological control of diapause in flies. In *The evolution of insect life cycles*, ed. F. Taylor and R. Karban, chapter 3. New York: Springer-Verlag.

Vinogradova, E. B. and Marchenko, M. I. 1984. The use of temperature parameters of fly growth in the medicolegal practice. *Sud. Med. Exp.*, **27**: 16–19.

Wall, R., French, N. and Morgan, K. L. 1992. Effects of temperature on the development and abundance of the sheep blowfly *Lucilia sericata* (Diptera: Calliphoridae). *Bull. Entomol. Res.*, **82**: 125–131.

Wall, R., Langley, P. A. and Morgan, K. L. 1991. Ovarian development and pteridine accumulation for age determination in the blowfly *Lucilia sericata. J. Insect Physiol.*, **37**: 863–868.

Wall, R., Langley, P. A. Stevens, J. and Clarke, G. M. 1990. Age-determination in the Old-World screw-worm fly *Chrysomya bezziana* by pteridine fluorescence. *J. Insect Physiol.*, **36**: 213–218.

Wallman, J. F. 1999. *Systematics and thermobiology of carrion-breeding blowflies (Diptera: Calliphoridae).* PhD Dissertation, Dept. of Environmental Biology, University of Adelaide, Australia.

Waterhouse, D. F. 1947. *The relative importance of live sheep and of carrion as breeding grounds for the Australian sheep blowfly* Lucilia cuprina. C.S.I.R.O. (Australia), Bulletin 217. Canberra: CSIRO.

Webster, F. M. 1890. *Insect life*, **5**: 356–358, 370–372.

Wells, J. D., Byrd, J. H. and Tantawi, T. I. 1999. Key to third-instar Chrysomyinae (Diptera: Calliphoridae) from carrion in the continental United States. *J. Med. Entomol.*, **36**: 638–641.

Wells, J. D., Introna, Jr., F., Di Vella, G., Campobasso, C. P., Hayes, J. and Sperling, F. A. H. 2001. Human and insect mitochondrial DNA analysis from maggots. *J. Forensic Sci.*, **46**: 685–687.

Wells, J. D. and King. J. 2001. Incidence of precocious egg development in flies of forensic importance (Calliphoridae). *Pan-Pacific Entomol.*, **77**: 235–239.

Wells, J. D. and Kurahashi, H. 1994. *Chrysomya megacephala* (Fabricius) (Diptera: Calliphoridae) development: rate, variation and the implications for forensic entomology. *Jap. J. Sanit. Zool.*, **45**: 303–309.

Wells, J. D. and LaMotte, L. R. 1995. Estimating maggot age from weight using inverse prediction. *J. Forensic Sci.*, **40**: 585–590.

Wigglesworth, V. B. 1967. *The principles of insect physiology.* London: Methuen.

Wijesundara, D. P. 1957. The life-history and bionomics of *Chrysomyia megacephala* (Fab.). *Ceylon J. Sci.*, **25**: 169–185.

Williams, H. 1984. A model for the aging of fly larvae in forensic entomology. *Forensic Sci. Int.*, **25**: 191–199.

Williams, H. and Richardson, A. M. M. 1984. Growth energetics in relation to temperature for larvae of four species of necrophagous flies (Diptera: Calliphoridae). *Aust. J. Ecol.*, **9**: 141:152.

Williams, K. L., Browne, L. B., Gerwin, A. C. M. van, Van Gerwin, A. C. M., and Barton-Browne, L. 1977. Ovarian development in autogenous and anautogenous *Lucilia cuprina* in relation to protein storage in the larval fat body. *J. Ins. Physiol.*, **23**: 659–664.

Xue, W. and Chao, C. (eds.) 1996. *Flies of China*, vols l. and 2. Shenyang: Liaoning Science and Technology Press. (In Chinese.)

Zdarek, J. 1985. Regulation of pupariation in flies. *Comprehensive insect physiology*, vol 8, pp. 301–333.Oxford: Pergamon.

Zdarek, J. and Fraenkel, G. 1972. The mechanism of puparium formation in flies. *J. Exp. Zool.*, **179**: 315–324.

Zenovjeva, K. B. 1978. Maternal induction of the larval diapause in *Lucilia hirsutula* Gr., *L. illustris* Meig., *Calliphora uralensis* Vill. (Diptera, Calliphoridae). In *Photoperiodic reactions of the insects*, ed. V. A. Zaslovsky, pp. 80–94. Leningrad: Nauka.

Zenovjeva, K. B. and Vinogradova, E. B. 1972. The control of seasonal development in parasites of flesh and blowflies. II. Ecological control of winter adaptations in *Calliphora vicina* R.-D. (Diptera, Calliphoridae). In *Host–parasite relationships in insects.* V. A. Zaslavsky, ed. pp. 90–9. Leningrad: Nauka.

Zumpt, F. 1965. *Myiasis in man and animals in the Old World.* London: Butterworths.

PART 2

"A tiny fly can choke a big man."

Solomon Ibn Gabirol (1021?–1058?)

Preface to Part II

This book represents more than "just" the first comprehensive treatment of the scientific and legal aspects of forensic entomology. For me, as a law professor and a litigator, it is also the vehicle for summarizing and analyzing the law of scientific evidence – a topic of great personal and professional interest to me because of my unconventional career path, which took me from a Master of Science degree in entomology to a Juris Doctor and Master of Laws.

My work on this book required me to move far beyond the familiar borders of United States law. I delved into the case law and statutory law of other major nations of the world as well, to make this truly a global resource for the legal practitioner and his or her scientific experts who face the complex task of evaluating and utilizing forensic entomology evidence and other scientific evidence in litigation. As the world of forensic science advances, the legal field is ever obliged to play "catch up" to handle the latest developments appropriately and ensure fairness and justice as the science enters the courtroom. I have endeavored to set forth the proper legal framework within which these issues can be addressed, worldwide.

My esteemed co-author, mentor, and friend, Professor Greenberg, has made this volume the definitive work on forensic entomology from the scientific perspective. It has been my privilege to have my words appear alongside his. It is my most sincere hope that my contributions concerning the law of scientific evidence in general, and forensic entomology in particular, will be judged worthy of this honor.

Professor John Charles Kunich

6

The Law of Scientific Evidence

Simply stated, the purpose of forensic entomology is to assist in the process of proving who did what to whom, when. Entomological evidence, facts, principles, and analysis may directly aid police officers and other investigators in their crime-solving activities. Ultimately, though, this would be of little or no practical value unless the entomological materials and conclusions to be drawn therefrom were also legally admissible as evidence in a court of law. It is in court where the science truly meets the law and may influence the outcome of a given case.

Even in that large majority of cases that are plea-bargained or otherwise disposed of prior to trial, issues of admissibility of the entomological evidence are of central importance. Unless there is a reasonable likelihood that the evidence would be admissible in the event of a litigated trial, attorneys will discount it and it will not be a factor they need to consider in weighing their various settlement options. Entomological facts become a factor in influencing the results of legal actions in direct proportion to their acceptance as relevant, material, and competent evidence in court.

Although forensic entomology is usually considered primarily a tool in the conduct of criminal proceedings, it is equally applicable in the civil law context. Particularly in wrongful death actions, this type of evidence may be highly probative of the time and circumstances of the fatality in question, and thus be a key factor in the case. Additionally, forensic entomology evidence can be pivotal in civil actions concerning spoiled food or food-related products. Therefore, the following evidentiary discussion should be considered pertinent in all types of litigation, both criminal and civil. The only difference of significance for our purposes is the standard of proof. In criminal actions, of course, the prosecution has the burden of establishing every element of each charge beyond a reasonable doubt, whereas in civil litigation the plaintiff need only prove his or her case by a preponderance of the evidence, i.e., that it is more likely than not that the plaintiff's position on every element of each claim is true. For all intents and purposes, the rules of evidence are identical for both criminal and civil cases.

In this chapter we will examine the essentials of the law of scientific evidence in American courts of law, and will compare the American approach with that of several other nations. Although we will delve into the intricacies of the law from an attorney's perspective, non-lawyers who are interested in forensic entomology should also profit from this analysis, because the law establishes the framework within which the field operates, including the ground rules under which the evidence may be used in actual litigation. It is important that attorneys, scientists, and investigators alike have an understanding of the legal framework that governs the use of this and all other scientific evidence at trial. Without such a foundation, there is the risk that a great deal of time and effort might be expended without any benefit in terms of the final outcome of any given case. And influencing the outcome of a case is the reason why the field of forensic entomology exists in the first place. With that in mind, we will now examine the law governing the admissibility of evidence in general, with emphasis on scientific evidence, and then in the following chapter move on to specific issues relevant to forensic entomology in particular.

General Rules Governing Admissibility of Evidence

The Federal Rules of Evidence control the admissibility of evidence in all federal trials within the United States. Additionally, every State has its own rules, which are often closely patterned after the federal rules, and which are used in trials conducted in State courts within each jurisdiction. Because the Federal Rules either apply directly or are the model for the State rules in virtually every courtroom in the United States, we will focus on them.

Obviously, for actions taking place in State court, it is essential to be aware of any deviations that may exist from the Federal Rules of Evidence with regard to expert opinion testimony, admissibility of evidence in general, and the applicable standard for evaluating the admissibility of scientific evidence. It is beyond the scope of this book to address the evidentiary rules of all State jurisdictions, but the principles discussed herein should be instructive, if not directly applicable, in all cases.

There are a few rules that form the framework within which the vast majority of evidentiary issues are determined at trial. Practitioners must be familiar with these rules and gear their litigation tactics to satisfying their requirements. It may be useful to conceptualize the rules of evidence as a filter that controls the content of material entering a courtroom. The judge uses this filter to regulate the flow of information to the trier of fact, which is the jury in trials by jury and the judge in trials by judge alone. The facts, as determined by the trier of fact, are of course the fuel which powers the legal engine, i.e., the applicable rules of law, to arrive at the final outcome in every trial. Understood in this manner, the rules of evidence play a direct and often pivotal role in determining the result in court; evidence must pass through this evidentiary filter if it is to be a factor in the case. Thus, any variables that may affect the tendency of information either to pass successfully through this filter or be trapped and rejected by it must be understood and thoroughly incorporated into both pretrial preparation and in-court tactics.

We begin with the basic and very liberal presumption embodied in the rules, that all relevant evidence is admissible. Federal Rule of Evidence 402 states:

Relevant Evidence Generally Admissible; Irrelevant Evidence Inadmissible. All relevant evidence is admissible, except as otherwise provided by the Constitution of the United States, by Act of Congress, by these rules, or by other rules prescribed by the Supreme Court pursuant to statutory authority. Evidence which is not relevant is not admissible.

Of course, the word "except" is very important in this rule, and that is the battleground where the legal war over the law of evidence is fought. The rules of evidence themselves are replete with exceptions to the general admissibility principle, and additional exceptions have been carved out by the courts on Constitutional grounds. Nonetheless, it is significant that the general presumption is in favor of admissibility of relevant evidence. It is easy to lose sight of this overarching philosophy of inclusion when entangled in the thicket of exceptions and counter-exceptions that criss-cross the rules, and if litigators and judges keep it in mind it can serve as a useful navigational reference point, an evidentiary Polaris.

This Rule 402 principle leads directly to the question of what evidence is relevant. Federal Rule of Evidence 401 formally defines the term, and is similarly expansive in its tendency to allow, rather than to exclude, evidence:

Definition of "Relevant Evidence." "Relevant evidence" means evidence having any tendency to make the existence of any fact that is of consequence to the determination of the action more probable or less probable than it would be without the evidence.

Rule 401 is, on its face, a very inclusive standard. So long as the evidence has "any tendency" to make "any fact of consequence" in influencing the result of the case more likely or less likely, it is relevant. Taken together with Rule 402, this rule provides a strong presumption in favor of admissibility, again of course absent an applicable exception.

These exceptions are too numerous to review in detail here, but Federal Rule of Evidence 403 deserves special attention, both because of its importance to the admissibility of forensic entomology evidence and because it so directly empowers the trial judge to exclude relevant evidence. It reads as follows:

Exclusion of Relevant Evidence on Grounds of Prejudice, Confusion, or Waste of Time. Although relevant, evidence may be excluded if its probative value is substantially outweighed by the danger of unfair prejudice, confusion of the issues, or misleading the jury, or by considerations of undue delay, waste of time, or needless presentation of cumulative evidence.

Rule 403 gives the trial judge enormous discretion to refuse to admit evidence that in some way seems inappropriate. In Chapter 7, we will examine this issue in considerable detail, but it is wise to keep it in mind from the outset as we discuss the various factors that influence the admissibility of evidence. Even if it appears that an item of evidence has swept through the primary evidentiary filters, Rule 403 is always there as a form of failsafe, potentially ready to screen out otherwise admissible evidence. As we shall see, there are certain factors that affect the tendency of judges to exercise their Rule 403 power, and knowledgeable litigators can fold these considerations into their trial tactics.

As a final basic evidentiary principle, we must note the preeminent role played by the trial judge (the "court") in all such matters. Federal Rule of Evidence 104 governs:

Preliminary Questions. (a) *Questions of admissibility generally.* Preliminary questions concerning the qualification of a person to be a witness, the existence of a privilege, or the admissibility of evidence shall be determined by the court, subject to the provisions of subdivision (b). In making its determination it is not bound by the rules of evidence except those with respect to privileges. (b) *Relevancy conditioned on fact.* When the relevancy of evidence depends upon the fulfillment of a condition of fact, the court shall admit it upon, or subject to, the introduction of evidence sufficient to support a finding of the fulfillment of the condition. (c) *Hearing of jury.* Hearings on the admissibility of confessions shall in all cases be conducted out of the hearing of the jury. Hearings on other preliminary matters shall be so conducted when the interests of justice require, or when an accused is a witness and so requests. (d) *Testimony by accused.* The accused does not, by testifying upon a preliminary matter, become subject to cross-examination as to other issues in the case. (e) *Weight and credibility.* This rule does not limit the right of a party to introduce before the jury evidence relevant to weight or credibility.

Under Rule 104, it is the trial judge who acts as gatekeeper for all evidence at trial. As provided in this rule, key evidentiary issues such as the qualification of an expert witness or the admissibility of a particular type of scientific evidence will usually be litigated with the jury out of the courtroom, so as not to taint them with extraneous or inadmissible information. This allows the judge to consider otherwise inadmissible evidence in making his or her determinations, without fear that the jury will be exposed to such matters. However, note also that subdivision (e) of this rule allows the litigants to attack or bolster evidence that has been ruled admissible, in an effort to diminish or increase its importance in the minds of the jurors. Therefore, even after an item of evidence has been admitted into evidence, there are important tactical issues to consider as to how best to affect the amount of weight the jury will accord to the evidence when the jurors are in the deliberation room deciding the result of the case. Admissible evidence can be effectively neutralized or magnified for purposes of how the jurors will use it in their decision making, depending on how the evidence is presented, undermined, or supported in the courtroom.

With this very basic framework as our general frame of reference, let us now examine the specific issues attendant to scientific evidence and the expert opinion evidence that typically conveys it to the jury. We will encounter some additional rules of evidence as we do so.

Expert Opinion Evidence

The main practical application of all scientific evidence in court is to serve as the material that enables an expert witness to testify and offer an opinion on a matter of significance to the case. Without the assistance of a properly qualified expert witness, it is unlikely any attorney could persuade a judge to admit scientific data into evidence, and even if it were so admitted, there is no way the information would be comprehensible to the non-scientists who almost always constitute the judge and jury. It is expert opinion testimony that digests the complex, foreign material of scientific information and presents it to the jurors in a form that they can understand and use in rendering their decision.

There are certain matters about which any witness can offer an opinion at trial. As stated in Federal Rule of Evidence 701, a non-expert witness can opine as to matters that are rationally based on the witness's perceptions and helpful to a clear understand of the testimony or the determination of a fact in issue. This type of opinion testimony reaches as far as phenomena within the common ken of the general populace, e.g., the approximate speed of a passing vehicle, the value of an item of the witness's personal property, whether a person was intoxicated, or the height of an individual. However, where, as in the case of forensic entomology evidence, the subject matter is beyond the realm of common knowledge, a witness must be properly qualified as an expert at trial before being allowed to offer opinions based thereon.

The significance of a witness being qualified as an expert, other than bestowing a certain perceived degree of additional prestige to everything he or she says in court, is that the expert witness has special latitude to express opinions from the witness stand. Federal Rule of Evidence 702 provides as follows:

> **Testimony by Experts.** If scientific, technical, or other specialized knowledge will assist the trier of fact to understand the evidence or to determine a fact in issue, a witness qualified as an expert by knowledge, skill, experience, training, or education, may testify thereto in the form of an opinion or otherwise, if (1) the testimony is based upon sufficient facts or data, (2) the testimony is the product of reliable principles and methods, and (3) the witness has applied the principles and methods reliably to the facts of the case.[1]

This opinion testimony may even include the expert's view on an "ultimate issue" in the trial that is to be decided by the judge or jury, except as to the mental state of the defendant in a criminal case.[2]

Rule 702, which has analogues in every State, does not set a particularly high standard. It requires simply that, to qualify as an expert, the witness must possess "knowledge, skill, experience, training, or education." This expertise

[1] Federal Rules of Evidence 701, 702, and 703 were amended, effective on December 1, 2000. With regard to Rule 702, the former rule read as follows:

> **Testimony by Experts.** If scientific, technical, or other specialized knowledge will assist the trier of fact to understand the evidence or to determine a fact in issue, a witness qualified as an expert by knowledge, skill, experience, training, or education, may testify thereto in the form of an opinion or otherwise.

[2] See Federal Rule of Evidence 704.

must, of course, be related to the specific subject matter about which the witness will testify in the case at hand. A court would not allow an expert witness qualified as a chemist to opine as to a psychological syndrome, for example. The trial court judge enjoys wide discretion in deciding whether a given witness is so qualified.

Criminal trials in England use a very similar approach. In England, the differences between expert and lay witnesses are that: (1) experts are permitted to testify upon matters of opinion as well as fact; (2) experts are regarded as the living embodiment of a corpus of learning, and thus are not constrained by the hearsay rule from adopting the views of other experts in the field; (3) expert witnesses may be paid; and (4) the expert's evidence can be more easily admitted in documentary form.[3]

Rule 702 does not require an expert witness to have any particular credentials, degrees, professional status, or the like. The idea is simply to be helpful, that is, to assist the judge and the jury in understanding facts of importance to the outcome of the case. If a witness has sufficient knowledge, experience, etc., to be able to help these triers of fact make their decisions at trial, there is no legal impediment to the judge recognizing that person as an expert witness and allowing him or her to testify accordingly. Naturally, however, a witness who is fortunate enough to have earned advanced degrees, has published widely in the field, and has received various professional honors will tend to be more influential with the judge and jury than a witness with less impressive indicia of expertise. As we have seen during our discussion of Rule 104(e), the fact that a piece of evidence has been admitted or a person has been qualified as an expert does not in any way preclude the introduction of further evidence that might affect the weight the jury will give that evidence or that expert's opinion. Such issues of weight and credibility are where many trials are won or lost.

The primary legal difference between lay and expert opinion is that opinion testimony by a lay witness must be "rationally based on the perception of the witness."[4] In contrast, an expert witness need not have been an eye-witness, and indeed it would be extraordinarily rare if he or she were. In lieu of the usual way direct means by which a witness obtains the raw materials from which to form an opinion, Federal Rule of Evidence 703 sets forth the types of materials that may properly substitute for eye-witness perception in forming the basis of an expert witness's opinion at trial:

Bases of Opinion Testimony by Experts. The facts or data in the particular case upon which an expert bases an opinion or inference may be those perceived by or made known to the expert at or before the hearing. If of a type reasonably relied upon by experts in the particular field in forming opinions or inferences upon the subject, the

[3] See Peter Alldridge, *Scientific Expertise and Comparative Criminal Procedure,* 3 Int'l J. Evid. & Proof 141, 149 (1999); C. TAPPER, CROSS AND TAPPER ON EVIDENCE, 8th ed. 543 et seq. (1995).

[4] Federal Rule of Evidence 701(a).

facts or data need not be admissible in evidence in order for the opinion or inference to be admitted. Facts or data that are otherwise inadmissible shall not be disclosed to the jury by the proponent of the opinion or inference unless the court determines that their probative value in assisting the jury to evaluate the expert's opinion substantially outweighs their prejudicial effect.[5]

Therefore, an expert witness has much broader latitude in terms of the facts that may be relied on in forming opinions. These legal advantages, coupled with the formidable, albeit intangible benefit of being officially anointed as an "expert witness" by the judge in the presence of the jury, makes the use of expert testimony a potentially powerful weapon in a litigator's arsenal. This is a poorly kept secret, and as attorneys have seen the gains to be won with experts in litigation, the use of experts has become virtually a matter of course in a large percentage of cases.[6]

Because we are concerned with expert scientific testimony, we will now focus on the particular legal issues attendant to science in the courtroom. As we shall see, the relationship between science and the law in the courts has been an uneasy one.

Is Scientific Evidence Hazardous to the Court's Health?

Underlying the law of scientific evidence is a perception, by some judges and by the drafters of the rules of evidence, that this type of evidence carries with it an unusual potential for abuse. The non-scientists who typically become attorneys, judges, and legislators tend to be uncomfortable with science. They did not seek out nor particularly excel in science classes in college. They do not work with science themselves on a regular basis in their professional careers. And they believe that the non-scientists who generally populate their juries share the same preconceptions and deficiencies in the area of science.

Unfamiliarity in this instance breeds not contempt, but rather a perception that juries will often give too much deference to scientific expert testimony and assign too much weight to scientific evidence. The view is generally that, due to lack of scientific understanding on the part of juries and the halo effect of expert witness status, scientific evidence will enjoy "a posture of mystic infallibility in

[5] This is the amended version of Rule 703, effective December 1, 2000. Previously, Rule 703 read as follows:

> **Bases of Opinion Testimony by Experts.** The facts or data in the particular case upon which an expert bases an opinion or inference may be those perceived by or made known to the expert at or before the hearing. If of a type reasonably relied upon by experts in the particular field in forming opinions or inferences upon the subject, the facts or data need not be admissible in evidence.

[6] See Gross, *Expert Evidence*, Wisc. L. Rev. 1113, 1118–19 (1991), citing a Rand Corporation study of the use of experts in trials in California courts of general jurisdiction during the early 1980s. Expert witnesses appeared in 86% of the trials studied, with an average of 3.3 experts per trial.

the eyes of the jury."[7] Non-scientist juries are assumed to lack the acumen to perceive flaws in scientific evidence, with the resultant potential for an uncorrected erroneous result at trial.

A lay jury will lack the very training, experience, and education that enables a scientific witness to qualify as an expert. With the jury unarmed in this manner, the law has presumed that there is a danger of blind acceptance of whatever evidence is shrouded in the hallowed garb of science and ceremoniously placed before the jury. This perceived risk has resulted in an array of safeguards designed to ensure the reliability and accuracy of scientific evidence.

There are two principal ways in which scientific evidence may be inaccurate. The underlying scientific theory and principles may be erroneous, or "junk science" as the current catch-phrase puts it. Or, the scientific theory may be valid but the person or persons applying it may have committed a significant error in the case at hand. To deal with these twin risks, most American jurisdictions today have set forth, whether explicitly or implicitly, a four-part test which the party seeking to introduce scientific evidence must pass:

1. Any sample tested must be shown to have been uncontaminated, with appropriate proof of chain of custody to establish that the sample was in fact the actual sample taken from the scene of the incident in question.
2. The scientific theory and technique must be shown to be valid, i.e., it measures what it purports to measure.
3. The theory and technique were properly applied in the case at hand.
4. The test results were properly analyzed and interpreted.

In this chapter we will focus on the second part of this test, because it is not tied to the vagaries of application in any particular case but rather is a global issue that will be equally applicable in any trial involving a particular type of scientific evidence. However, in a subsequent chapter we will revisit the other prongs of the test as well, because they can be pivotal in determining the admissibility or weight of scientific evidence in every case.

Courts may use any of the following four methods to establish, for purposes of a given case, the validity of a particular scientific theory or method:

1. Judicial notice.
2. Stipulation by all the parties to the case.
3. Legislative recognition of the validity of the method/theory.
4. In-court presentation of evidence, including expert testimony.

The first three methods, it should come as no surprise, do not generate much controversy. By their very nature, they deal with scientific principles or scientific techniques that are already firmly rooted in the modern judicial system as well established, time-tested weapons in the court's arsenal of truth-finding. It

[7] *United States v. Addison*, 498 F. 2d 741, 744 (D.C. Cir. 1974).

is the fourth method where all the sound and fury are to be found. It is here that new technologies and novel scientific theories pass through the refiner's fire and either are accepted or found wanting. Additionally, even the vast majority of "traditional" forms of scientific evidence must also undergo this case-by-case determination of admissibility, and will continue to do so until and unless they eventually gain formal recognition by the legislature or become an accepted subject for judicial notice.

This process of litigating the admissibility of frequently-encountered scientific methods may seem wasteful and duplicative, requiring as it does a great deal of time, effort, and money. In each case where such evidence is in issue, experts must be employed and brought in to testify. Attorneys for both sides must conduct direct- and cross-examination of each expert witness. Judges must evaluate often-voluminous quantities of documents proffered by both sides in an effort to prove or disprove the worthiness of the science and the technique at issue. This is in fact an expensive, resource-intensive enterprise. But in that substantial percentage of cases in which judicial notice, legislative codification, and mutual stipulation are not available, this is the only remaining way for the admissibility of the evidence to be adjudicated.

We will now turn our attention to the legal standards used by the courts as they struggle through this process of evaluating the validity/reliability of such scientific theories and techniques. These tests are the legal hurdles over which litigators and their expert witnesses must leap as a prerequisite to offering expert opinion testimony or other scientific evidence in court.

Legal Standards for Evaluating Scientific Evidence

FRYE

For about 70 years, the leading case governing the standard for admissibility of scientific evidence was *Frye v. United States.*[8] The test enunciated in *Frye* was the one used by no fewer than 45 States and most of the federal circuits, as recently as the late 1970s.[9] Even today, many State courts continue to use the *Frye* test, despite more recent developments to be discussed shortly.[10]

In the *Frye* case, the defendant had been given, pre-trial, a systolic blood pressure test (an early form of polygraph or "lie detector" test), which measured his blood pressure as he was asked a series of questions. Defendant Frye had a witness who was prepared to testify that changes (or lack thereof) in Frye's blood pressure during questioning indicated that he was being truthful in his responses when he denied committing the crime of second-degree

[8] 293 F. 1013 (D.C. Cir. 1923).

[9] Recent Developments, 64 Cornell L. Rev. 875, 878–79 (1979); Note, 40 Ohio St. L.J. 757, 769 (1979). A notable exception was the Second Circuit, which abandoned the *Frye* test in *United States v. Williams*, 583 F.2d 1194, 1197–99 (2d Cir. 1978), cert. denied, 439 U.S. 1117 (1979).

[10] *From* Frye *to* Daubert: *Is a Pattern Unfolding?*, 35 Jurimetrics J. 191, 193 (1995).

murder. The witness was ready to testify in support of the scientific theory and also his instrumentation. Although such evidence would have satisfied the authentication requirement applicable to evidence in general, the trial court judge refused to allow it to go before the jury after hearing the witness testify in a session outside the presence of the jurors. Frye then asked to be administered a similar test in court, in the presence of the jury, but the judge denied his request. Frye was convicted. On appeal, he argued that the trial judge's refusal to allow expert testimony on the results of his test was reversible error. However, the Court of Appeals for the District of Columbia affirmed the exclusion of the evidence and upheld Frye's conviction.

The Court of Appeals based its decision on the fact that the foundation for the evidence was incomplete, in that the witness did not testify that the technique had gained general acceptance in the applicable scientific community, i.e., the fields of psychology and physiology. The Court stated:

> Just when a scientific principle or discovery crosses the line between the experimental and demonstrable stages is difficult to define. Somewhere in this twilight zone the evidential force of the principle must be recognized, and while the courts will go a long way in admitting expert testimony deduced from a well-recognized scientific principle or discovery, the thing from which the deduction is made must be sufficiently established to have gained general acceptance in the particular field in which it belongs. We think the systolic blood pressure deception test has not yet gained such standing and scientific recognition among physiological and psychological authorities as would justify the courts in admitting expert testimony deduced from the discovery, development, and experiments thus far made.[11]

Interestingly, the Court did not cite any authority in support of the test it announced, nor did it set forth any policy rationale for it.

The *Frye* "general acceptance" test certainly did not provide a clear, easily used standard, despite its widespread adoption by courts throughout the United States.[12] Its most glaring defect is that it appears to call for some type of vote or poll among scientists in a given field as the gauge of a technique's validity, which, as applied by some courts, amounted to asking the very specialists who had developed a certain technique and had a personal stake in its acceptance to pass judgment on their own bread and butter. Certainly, although the degree of acceptance by scientific peers is one index of validation, the real question is whether the technique or hypothesis has been borne out by repeated experimentation and replication under controlled scientific conditions. The test should not turn on a counting of heads – even educated heads – or a show of

[11] *Frye v. United States*, 293 F. 1013, 1014 (D.C. Cir. 1923).

[12] See, e.g., Jay P. Kesan, *Note: An Autopsy of Scientific Evidence in a Post-Daubert World*, 84 Geo. L.J. 1985, 1991 (1996); Paul C. Giannelli, *The Admissibility of Novel Scientific Evidence:* Frye v. United States, *a Half-Century Later*, 80 Colum. L. Rev. 1197, 1208 (1980).

hands, but, as one court put it, on whether the theory is supported by "solid empirical research."[13]

Other critics argued that the *Frye* test reflects an unproven, patronizing, paternalistic assumption that non-scientist jurors and judges are incapable of evaluating and properly weighing scientific testimony.[14] Contrary to the bias inherent in the *Frye* test, the evidence from a number of studies indicates that "[t]he image of a spellbound jury mesmerized by . . . a forensic expert is more likely to reflect . . . fantasies than the . . . realities of courtroom testimony.:[15] These critics argued that the *Frye* test supplanted detailed analysis by the court with mere scholarly peer review and publication, thereby allowing nonjudicial persons to usurp the judge's role and make what is essentially a judicial policy decision for the court.[16] In application, the test amounted to abdication of judicial responsibility, with a vaguely-defined group of scientists sitting in place of the judge as the arbiter of admissibility of scientific evidence.

A further problem was that both the method of analysis and the results produced by courts purporting to use the seemingly straightforward *Frye* standard were far from uniform. Some courts did little more than "count noses" to determine general acceptance within the scientific community,[17] while others conducted in-depth analysis.[18]

Even courts that adhere to the *Frye* standard and support its utility as a filter for unreliable pseudo-scientific evidence have had to wrestle with (or avoid) the unanswered questions it dumped into the laps of trial judges. For example, what percentage of experts from a given field of study must accept a scientific technique or principle before it can be deemed "generally accepted," and how is such a poll/vote to be determined?[19] Also, how broadly or narrowly is the applicable field to be defined, i.e., how far down the ladder of specialization do we go in arriving at the appropriate pool of scientists to be "polled"?[20] Which

[13] *State v. York*, 564 A.2d 389, 390–91 (Me. 1989); see also *United States v. Stifel*, 433 F.2d 431, 438 (6th Cir. 1970); Black, *A Unified Theory of Scientific Evidence*, 56 Fordham L. Rev. 595, 625 (1988).

[14] Edward J. Imwinkelried, *The Standard for Admitting Scientific Evidence: A Critique from the Perspective of Juror Psychology*, 28 Vill. L. Rev. 554, 567–68 (1983).

[15] Rogers and Ewing, *Ultimate Opinion Proscriptions: A Cosmetic Fix and a Plea for Empiricism*, 13 Law & Hum. Behav. 357, 363 (1989).

[16] See David L. Faigman, *Making the Law Safe for Science: A Proposed Rule for the Admissibility of Expert Testimony*, 35 Washburn L.J. 401, 405 (1996).

[17] See Barry C. Scheck, *DNA and* Daubert, 15 Cardozo L. Rev. 1959, at 1959 (1994).

[18] See, e.g., *In re Agent Orange Product Liability Litigation*, 611 F. Supp. 1223 (E.D.N.Y. 1985).

[19] See James E. Starrs, Frye v. United States *Restructured and Revitalized: A Proposal to Amend Federal Evidence Rule* 702, 26 Jurimetrics 249, 250 (1986).

[20] As alluded to previously, if the field is too narrowly defined, the *Frye* test amounts to asking people whether the subject matter which they have either developed themselves, or rely on for their livelihood, or both, has validity. It is not surprising that such a group tends to give its enthusiastic endorsement of whatever field of endeavor is in question.

facets of the testimony or underlying rationale were required to be "generally accepted"?[21] Despite such nagging questions, many courts for many years have extolled the virtues of the *Frye* test, even going so far as to supply the policy bases which the *Frye* court itself left out in its brief opinion. See, e.g., *People v. Kelly*[22], in which the California Supreme Court hypothesized:

[T]he *Frye* test... may well promote a degree of uniformity of decision. Individual judges whose particular conclusions may differ regarding the reliability of particular scientific evidence, may discover substantial agreement and consensus in the scientific community... The primary advantage, however, of the *Frye* test lies in its essentially conservative nature. For a variety of reasons, *Frye* was deliberately intended to interpose a substantial obstacle to the unrestrained admission of evidence based upon new scientific principles.[23]

The number of States that follow the *Frye* test has dramatically fallen since the 1990s, particularly in the wake of the 1993 United States Supreme Court decision in *Daubert v. Merrell Dow Pharmaceuticals, Inc.*[24] Still, as recently as 1996, 21 states continued to adhere to *Frye*.[25] In the federal courts, of course, *Daubert* immediately supplanted *Frye*, which previously had been accepted by a majority of the federal circuits.[26] Under the American judicial system,

[21] See examples described in Andre A. Moessens, *Admissibility of Scientific Evidence – An Alternative to the* Frye *Rule*, 25 Wm. & Mary L. Rev. 545, 553 (1984).

[22] 549 P.2d 1240 (Cal. 1976). [23] Id., at 1244–45. [24] 509 U.S. 579 (1993).

[25] The states employing the *Frye* test as of 1996 were: Alabama, *Ex parte Perry*, 586 So. 2d 242, 247 (Ala. 1991); Alaska, *Harmon v. State*, 908 P.2d 434, 439 n.5 (Alaska App. 1995); Arizona, *State v. Boles*, 905 P.2d 572, 578 (Ariz. App. 1995); California, *People v. Leahy*, 882 P.2d 321, 324–31 (Cal. 1994); Colorado, *Lindsey v. People*, 892 P.2d 281, 288 (Colo. 1995); Connecticut, *State v. Esposito*, 670 A.2d 301, 316 (Conn. 1996); Florida, *Hayes v. State*, 660 So.2d 257, 262 (Fla. 1995); Illinois, *People v. Watson*, 629 N.E.2d 634, 640–41 (Ill. App. 1994); Kansas, *State v. Hill*, 895 P.2d 1238, 1244–46 (Kan. 1995); Maryland, *Armstead v. State*, 673 A.2d 221, 225 n.4, 228 (Md. 1996); Michigan, *People v. Lee*, 537 N.W. 2d 233, 249 n.17 (Mich. App. 1995); Minnesota, *State v. Klawitter*, 518 N.W.2d 577, 588 n.1 (Minn. 1994); Mississippi, *Materials Transp. Co. v. Newman*, 656 So.2d 1199, 1203–04 (Miss. 1995); Missouri, *State v. Funke*, 903 S.W.2d 240, 244 (Mo. App. 1995); Nebraska, *State v. Carter*, 524 N.W.2d 763, 779 (Neb. 1994); New Hampshire, *State v. Vandebogart*, 652 A.2d 671, 677–78 (N.H. 1995); New Jersey, *State v. Williams*, 599 A.2d 960, 963 (N.J. Super. L. Div. 1991); New York, *People v. Yates*, 637 N.Y.S.2d 625, 625 (N.Y. Sup. Ct. 1995); North Dakota, *City of Fargo v. McLaughlin*, 512 N.W.2d 700, 705 (N.D. 1994); Pennsylvania, *Commonwealth v. Crews*, 640 A.2d 395, 399 (Pa. 1994); and Washington, *State v. Russell*, 882 P.2d 747, 761 (Wash. 1994).

[26] Seven circuits had interpreted the Federal Rules of Evidence to embody *Frye*: *Christophersen v. Allied-Signal Corp.*, 939 F.2d 1106, 1110 (5th Cir. 1991) (per curiam); *United States v. Two Bulls*, 918 F.2d 56, 60 (8th Cir.) reh'g granted, vacated en banc, and remanded, 925 F.2d 1127 (8th Cir. 1991); *United States v. Gillespie*, 852 F.2d 475, 480 (9th Cir. 1988); Kropinski v. World Plan Executive Council – U.S., 853 F.2d 948, 956 (D.C. Cir. 1988); *United States v. Metzger*, 778 F.2d 1195, 1203 (6th Cir. 1985); *United Ellis v. Int'l Playtex, Inc.*, 745 F.2d 292, 303–04 (4th Cir. 1984); and *United States v. Smith*, 869 F.2d 348, 350 (7th Cir. 1989). Two other circuits had held that the *Frye* test was incompatible with the Federal Rules: *United States v. Jakobetz*, 955 F.2d 786, 794 (2d Cir.), cert. denied, 113 S. Ct. 104 (1992); and *United States v. Downing*, 753 F.2d 1224, 1237 (3d Cir. 1985). See Michael J. Saks, *Merlin and Solomon: Lessons from the Law's Formative Encounters with Forensic Identification Science*, 49 Hastings L. J. 1069, 1077 (1998).

because the Supreme Court based its decision on the Federal Rules of Evidence (which are not controlling authority in State proceedings), State courts remain free to use *Frye*, while the federal courts are bound to follow *Daubert*.

It should be noted that, even before *Daubert*, some courts used tests other than the *Frye* "general acceptance" approach. For example, some favored a more liberal "substantial acceptance" test, under which evidence would be admitted if it was accepted by a significant minority of experts in the field.[27] This approach was thought by some to be more appropriate in light of the way scientific knowledge advances and evolves. Other courts used what might be called "multifactor reliability tests" which required the judge to examine the scientific validity of the expert's methods, under the theory that this would be indicative of evidentiary reliability.[28] One "multifactor" court set forth a five-part test to evaluate the reliability of scientific evidence, involving: (1) potential rate of error; (2) existence and maintenance of standards; (3) the level of care with which the technique has been used and whether it lends itself to abuse; (4) any analogy that could be drawn between the proffered evidence and other scientific evidence routinely admitted into evidence; and (5) the presence of failsafe characteristics.[29]

It is interesting to note that, at the same time that *Frye* was losing some of its support within the United States, it was gaining adherents in other nations. Until about 1990, most common law nations, including England, Canada, Australia, and New Zealand, uniformly subjected expert scientific testimony to a liberal "relevancy" test for admissibility.[30] Although the rules regarding admissibility of scientific evidence are in flux in many of the British Commonwealth countries, the trend in recent years has been towards some stricter form of *Frye*-like judicial scrutiny based on the general acceptance test.[31] None of these nations, however, has developed a uniform test for admissibility to replace the relevancy standard.[32] We will examine the legal situation in these nations in more detail near the end of this chapter.

DAUBERT

Many years of controversy, coupled with questions as to the continued viability of *Frye* in the federal courts in the aftermath of the adoption of the Federal

[27] See, e.g., *United States v. Torniero*, 735 F.2d 725 (2nd Cir. 1984), cert. denied 469 U.S. 1110 (1985); *United States v. Gould*, 741 F.2d 45 (4th Cir. 1984).

[28] See, e.g., *United States v. Downing*, 753 F.2d 1224, 1237–38 (3rd Cir. 1985). This decision was important as an influence in *Daubert* itself. The *Downing* court recognized that novel scientific evidence might be helpful to the jury, and required the trial judge to conduct a preliminary inquiry focusing on: (1) the soundness and reliability of the process or technique used; (2) the possibility that admitting the evidence would overwhelm, confuse, or mislead the jury; and (3) the connection between the scientific research or test result to be presented and particular disputed factual issues in the case at hand.

[29] *United States v. Williams*, 443 F. Supp. 269, 273 (S.D.N.Y. 1977), aff'd, 583 F.2d 1194, 1196–99 (2nd Cir. 1978), cert. denied, 439 U.S. 1117 (1979).

[30] David E. Bernstein, *Junk Science in the United States and the Commonwealth*, 21 Yale J. Int'l L. 123, 138–9 (1996). [31] Id. [32] Id.

Rules of Evidence, had set the stage for the Supreme Court to address the issue. One key factor was the extensive litigation during the late 1970s and 1980s alleging that the morning sickness anti-nausea drug Bendectin causes birth defects. Dozens of Bendectin cases resulted in evidentiary rulings that highlighted the need for renewed attention on the issue of scientific evidence.[33] In these and other cases, scientific evidence was called into question or even discredited, giving rise to considerable controversy concerning the intrusion of "junk science"[34] into the sanctuary of the courtroom. Finally, in *Daubert,* one of the Bendectin cases, the Court directly dealt with the standard for admitting expert testimony, particularly of a novel scientific nature, in a federal trial.

In *Daubert,* the case centered around two minor children born with serious birth defects. The children and their parents brought suits alleging the birth defects were caused by the mothers' ingestion of Bendectin, an anti-nausea drug, during pregnancy.[35] During the battle of the expert witnesses at trial, the plaintiffs sought to counter the defense experts with the testimony of eight of their own well-credentialed experts. Plaintiffs' expert testimony was based on animal-cell studies, live-animal studies, and chemical-structure analysis, as well as re-calculations of data in previously published studies that had found no causal link between Bendectin and birth defects. The trial court refused to accept this evidence, holding that expert opinion based on these foundations rather than on epidemiological evidence is not admissible to prove causation of the birth defects. The re-analysis of data in previous studies was rejected because it was unpublished and not subjected to the normal peer review process, but rather was generated solely for use in litigation. As a result, the court concluded that plaintiffs had provided an insufficient foundation to allow testimony from their experts at trial, and thus plaintiffs could not meet their burden of proving causation at the trial court level.

The Supreme Court held that the *Frye* test was superseded (within the federal court system) by the adoption of the Federal Rules of Evidence.[36] The Court noted that the Federal Rules of Evidence established a liberal basic

[33] Several juries in these cases, persuaded in part by expert scientific testimony, found for the plaintiffs, awarding tens of millions of dollars in damages, despite the existence by this time of overwhelming epidemiological evidence that Bendectin does not cause birth defects. See, e.g., *Ealy v. Richardson-Merrell, Inc.*, 897 F.2d 1159 (D.C. Cir.), cert. denied, 498 U.S. 950 (1990); *Richardson v. Richardson-Merrell, Inc.*, 857 F.2d 823 (D.C. Cir. 1988), cert. denied, 493 U.S. 882 (1989). In reaction, many courts began excluding evidence in Bendectin cases, creating precedents favoring stricter scrutiny of scientific evidence. See Michael D. Green, *Expert Witnesses and Sufficiency of Evidence in Toxic Substances Litigation: The Legacy of Agent Orange and Bendectin Litigation,* 86 Nw. U. L.Rev. 643, 671–80 (1992).

[34] This term, "junk science," was popularized by Peter Huber, who defined it as "jargon-filled, serious-sounding deception." See PETER W. HUBER, GALILEO'S REVENGE: JUNK SCIENCE IN THE COURTROOM 2, 3 (1991); Paul C. Giannelli, *"Junk Science": The Criminal Cases,* 84 J. Crim. L. & Criminology 105, 107–08 (1993).

[35] *Daubert,* 509 U.S. at 582–83.

[36] Id., at 589.

standard for relevant evidence, i.e., evidence which has "any tendency to make the existence of any fact that is of consequence to the determination of the action more probable or less probable than it would be without the evidence."[37] Also, the Court cited Rule 702, which as we have seen governs expert testimony, and pointed out that the rule says nothing at all about any "general acceptance" requirement, let alone establishes it as an absolute prerequisite to admissibility. In rejecting the *Frye* test, the Court stated, rather bluntly, "That austere standard, absent from and incompatible with the Federal Rules of Evidence, should not be applied in federal trials."[38]

The *Daubert* Court did note that the Federal Rules of Evidence, particularly Rule 702, contemplate some judicial involvement in regulating the topics and theories about which an expert may testify. The Court mentioned the following general requirements:

> The subject of an expert's testimony must be "scientific . . . knowledge." . . . [I]n order to qualify as "scientific knowledge," an inference or assertion must be derived by the scientific method. Proposed testimony must be supported by appropriate validation.[39]

The Court briefly discussed the distinction scientists usually draw between "validity" (the extent to which a principle supports what it purports to support) and "reliability" (the extent to which application of a principle produces consistent, replicable results).[40] The Court stated that, for the purposes at hand, the focus was on evidentiary reliability, i.e., the trustworthiness of the evidence, and in a case involving scientific evidence, evidentiary reliability will be predicated on *scientific validity*.

Succinctly stated, the Court enunciated the following test for federal judges to apply when faced with a proffer of scientific evidence:

> [T]he trial judge must determine at the outset, pursuant to Rule 104(a), whether the expert is proposing to testify to (1) scientific knowledge that (2) will assist the trier of fact to understand or determine a fact in issue. This entails a preliminary assessment of whether the reasoning or methodology properly can be applied to the facts in issue.[41]

Naturally, this standard is not as simple as it appears. Indeed, the *Daubert* Court itself recognized that "[m]any factors will bear on the inquiry, and we do not presume to set out a definitive checklist or test."[42] The Court went on to enumerate several factors to be considered in gauging the admissibility of scientific evidence:

[37] Fed. Rule of Evid. 401.

[38] *Daubert*, 509 U.S. at 589. The Court's characterization of the *Frye* test as "austere" is a bit surprising in light of the criticisms *Frye* absorbed for permitting too much "junk science" in the courtroom. Apparently the comment is attributable to the specific result at the trial court level in the *Daubert* case itself, and was not reflective of any data concerning the general liberality or restrictiveness of *Frye* as an evidentiary standard. [39] Id., at 590. [40] Id.

[41] Id., at 592–93. [42] Id., at 593.

Ordinarily, a key question to be answered in determining whether a theory or technique is scientific knowledge that will assist the trier of fact will be whether it can be (and has been) tested . . . Another pertinent consideration is whether the theory or technique has been subjected to peer review and publication . . . [S]ubmission to the scrutiny of the scientific community is a component of "good science," in part because it increases the likelihood that substantive flaws in methodology will be detected. The fact of publication (or lack thereof) in a peer-reviewed journal will thus be a relevant, though not dispositive, consideration in assessing the scientific validity of a particular technique or methodology . . . Additionally, in the case of a particular scientific technique, the court ordinarily should consider the known or potential rate of error . . . and the existence and maintenance of standards controlling the technique's operation. Finally, "general acceptance" can yet have a bearing on the inquiry . . . Widespread acceptance can be an important factor in ruling particular evidence admissible, and a known technique that has been able to attract only minimal support within the community may properly be viewed with skepticism. The inquiry envisioned by Rule 702 is, we emphasize, a flexible one. The overarching subject is the scientific validity – and thus the evidentiary relevance and reliability – of the principles that underlie a proposed submission. (Citations omitted.)[43] [Citations omitted.]

It should be noted that one key aspect of this analysis is a relevance requirement, which is both obvious and easy to overlook in the clutter of other issues to be considered. For evidence to "assist the trier of fact," it must be relevant to some fact in issue in the case at hand, i.e., it must "fit" the case. The *Daubert* Court took this concept of "fit" from *United States v. Downing,*[44] in which it was held that as a precondition for admission of expert testimony regarding the fallibility of eyewitness identifications in cross-racial situations or under stressful conditions, there must be a detailed offer of proof establishing that, in the case before the court, eyewitness identifications were in fact cross-racial or made under stressful conditions. Thus, proffered expert opinion must be shown to relate to facts or conditions that exist in the case before the court.[45]

The *Daubert* Court concluded by responding to those critics of a more relaxed rule for admitting scientific evidence, which the Court believed it was setting forth. Unswayed by arguments that "befuddled juries are confounded by absurd and irrational pseudoscientific assertions," the Court found such arguments "overly pessimistic about the capabilities of the jury, and of the adversary system generally."[46] Further,

Vigorous cross-examination, presentation of contrary evidence, and careful instruction on the burden of proof are the traditional and appropriate means of attacking shaky but admissible evidence. Additionally, in the event the trial judge concludes that the scintilla of evidence presented supporting a position is insufficient to allow a reasonable juror to conclude that the position more likely than not is true, the court remains free to direct a

[43] Id., at 593–94. [44] 753 F.2d 1224, 1242 (3d Cir. 1985).

[45] G. Michael Fenner, *The* Daubert *Handbook: The Case, Its Essential Dilemma, and Its Progeny,* 29 Creighton L. Rev. 939, 1005 (1996). [46] *Daubert,* 509 U.S. 579, at 595.

judgment, Fed. Rule Civ. Proc. 50(a), and likewise to grant summary judgment, Fed. Rule Civ. Proc. 56. These conventional devices, rather than wholesale exclusion under an uncompromising "general acceptance" test, are the appropriate safeguards where the basis of scientific testimony meets the standards of Rule 702.[47]

Interestingly, on remand, the Ninth Circuit Court of Appeals applied the new *Daubert* test but nonetheless concluded that the plaintiffs' evidence was inadmissible. The court found it very significant that plaintiffs' experts were not proposing to testify about matters "growing naturally and directly out of research they have conducted independent of the litigation," but rather developed their opinions expressly for purposes of testifying. The court stated,

> That an expert testifies for money does not necessarily cast doubt on the reliability of his testimony, as few experts appear in court as an eleemosynary gesture. But in determining whether proposed expert testimony amounts to good science, we may not ignore the fact that a scientist's normal workplace is the lab or the field, not the courtroom or the lawyer's office ... If the proffered expert testimony is not based on independent research, the party proffering it must come forward with other objective, verifiable evidence that the testimony is based on "scientifically valid principles."[48]

Whereas the *Frye* test came to be known as the "general acceptance" standard, the *Daubert* test has been called the "relevancy plus reliability" standard. This new standard has been adopted by many State courts and, obviously, applies in all federal courts, at least in the civil context and at least in cases involving novel scientific evidence. However, there is no shortage of controversy and unresolved issues under *Daubert*, just as under *Frye*, including the threshold issue of the types of cases in which the *Daubert* test should be applied.[49]

For example, and somewhat surprising in light of the SupremeCourt's belief that it was liberalizing the standard for admissibility from that which prevailed under the presumably more restrictive and "austere" *Frye* test, there is no clear answer as to whether *Daubert* is resulting in more evidence being admitted, or

[47] Id., at 596. The Rule 50 power mentioned was employed at the lower court level and upheld by the Supreme Court in *Weisgram v. Marley Co.*, 120 S. Ct. 1011; 2000 U.S. LEXIS 1011; 145 L. Ed. 2d 958 (Feb. 22, 2000).

[48] 43 F.3d 1311, 1317–18 (9th Cir. 1995). *See also Wilson v. Merrell-Dow Pharmaceuticals, Inc.*, 160 F.3d 625, 629–31 (10th Cir. 1998) (upholding exclusion of Bendectin evidence under the *Daubert* test).

[49] The Supreme Court contributed to this confusion with its own lack of clarity as to whether the standard should apply in contexts other than federal civil cases involving novel scientific evidence. The Court stated, "although the *Frye* decision itself focused exclusively on 'novel' scientific techniques, we do not read the requirements of Rule 702 to apply specially or exclusively to unconventional evidence. *Daubert*, 509 U.S. at 592 n.11. This appears to imply that the test should be used in a broader context than novel scientific evidence alone. Also, because the *Frye* standard, which the Court overruled, arose in a criminal case, it seems likely that the Court intended the new test to apply in the criminal as well as civil context, although the Court did not explicitly so state.

less.[50] In the immediate aftermath of *Daubert*, several commentators considered it stricter than *Frye*, and prognosticated that the decision would decrease the admissibility of scientific evidence. Conversely, other legal scholars pointed out that the testimony at issue in *Daubert* was "novel" and outside the scientific mainstream, and concluded that the new mode of analysis under *Daubert* would widen the types of evidence admissible in court.[51]

As another example of post-*Daubert* confusion, some courts concluded that the *Daubert* test only applies to scientific evidence, and not to the other types of expert testimony provided for by Federal Rule of Evidence 702, i.e., "technical, or other specialized knowledge."[52] For example, in *United States v. Starzecpyze1*,[53] the court declined to apply *Daubert* to forensic document examination (FDE) evidence. The court discussed the "lack of critical scholarship" in articles on FDE and the fact that although FDEs are accepted within their own community, this community is devoid of financially disinterested parties such as academics. The court concluded that "the testimony at the *Daubert* hearing firmly established that forensic document examination, despite the existence of a certification program, professional journals and other trappings of science, cannot, after *Daubert*, be regarded as 'scientific . . . knowledge.'" However, in a surprise twist, the court still deemed FDE evidence eligible for consideration as "technical, or other specialized knowledge" under Rule 702, irrespective of the *Daubert* test. Such non-scientific, or "skilled" testimony, can be admissible under Rule 702 and the general relevancy rules (Rules 104(a) and 403) without reference to the *Daubert* standard, so long as it satisfies the requirements of the Rules themselves.[54]

[50] See Jonathon Hoffman, *A Briefcase and an Opinion: Post-*Daubert *Expert Testimony – A Major Shift*, 22 Prod. Safety & Liab. Rep. (BNA) 379 (Apr. 8, 1994) (including a survey concluding that two-thirds of post-*Daubert* products liability cases which cited *Daubert* excluded expert testimony.)

[51] See Jay P. Kesan, *Note, An Autopsy of Scientific Evidence in a Post-*Daubert *World*, 84 Geo. L. J. 1985, 2013 (1996) (stating that "as one might reasonably expect from a more liberal admissibility requirement, post-*Daubert* courts are admitting more scientific evidence in civil and criminal cases.") *See also McKnight ex rel Ludwig v. Johnson Controls, Inc.*, 36 F.3d 1396 (8th Cir. 1994) (wherein opposing sides argued the opposite viewpoints as to whether *Daubert* makes expert testimony more or less readily admissible.)

[52] The Supreme Court opened the door for this interpretation by noting that its discussion of Rule 702 was limited to the scientific context because that was the type of evidence at issue in the case before it, while also recognizing that Rule 702 applies to technical or other specialized knowledge as well as scientific knowledge. *Daubert*, 509 U.S. at 590 n.8.

[53] 880 F. Supp. 1027 (S.D.N.Y. 1995).

[54] See also *Iacobelli Constr. V. County of Monroe*, 32 F.3d 19 (2nd Cir. 1994) (dealing with evidence from a geotechnical consultant and an underground-construction consultant and holding the *Daubert* standard inapplicable to such non-scientific evidence), and *United States v. Webb*, 115 F.3d 711, 716 (9th Cir. 1997) (holding that the *Daubert* test only applies to scientific and not technical or specialized expert knowledge); *McKendall v. Crown Control Corp.*, 122 F.3d 803 (9th Cir.1997) (same); *United States v. Jones*, 107 F.3d 1147, 1157 (6th Cir. 1997) (recognizing that a "lack of empirical evidence in the field of handwriting analysis" precluded admissibility under *Daubert*, but admitting this same evidence as specialized knowledge that would be helpful to the jury.)

This post-*Daubert* distinction between "scientific" evidence and "technical" or "specialized" expert opinion was also deemed important in *United States v. Bynum*.[55] The Court in *Bynum* stated:

> Scientific knowledge is generated through the scientific method – subjecting testable hypotheses to the crucible of experiment in an effort to disprove them. An opinion that defies testing, however defensible or deeply held, is not scientific.[56]

Under this view, it was methodology that is the key to separating scientific knowledge from technical or specialized knowledge. Simply stated, it is not so much what the conclusions are as it is how they are reached.[57]

The reasoning that led these courts, in the immediate aftermath of *Daubert*, to apply a different evidentiary standard to scientific knowledge as opposed to technical or specialized knowledge was, entirely properly, subject to much criticism. For instance, scientific evidence does not seem to be any more prone to misunderstanding or misuse than any other type of evidence that derives from a specialized field of expertise. And even assuming that a different standard should apply for scientific evidence, the line separating such information from other technical or specialized knowledge is anything but bright. The factors, for example, that spurred the *Starzecpyzel* court to hold forensic document examiner evidence non-scientific, are very subjective and amenable to the opposite conclusion, depending upon the eye of the beholder/judge in any given case. As a result, some courts reached quite confusing decisions, as in *United States v. Velasquez*,[58] in which the court opined that the *Daubert* test for scientific expert testimony does not apply to the field of handwriting analysis, yet, out of an "exercise of caution," applied *Daubert* anyway and concluded that the handwriting analysis evidence was sufficiently reliable to be admissible.

The Supreme Court resolved much of this confusion in *Kumho Tire Co. v. Carmichael*.[59] There, the Court held that a trial judge has the same gatekeeping function under Rule 702 in all instances of expert testimony, including testimony based not upon "scientific" knowledge but rather upon "technical" or "other specialized knowledge."[60] The Court also held that in all such cases "a

[55] 3 F.3d 769 (4th Cir. 1993). [56] Id., at 773.

[57] See Edward J. Imwinkelried, *The Daubert Decision*, Trial 60, 63 (Sept. 1993). See also Edward J. Imwinkelried, *The Next Step After* Daubert: *Developing a Similarly Epistemological Approach to Ensuring the Reliability of Nonscientific Expert Testimony*, 15 Cardozo L. Rev. 2271, 2281 (April 1994). [58] 64 F.3d 844, 850–51 (3rd Cir. 1995).

[59] 526 U.S. 137, 119 S.Ct. 1167, 143 L.Ed.2d 238 (1999). In *Kumho Tire*, the plaintiffs claimed that a manufacturing defect caused a tire to blow out, which, in turn, resulted in numerous injuries and one death. Plaintiffs' expert, who had a masters degree in mechanical engineering and 10 years work experience at Michelin America, Inc., as well as prior consulting experience in other tire blowout cases, gave his opinion that a manufacturing defect or design defect caused the plaintiffs' injuries. He based his opinion upon the combination of his knowledge of tire failures, a personal theory of the cause of tire failures, and his inspection of the tire at issue. There was some question as to the reliability of the expert's methodology. 119 S.Ct. 1171–72, 143 L.Ed.2d 247–48. [60] 119 S.Ct. 1174, 143 L.Ed.2d 251.

trial court should consider the specific factors identified in Daubert where they are reasonable measures of the reliability of expert testimony."[61]

The Court looked to the express language of Rule 702 as well as to the policy rationale underlying it, and reasoned as follows:

> This language makes no relevant distinction between "scientific" knowledge and "technical" or "other specialized" knowledge. It makes clear that any such knowledge might become the subject of expert testimony. In Daubert, the Court specified that it is the Rule's word "knowledge," not the words (like "scientific") that modify that word, that "establishes a standard of evidentiary reliability." Hence, as a matter of language, the Rule applies its reliability standard to all "scientific," "technical," or "other specialized" matters within its scope . . . Neither is the evidentiary rationale that underlay the Court's basic Daubert "gatekeeping" determination limited to "scientific" knowledge. Daubert pointed out that Federal Rules 702 and 703 grant expert witnesses testimonial latitude unavailable to other witnesses on the "assumption that the expert's opinion will have a reliable basis in the knowledge and experience of his discipline.". . . The Rules grant that latitude to all experts, not just to "scientific" ones. Finally, it would prove difficult, if not impossible, for judges to administer evidentiary rules under which a gatekeeping obligation depended upon a distinction between "scientific" knowledge and "technical" or "other specialized" knowledge. There is no clear line that divides the one from the others. Disciplines such as engineering rest upon scientific knowledge. Pure scientific theory itself may depend for its development upon observation and properly engineered machinery. And conceptual efforts to distinguish the two are unlikely to produce clear legal lines capable of application in particular cases.[62]

The *Kumho Tire* opinion went on to hold that "a trial judge determining the 'admissibility of an engineering expert's testimony' may consider several more specific factors that Daubert said might 'bear on' a judge's gatekeeping determination," as deemed appropriate by the judge in any given case.[63] Such additional factors, in "technical" or "other specialized" knowledge cases as well as "scientific" ones, may include whether a certain theory or technique can be or has been tested; whether it has been subjected to peer review and publication; whether the technique features a high known or potential rate of error and whether there are standards controlling the technique's operation; and whether the theory or technique enjoys general acceptance within a relevant scientific community.[64] The Court emphasized that the *Daubert* factors do not constitute a "definitive checklist or test" and that under *Daubert* the Rule 702 inquiry is "a flexible one,"[65] stating:

> We agree with the Solicitor General that "the factors identified in Daubert may or may not be pertinent in assessing reliability, depending on the nature of the issue, the expert's particular expertise, and the subject of his testimony." The conclusion, in our view, is that we can neither rule out, nor rule in, for all cases and for all time the

[61] 119 S.Ct. 1176, 143 L.Ed.2d 252.
[62] 119 S.Ct. 1174, 143 L.Ed.2d 250.
[63] 119 S.Ct. 1175, 143 L.Ed.2d 251.
[64] 119 S.Ct. 1175, 143 L.Ed.2d 251.
[65] Id.

applicability of the factors mentioned in Daubert, nor can we now do so for subsets of cases categorized by category of expert or by kind of evidence. Too much depends upon the particular circumstances of the particular case at issue.[66]

Thus, the *Kumho Tire* decision stressed the general applicability of the *Daubert* principles, and somewhat clarified the considerable discretion vested in trial judges.[67] In so supporting the latitude enjoyed by trial judges in their rulings on evidentiary matters, including the admissibility of expert opinion testimony of all types, the Court reiterated its post-*Daubert* holding in *General Electric Co. v. Joiner.*[68] The *Joiner* decision held that the deferential abuse-of-discretion standard is appropriate for use by Courts of Appeals in reviewing trial judge decisions to admit or exclude expert scientific testimony at trial.[69] The trial court in *Joiner* had concluded that the studies upon which the expert witnesses relied were insufficient to support their opinion testimony, and the Supreme Court held that the resultant decision to exclude the experts' testimony was not an abuse of discretion.[70]

Post-*Daubert* Courts of Appeals have interpreted its reliability rule to mean that a reliable expert opinion must be based on scientific, technical, or other specialized knowledge and not on belief or speculation, and that inferences must be derived using scientific or other valid methods. Reliability of specialized knowledge and methods for applying it to various circumstances may be indicated by testing, peer review, evaluation of rates of error, and general acceptability.[71] In reliance on the guidance in *Kumho Tire*, the *Daubert* analysis has been applied to such evidence as handwriting analysis,[72] the reliability of eyewitness identification,[73] statistical analysis of disparate impact in an employment discrimination case,[74] radiation dose-effect analysis,[75] the carcinogenicity of a milk additive,[76] the behavioral patterns of typical child

[66] 119 S.Ct. 1175, 143 L.Ed.2d 251–2 (citations omitted). The Court went so far as to hold that "some of Daubert's questions can help to evaluate the reliability even of experience-based testimony." 119 S.Ct. 1176, 143 L.Ed.2d 252.

[67] See generally, Kimberly M. Hrabosky, Note: Kumho Tire v. Carmichael: *Stretching* Daubert *Beyond Recognition*, 8 Geo. Mason L. Rev. 203 (1999).

[68] 522 U.S. 136 (1997).

[69] Id. at 141–43. See also *United States v. Gilliard*, 133 F.3d 809 (11th Cir.1998).

[70] Id. at 146–47.

[71] See *Oglesby v. General Motors Corp.*, 190 F.3d 244, 250 (4th Cir. 1999).

[72] See *United States v. Paul*, 175 F.3d 906, 909–12 (11th Cir. 1999).

[73] See *United States v. Hall*, 165 F.3d 1095, 1100–07 (7th Cir. 1999) (upholding exclusion of expert testimony as to the reliability of eyewitness identification, under *Daubert* factors).

[74] See *Munoz, et al., v. Peters*, 200 F.3d 291, 300–02 (5th Cir. 2000) (upholding refusal to allow expert testimony due to lack of reliability).

[75] See *In re TMI Litigation*, 193 F.3d 613 (3rd Cir. 1999) (extensive discussion of various examples of expert testimony concerning the health effects of the Three Mile Island nuclear power plant incident).

[76] See *National Bank of Commerce v. Associated Milk Producers, Inc.*, 191 F.3d 858, 862–65 (8th Cir. 1999) (noting that under *Daubert* a trial court's analysis of the proffered evidence is limited solely to principles and methodology, not on the conclusions that they generate. A court is not free to choose between the conflicting views of experts whose principles and methodology are reliable and relevant).

molesters,[77] the possibility of a causal link between silicone breast implants and cancer,[78] toxicological evidence as to dosage and effect of cocaine ingestion,[79] the typical behavior of street gang members,[80] the existence of a design defect in machinery,[81] the nature, organization, rules, jargon, and structure of organized crime,[82] the occurrence of seatbelt failure and its injury-related ramifications,[83] and the content validity of standardized tests,[84] among others.

[77] See *United States v. Romero*, 189 F.3d 576, 584–87 (7th Cir. 1999) (upholding admissibility of such expert testimony, although noting potential problems with "profiling" types of evidence).

[78] See *Allison v. McGhan Medical Corp.*, 184 F.3d 1300, 1310–22 (11th Cir. 1999) (upholding exclusion of testimony by three experts after extensive discussion of *Daubert* factors). The court stated:

> While meticulous Daubert inquiries may bring judges under criticism for donning white coats and making determinations that are outside their field of expertise, the Supreme Court has obviously deemed this less objectionable than dumping a barrage of questionable scientific evidence on a jury, who would likely be even less equipped than the judge to make reliability and relevance determinations and more likely than the judge to be awestruck by the expert's mystique. Also, a judge may enlist outside experts to assist in this sometimes very difficult decision. Using independent court-appointed experts may serve to quell the pseudo-scientist criticism.

Id. at 1310. See also *Curtis v. M&S Petroleum, Inc.*, 174 F.3d 661, 668–71 (5th Cir. 1999) (holding the trial court improperly excluded expert testimony as to the health effects of exposure to benzene; the Fifth Circuit found an adequate foundation in the record to support the expert's opinion testimony); *Kennedy v. Collagen Corp.*, 161 F.3d 1226, 1227–31 (9th Cir. 1998) (same, as to health effects of a particular type of injection for wrinkle removal in skin).

[79] See *Ruiz-Troche v. Pepsi-Cola*, 161 F.3d 77, 80–87 (1st Cir. 1998) (reversing the exclusion of expert testimony and related evidence; detailed discussion of *Daubert* and its meaning).

[80] See *United States v. Hankey*, 203 F.3d 1160 (9th Cir. 2000) (affirming the admission of gang-expert police testimony on rebuttal to counter defense claims). The Ninth Circuit noted that the trial judge had probed extensively into the expert's experience-based qualifications, and found this all that could have been required given the non-scientific nature of the evidence. In emphasizing the flexibility of the inquiry under *Daubert* and *Kumho Tire*, the court stated, "The Daubert factors (peer review, publication, potential error rate, etc.) simply are not applicable to this kind of testimony, whose reliability depends heavily on the knowledge and experience of the expert, rather than the methodology or theory behind it." Id.

[81] See *Padillas v. Stock-Gamco, Inc.*, 186 F.3d 412, 416–18 (3rd Cir. 1999) (reversing the trial court's summary judgment exclusion of expert evidence because the court failed to use proper procedure. The Third Circuit stated that the district court's analysis of the proferred evidence did not establish that the expert lacked "good grounds" for his opinions, but rather that they were "insufficiently explained and the reasons and foundations for them inadequately and perhaps confusingly explicated." If the district court was concerned with the factual dimensions of the expert evidence, it should have held an in limine hearing to assess the admissibility of the evidence, thereby giving the proponent of the evidence an opportunity to respond to the court's concerns. Id. at 418. See also *Jaurequi v. Carter Manufacturing Co.*, 173 F.3d 1076. 1081–85 (8th Cir. 1999) (upholding exclusion of alternative-design/failure to warn expert testimony as irrelevant and insufficiently reliable).

[82] See *United States v. Tocco*, 200 F.3d 401, 418 (6th Cir. 2000) (upholding the admission of expert testimony by an FBI agent).

[83] See *Clark v. Takata Corp.*, 192 F.3d 750, 755– (7th Cir. 1999) (upholding rejection of expert testimony as unreliable due to faulty methodology and unsupported conclusions).

[84] See *Bryant, et al., v. City of Chicago*, 200 F.3d 1092, 1098 (7th Cir. 2000).

In *Weisgram v. Marley Co.,*[85] a unanimous Supreme Court offered some practical advice for litigants in the post-*Daubert* era as follows:

> Since Daubert, . . . parties relying on expert evidence have had notice of the exacting standards of reliability such evidence must meet. It is implausible to suggest, post-Daubert, that parties will initially present less than their best expert evidence in the expectation of a second chance should their first try fail. We therefore find unconvincing Weisgram's fears that allowing courts of appeals to direct the entry of judgment for defendants will punish plaintiffs who could have shored up their cases by other means had they known their expert testimony would be found inadmissible. In this case, for example, although Weisgram was on notice every step of the way that Marley was challenging his experts, he made no attempt to add or substitute other evidence.[86]

The Court appeared to recognize the uncertainty that still surrounds the admissibility of expert testimony, both at trial and on appeal, and cautioned litigants to do everything feasible at the trial level to supplement expert testimony with other supporting evidence.[87] In other words, buy "insurance" against unfavorable judicial assessment of your expert testimony, lest your case become unsupportable when the key evidence is ruled inadmissible.

Whatever formal doctrinal framework courts assert they are following, in practice they tend to weigh the admissibility of scientific evidence by considering whatever factors seem appropriate to them, in a realistic approach to handling the evidence before them in the instant case. *Daubert* and *Kumho Tire* have provided a general framework within which courts must operate, but those decisions, as we have seen, stressed the great discretion vested in trial judges in ruling on expert/scientific evidence.[88] In chapter 7, we will examine the practical factors most frequently deemed important by the courts, and then we will evaluate the admissibility of forensic entomology evidence in light of those factors as well as under the *Frye* and *Daubert* tests.

Earlier, we alluded to the proposed amendment to Federal Rule of Evidence 702. The Advisory Committee's Note accompanying the proposed amendment states that the Rule is being changed in response to *Daubert* and "the many cases applying *Daubert,* including *Kumho Tire.*" The Note states, "In *Daubert* the Court charged trial judges with the responsibility of acting as gatekeepers to exclude unreliable expert testimony, and the Court in *Kumho* clarified that

[85] 120 S. Ct. 1011; 2000 U.S. LEXIS 1011; 145 L. Ed. 2d 958 (Feb. 22, 2000). In *Weisgram,* plaintiffs used three expert witnesses to establish the existence of a defect in a heater and its causal link to a fatal fire. Although allowed by the trial court, the Eighth Circuit ruled that the expert testimony was unreliable under *Daubert,* and, absent this key evidence, the defendant was entitled to judgment as a matter of law because the remaining evidence was insufficient to constitute a submissible case. See 169 F.3d 514, 517 (8th Cir.1999). The Supreme Court affirmed. [86] Id. at *27–28. (Citations omitted.) [87] Id.

[88] See generally, Michael H. Graham, *The Expert Witness Predicament: Determining "Reliable" Under the Gatekeeping Test of* Daubert, Kumho, *and Proposed Amended Rule 702 of the Federal Rules of Evidence,* 54 U. Miami L. Rev. 317 (2000).

this gatekeeper function applies to all expert testimony, not just testimony based in science." Additionally,

> The amendment affirms the trial court's role as gatekeeper and provides some general standards that the trial court must use to assess the reliability and helpfulness of proffered expert testimony. Consistently with *Kumho*, the Rule as amended provides that all types of expert testimony present questions of admissibility for the trial court in deciding whether the evidence is reliable and helpful. Consequently, the admissibility of all expert testimony is governed by the principles of Rule 104(a). Under that Rule, the proponent has the burden of establishing that the pertinent admissibility requirements are met by a preponderance of the evidence. *Daubert* set forth a non-exclusive checklist for trial courts to use in assessing the reliability of scientific expert testimony. . . . No attempt has been made to "codify" these specific factors. *Daubert* itself emphasized that the factors were neither exclusive nor dispositive. Other cases have recognized that not all of the specific Daubert factors can apply to every type of expert testimony. . . . The standards set forth in the amendment are broad enough to require consideration of any or all of the specific *Daubert* factors where appropriate.[89]

METHODS OF ANALYSIS IN NATIONS OTHER THAN THE UNITED STATES

As mentioned earlier, most nations other than the United States have not articulated a specific test for determining the admissibility of scientific evidence. To some extent, the American tests have influenced the law in other nations.[90] However unsettled and shifting the current legal situation may be, it should nonetheless be useful to discuss some of the methods that are employed around the world.

In England and Wales, it is essentially left for the trial judge to decide, in each case, whether the subject matter area in question has developed to such a point that a person with appropriate qualifications can give expert evidence about it.[91] There is no special test or threshold issue standard governing admissibility; all that is required is that the evidence meet the usual requirements of relevancy and helpfulness.[92] The determination of admissibility is treated as a finding of fact, not a finding of law, and as such has no generalized application to other cases. In other words, a holding by one trial court that a given scientific technique does or does not satisfy a particular test for admissibility has no status as legal authority beyond that specific case. Judges in other cases, or even the same judge in other cases, may freely reach the opposite conclusion as to admissibility and may use a completely different method of analysis. The issue must be taken up de novo in each case.[93] This approach, of course, is both

[89] Id. at 348.

[90] Sophia I. Gatowski, et al., *The Diffusion of Scientific Evidence: A Comparative Analysis of Admissibility Standards in Australia, Canada, England, and the United States, and Their Impact on Social and Behavioural Sciences*, 4 Expert Evid. 86 (1996).

[91] See Peter Alldridge, *Scientific Expertise and Comparative Criminal Procedure*, 3 Int'l J. Evid. & Proof 141, 154 (1999). [92] Id. [93] Id. at 155.

highly flexible and also somewhat inefficient and unpredictable.[94] It has also led to a relative paucity of jurisprudence in the English courts that might illuminate the issue of admissibility of scientific evidence.

Traditionally, English common law held that any witness accredited as an expert may testify on the subject of his or her expertise, subject only to requirements of "relevancy" and "helpfulness."[95] This is an extremely liberal rule, which generated some scholarly criticism as "benevolent acquiescence,"[96] but remained basically unchanged for many years.[97] One commentator summarized the situation as follows:

> The English law has proceeded on this question not according to any established principles – the authorities are bereft of clear suggestions as to what these might be – but by gradually accepting particular fields into the legal fold over a period of time, to the effective, though not in theory absolute, exclusion of those which have not benefitted from this process.[98]

In England and Wales, a trial judge, choosing whatever standard seems appropriate, must decide whether a field of learning has developed to a sufficient degree that a person of proper qualifications may give testimony concerning it.[99] There has generally not been much judicial scrutiny as to whether

[94] See A.J.P. Kenny, *The Expert in Court,* 99 L.Q.R. 197 (1983).

[95] See Peter Alldridge, *Forensic Science and Expert Evidence,* 21 J. L. & Soc'y 136, 144 (1994). There is also a requirement, analogous to Rule 403, that the evidence must possess probative value that outweighs any prejudicial effect. *See* D. Carson, *Some Legal Issues Affecting Novel Forms of Expert Evidence,* 1 Expert Evid. 79 (1992).

[96] See, Ian R. Freckelton, *Science and the Legal Culture,* 2 Expert Evid. 107, 110 (1993); R. Coleman and H. Walls, *The Evaluation of Scientific Evidence,* 1974 Crim. L. Rev.

[97] See CAROL A.G. JONES, EXPERT WITNESSES: SCIENCE, MEDICINE, AND THE PRACTICE OF LAW 18–95 (1994). Jones outlines the history of the use of expert knowledge in the English courtroom as follows. A "community of witnesses-cum-jurors" decided cases until 1833. Id. at 23. According to Jones, "It was perfectly possible and routine for trials to proceed without any witnesses at all. The jurors were the witnesses. Only when the dual role was distilled into two separate roles (witness and juror) do we see the development of a distinct witness role in English courts of law." Id. The original community jurors were sent out around the country to collect information about land and land related matters. Id. "Their task was to supply facts from their own private knowledge and/or information which they had gathered from their investigations." Id. at 23–24. However, the community jurors would sometimes need specific knowledge, and specialist jurors were used to investigate and report back. Id. at 25. In 1670, the Bushell case , which ended the practice of imprisoning dissident juries, prompted a change in legal rule. Id. at 31. The law needed the opinion evidence of expert witnesses but had restricted the province of witnesses to evidence of fact because to do otherwise would be to let in a threat to judicial control. To get itself out of this quandary, it devised a deceptively simple solution: it made expert witnesses an exception to the rule forbidding witnesses to give opinion evidence. From this point onwards experts became special sorts of witnesses. Only experts could give evidence of opinion. This is now the main legal rule about expert evidence. Id. at 33. Thus the expert witness replaced the specialized juror and court-appointed expert in England. Id. at 35.

[98] T. HODGKINSON, EXPERT EVIDENCE: LAW AND PRACTICE 131–32 (1990).

[99] Peter Alldridge, *Recognising Novel Scientific Techniques: DNA As a Test Case,* 1992 Crim. L. Rev. 687, 692.

a given theory or technique is reliable or whether it is accepted in the relevant scientific community.[100] On the other hand, expert testimony has traditionally been allowed only if the subject matter is not within the common knowledge of the jurors.[101]

There have been some notorious examples in which this liberal approach to admissibility has led to miscarriages of justice. *Preece v. H.M. Advocate*[102] featured a prosecution of Preece for the murder of a female hitchhiker. The prosecution proffered the testimony of a scientist who had examined seminal stains on the woman's pants and identified them as belonging to blood type A. The expert also stated that Preece had blood type A, but failed to mention that the victim was of the same blood type and that the stains probably included some of her secretions. The expert claimed that he could distinguish between male and female secretions, but his method had never been reported in the scientific literature. Preece was convicted by the jury, but the conviction was reversed on appeal upon a finding that the expert had not acted objectively and that all of the evidence he presented was suspect.[103]

English statutory law, partially in response to the *Preece* controversy, now gives trial judges in criminal cases wide discretion to exclude evidence, including expert testimony. Pursuant to Section 78 of the Police and Criminal Evidence Act of 1984, a judge may exclude evidence if it appears from the totality of the circumstances that its admission would adversely affect the fairness of the proceedings.[104] In practice, however, judges have rarely exercised this power to keep out forensic or scientific evidence and have continued to admit even highly questionable evidence, sometimes with tragic consequences.[105]

For example, in *R. v. Robb*,[106] the defendant was convicted based on expert testimony that identified his voice from tapes of conversations. The expert himself had testified that the weight of informed opinion agreed that his technique was unreliable, and the expert failed to describe the criteria on which he relied in forming his opinions.[107] However, the appellate court upheld the conviction, holding that neither general acceptance nor reliability is a precondition to the admissibility of expert scientific testimony, although it would not allow a "quack, a charlatan, or an enthusiastic amateur" to testify as an expert.[108] The

[100] Ian R. Freckelton, *Science and the Legal Culture*, 2 Expert Evid. 107, 110 (1993).

[101] David E. Bernstein, *Junk Science in the United States and the Commonwealth*, 21 Yale J. Int'l L. 123, 167 (1996); Sophia I. Gatowski, et al., *The Diffusion of Scientific Evidence: A Comparative Analysis of Admissibility Standards in Australia, Canada, England, and the United States, and Their Impact on Social and Behavioural Sciences*, 4 Expert Evid. 86, 89 (1996).

[102] 1981 Crim. L. Rev. 783.

[103] Id. *See* J. K. Mason, *Expert Evidence in the Adversarial System of Criminal Justice*, 26 Med. Sci. & L. 8, 9 (1986); John Phillips, *A Winter's Tale–"The Slings and Arrows of Expert Evidence"*, 57 Law Inst. J. 710, 710–11 (1983).

[104] D. J. GEE and J. K. MASON, THE COURTS AND THE DOCTOR 61 (1990).

[105] David E. Bernstein, *Junk Science in the United States and the Commonwealth*, 21 Yale J. Int'l L. 123, 168 (1996).

[106] 93 Crim. App. 161 (1991).

[107] Id.

[108] Id. at 166.

court stated that expert testimony is admissible if the expert is qualified by academic training and practical experience, and is able to provide testimony with a "value significantly greater than that of the ordinary untutored layman."[109]

The extensive controversy surrounding the *Preece* and *Robb* cases, as well as the "Birmingham Six"[110] and "Maguire Seven"[111] cases, sparked a movement to reform the standards.[112] Perhaps in response, some English courts have excluded questionable expert testimony in DNA cases.[113] For the most part, however, English courts do not employ a stringent test for admissibility of scientific evidence akin to any of the prominent American standards, and probably are more liberal than courts in either Australia or Canada.[114]

Canada, which uses the same basic ad hoc approach as England and Wales in its trial courts, has set forth some overall guidance regarding scientific evidence. More so than most nations other than the United States, Canada has some specific standards by which such evidence can be evaluated, although the issue has only been considered in detail in very few Canadian cases.[115] The Supreme Court of Canada summarized its analytical paradigm in the *Mohan* case:

> Admission of expert evidence depends on the application of the following criteria: (a) relevance; (b) necessity in assisting the trier of fact; (c) the absence of any exclusionary rule; and (d) a properly qualified expert. Relevance is a threshold requirement to be decided by the judge as a question of law. Logically relevant evidence may be excluded if its probative value is overborne by its prejudicial effect, if the time required is not commensurate with its value or if it can influence the trier of fact out of proportion to its reliability. The reliability versus effect factor has special significance in assessing the admissibility of expert evidence. Expert evidence should not be admitted where there is a danger that it will be misused or will distort the fact-finding process, or will confuse the jury.[116]

[109] Id. [110] *R. v. McIlkenny,* 93 Crim. App. 287 (1991). [111] *R. v. Maguire,* 1992 Q.B. 936.

[112] See CAROL A. JONES, EXPERT WITNESSES: SCIENCE, MEDICINE, AND THE PRACTICE OF LAW 3 (1994) (cataloging miscarriages of justice attributed to expert scientific testimony); Peter Alldridge, *Recognizing Novel Scientific Techniques: DNA As a Test Case,* 1992 Crim. L. Rev. 687, 698 (arguing that "English law needs a test for the admissibility of novel scientific evidence which considers the evidence itself as well as the witness."); BEVERLY STEVENTON, THE ROYAL COMMISSION ON CRIMINAL JUSTICE, THE ABILITY TO CHALLENGE DNA EVIDENCE 37 (1993).

[113] See William Brown, *DNA Fingerprinting Back in the Dock,* New Sci., March 6, 1993, at 14; Jennifer Callahan, Survey: *The Admissibility of DNA Evidence in the United States and England,* 19 Suffolk Transnat'l L. Rev. 537 (1996).

[114] David E. Bernstein, *Junk Science in the United States and the Commonwealth,* 21 Yale J. Int'l L. 123, 170 (1996).

[115] Sophia I. Gatowski, et al., *The Diffusion of Scientific Evidence: A Comparative Analysis of Admissibility Standards in Australia, Canada, England, and the United States, and Their Impact on Social and Behavioural Sciences,* 4 Expert Evid. 86, 87 (1996).

[116] *R. v. Mohan,* 114 D.L.R. 4th 419, 89 C.C.C. 3d 402 (Sup. Ct. Can. 1994). This approach essentially adopts the Rule 403 standard used in American courts. *See* P.B. Limbert, *Beyond the Rule in Mohan: A New Model for Assessing the Reliability of Scientific Evidence,* 54 Univ. Toronto L. J. 65 (1996). The Canadian Supreme Court also addressed the issue of scientific evidence three times before *Mohan,* without establishing a general test, in *R. v. Abbey,* 68 C.C.C. 2d 394 (1982), in *R. v. Beland,* 2 S.C.R. 398 (1987), and in *R. v. Lavallee,* 55 C.C.C. 3d 97

This method reflects the recent trend for Canadian courts to scrutinize scientific evidence carefully, on an ad hoc basis, without settling on a uniform test to use in weighing the admissibility of scientific evidence generally.[117] This has been a matter of controversy within the Canadian legal community for a number of years.[118] The lower courts have wrestled with the issue, at times either expressly adopting or rejecting what they perceived to be the rule used in the United States.[119]

One of the more thoughtful lower court decisions was *R. v. Johnston*,[120] which dealt with DNA profiling. The court ruled that "the appropriate test is whether the evidence is relevant and helpful . . . The key to understanding 'helpful' in this context lies in equating it to 'reliable.'"[121] The court then outlined 14 factors that should be considered in assessing whether novel scientific evidence is "helpful."[122] Some other lower courts later used some or all of the *Johnston* factors as guides to their analysis.[123]

Footnote 116 (*cont.*)
(1990). Each case was decided on its facts, leaving lower courts without an overarching standard of admissibility.

[117] See JOHN SOPINKA, ET AL., THE LAW OF EVIDENCE IN CANADA 567 (1992); S. Rosalind Baker, *A Critical Approach to the Admissibility and Weight of DNA Evidence in Canada,* 20 C.R. 4th 212, 213 (1993) ("Currently, the test for the admissibility of novel scientific evidence in Canada is unclear."); *R. v. Lafferty,* 80 C.C.C. 3d 150, 151 (N.W.T. Sup. Ct. 1993), in which the court stated, "There is no specific test for admissibility of novel scientific evidence such as the test of general acceptance by the scientific community which is used in some United States jurisdictions."

[118] See Ronda Bessner, *The Road Not Taken: The Refusal of the Supreme Court of Canada to Articulate a Test for the Admissibility of Polygraph Evidence,* 60 C.R. 3d 55, 56 (1990) (criticizing the failure of the Canadian Supreme Court to settle the important legal question as to the standard trial courts should use in assessing the admissibility of evidence derived from scientific and quasi-scientific techniques such as polygraph, voice spectrograph, ballistics, or hypnosis); David Paciocco, *Evaluating Expert Opinion Evidence for the Purpose of Determining Admissibility: Lessons from the Law of Evidence,* 27 C.R. 4th 302, 310 (1994) ("[*Frye*] is not the law of Canada. Exactly what the law is, is not quite as clear . . ."); Marie Lussier, *Tailoring the Rules of Admissibility: Genes and Canadian Criminal Law,* 71 Can. B. Rev. 319, 337 (1992) ("[N]o clear position has yet been confirmed in Canadian jurisprudence with respect to the admissibility of novel scientific evidence . . .").

[119] See Peter Alldridge, *Scientific Expertise and Comparative Criminal Procedure,* 3 Int'l J. Evid. & Proof 141, 142–4 (1999).

[120] 69 C.C.C. 3d 395 (Ontario Ct., Gen. Div. 1992).

[121] Id. at 414.

[122] The factors are: (1) the potential rate of error; (2) the existence and maintenance of standards; (3) the care with which the scientific technique has been employed and whether it is susceptible to abuse; (4) whether there are analogous relationships with other types of scientific techniques that are routinely admitted into evidence; (5) the presence of fail-safe characteristics; (6) the expert's qualifications and stature; (7) the existence of specialized literature; (8) the novelty of the technique in its relationship to more established areas of scientific analysis; (9) whether the technique has been generally accepted by experts in the field; (10) the nature and breadth of the inference adduced; (11) the clarity with which the technique may be explained; (12) the extent to which basic data may be verified by the court and jury; (13) the availability of other experts to evaluate the technique; and (14) the probative significance of the evidence. Id. at 415.

[123] Sophia I. Gatowski, et al., *The Diffusion of Scientific Evidence: A Comparative Analysis of*

On a nationwide basis, however, the subsequent *Mohan* decision essentially established legal relevance and assistance to the trier of fact as the main factors for weighing the admissibility of expert testimony, and used a general balancing test patterned after Rule 403 as the measure of relevance.[124] Although *Mohan* analyzed the underlying validity of expert evidence, akin to the *Daubert* approach, and relied heavily on the issue of general acceptance in the scientific community, reflective of *Frye*, the Court cited no American case law in its opinion. In the aftermath of *Mohan*, lower Canadian courts have generally rejected *Frye* while at times employing some variant of the *Daubert* factors.[125]

Australia has also found its legal system enmeshed in lengthy controversies concerning scientific evidence.[126] The Australian courts moved from a liberal relevancy test to a *Frye*-like requirement that scientific evidence be generally accepted as reliable before it can be admitted in court. Indeed, the 1912 case of *R. v. Parker*[127] which held that fingerprint evidence could be admissible if the theory was "generally recognized by scientific men" actually pre-dated *Frye*. It was not until the 1950s, however, that Australian courts in general began using an "organized field of expertise" general acceptance test for evaluating scientific evidence, coupled with a requirement for a qualified expert in the applicable field.[128]

The leading Australian case on scientific evidence, *Clark v. Ryan*,[129] is often cited as allowing the admission of expert scientific testimony when it is necessary to assist the trier of fact, and when it complies with "scientific principles." However, the opinion does not provide any guidance as to what those scientific principles are, nor how to evaluate the proffered evidence with regard to scientific techniques and methods.[130] This rather vague field of expertise/general

Admissibility Standards in Australia, Canada, England, and the United States, and Their Impact on Social and Behavioural Sciences, 4 Expert Evid. 86, 88 (1996).

[124] See Peter Alldridge, *Scientific Expertise and Comparative Criminal Procedure*, 3 Int'l J. Evid. & Proof 141, 145 (1999).

[125] Id. at 147–48. See generally, Pierre Patenaude, Article: De L'Expertise Judiciaire Dans le Cadre du Proces Criminel et de la Recherche de la Verite: Quelques Reflexions, 27 Revue de Droit de L'Universite de Sherbrooke 1 (1996/1997) (summarizing the state of the law in Canada and Europe regarding expert scientific evidence).

[126] See generally, Ian Freckelton, *Judicial Attitudes Toward Scientific Evidence: The Antipodean Experience*, 30 U.C. Davis L. Rev. 1137 (1997) (containing a detailed analysis of the law of scientific evidence in Australia and New Zealand and criticizing the liberal admissibility standard, or lack thereof, in both nations).

[127] V.L.R. 152, 154 (1912).

[128] See *Transport Publishing Co. v. Literature Bd. Of Rev.*, 99 C.L.R. 111, 119 (1956); *Burger King Corp. v. Registrar of Trade Marks*, 128 C.L.R. 417 (1973); *Mattioli v. Parker*, Q.R. 499 (1973).

[129] *Clark v. Ryan*, 103 C.L.R. 486, 491 (1960) (declaring that scientific evidence is only admissible when it "so far partakes of the nature of a science as to require a course of previous habit, or study, in order to obtain a knowledge of it").

[130] See Sophia I. Gatowski, et al., *The Diffusion of Scientific Evidence: A Comparative Analysis of Admissibility Standards in Australia, Canada, England, and the United States, and Their Impact on Social and Behavioural Sciences*, 4 Expert Evid. 86, 88–89 (1996).

acceptance test continued to be used at least into the late 1970s.[131] Similar to American Rule 403, the trial judge in criminal trials also has long had discretion to exclude evidence if the probative value was substantially outweighed by prejudicial impact.[132]

Public concern in Australia erupted in the late 1970s as a result of some heavily publicized cases in which scientific/forensic evidence was perceived as improperly used. These very controversial "miscarriage of justice cases" focused attention on the deficiencies in the general acceptance/organized field of expertise test.[133] Most notable was the *Chamberlain* case.[134] Known in the media as the Dingo Baby case, it involved a woman, Alice Lynne Chamberlain, who, while on a camping trip, claimed that she saw her baby being carried off by a dingo. Neither her baby nor the dingo was ever found, though the baby's clothes were recovered. The government charged Chamberlain with murder, alleging that she had cut her baby's throat in the family car, later disposed of the body, and made it appear that the clothes had been damaged by a wild dog . At trial, both sides presented complex and contradictory scientific evidence relevant to the source of the cuts on the baby's clothes, the ability of a dingo to carry a baby, and what appeared to be blood stains on the baby's clothing and in Chamberlain's car.

The trial judge instructed the jury, in essence, that they could decide the case based on their own common sense and how convincing they found the experts, even if they did not comprehend the scientific evidence. The jury convicted Chamberlain of murder. On appeal, the court rejected a challenge to the jury instructions and affirmed the conviction, going so far as to focus on the personalities of the expert witnesses.[135] On appeal to the High Court, the majority again affirmed, despite a strong dissent criticizing the prosecution's scientific evidence.[136]

131 See *R. v. Gilmore,* 2 N.S.W.L.R. 935, 938–39 (1977) (finding that voice spectrography "is a recognized field in which a properly qualified expert can give admissible evidence."). *Gilmore* did not adopt *Frye* explicitly, although some commentators interpreted it in that way. See also ANDREW L.C. LIGERTWOOD, AUSTRALIAN EVIDENCE, Sect. 7.40, at 378 (2nd ed. 1993); Oliver P. Holdenson, *The Admission of Expert Evidence of Opinion as to the Potential Unreliability of Evidence of Visual Identification,* 16 Melb. U.L. Rev. 521, 541 (1988); Ian R. Freckelton, *Novel Scientific Evidence: The Challenge of Tomorrow,* 3 Austl. Bar. Rev. 243, 246–47 (1987); Ian R. Freckelton, *Expert Evidence and the Role of the Jury,* 12 Austl. Bar. Rev. 73, 87–91 (1994).

132 See Ian R. Freckelton, *Science and the Legal Culture,* 2 EXPERT EVID. 107, 109 (1993).

133 See Judy Bourke, *Misapplied Science: Unreliability in Scientific Test Evidence,* 10 Austl. Bar. Rev. 123, 132–34 (1993); Paul Gerber, *Playing Dice With Expert Evidence: The Lessons to Emerge From* Regina v. Chamberlain, 147 Med. J. Austl. 243 (1987); KEN CRISPIN, THE DINGO BABY CASE (1987); IAN R. FRECKELTON, THE TRIAL OF THE EXPERT 159–64 (1987); ROYAL COMMISSION OF INQUIRY INTO CHAMBERLAIN CONVICTIONS, COMMONWEALTH PARLIAMENT PAPER No. 192 (1987).

134 *R. v. Chamberlain,* 46 A.L.R. 493, 505 (1983).

135 *Chamberlain v. R.,* 72 F.L.R. 1, 30 (Fed. Ct. N. Terr. 1983).

136 *Chamberlain v. R.,* 153 C.L.R. 521 (1984).

There was a Royal Commission established to reevaluate the use of scientific evidence and to reassess the result in the *Chamberlain* case. In light of new evidence that the baby's clothes had been cut by a dingo's teeth, and based on detailed criticism of the government's scientific evidence, the Commission advised the Northern Territory Court of Appeal to free Mrs. Chamberlain, which it did.[137] The Commission's work also spurred a movement within Australia to adopt more stringent standards for the admissibility of scientific evidence, which was reflected in some judicial decisions from the territorial Supreme Courts.[138] In the opinion of some commentators, *Frye* was becoming the law of Australia, although most of the courts had not made that explicit.[139]

In early 1995, a new federal code of evidence was passed by the Australian parliament, setting forth rules governing the admissibility of scientific evidence. As in the United States, the Australian states are not required to follow the federal code rules, although they may adopt them.[140] Section 79 of the code now specifies that expert evidence is admissible if a person "has specialized knowledge based on the person's training, study or experience," and if the testimony "is wholly or substantially based on that knowledge."[141] The code also contains Section 135, which is modeled on American Rule 403. On its face, the code presents a very broad, liberal rule for the admissibility of expert testimony, although it remains to be seen how courts will construe the terms of the test.[142]

[137] See Ian R. Freckelton, *Of Blood, Babies and Bathwater*, 17 Alternative L.J. 10, 11 (1992).

[138] See Paul Gerber, *Playing Dice With Expert Evidence: The Lessons to Emerge From* Regina v. Chamberlain, 147 Med. J. Austl. 243 (1987). After the *Chamberlain* scandal, courts that applied the field of expertise test sometimes applied other, more stringent, tests as well. See, e.g., *R. v. Runjanjic*, 53 A. Crim. R. 362, 366 (1991) (an essential prerequisite to the admission of expert evidence of battered women's syndrome is that "it be accepted by experts competent in the field of psychology or psychiatry as a scientifically established facet of psychology."); *Casley-Smith v. F.S. Evans & Sons Party Ltd.*, 49 S.A. St. R. 314, 320 (1988) (holding that scientific evidence is admissible only if it is "sufficiently organised or recognized to be accepted as a reliable body of knowledge or experience."); *R. v. Lewis*, 29 A. Crim. R. 267 (S.C.N.T. 1987) (holding that the prosecution, when presenting novel scientific evidence, has a duty to "demonstrate its scientific reliability" in light of the "established universal view" regarding the reliability of the applicable technique. This test goes beyond *Frye* in apparently calling for universal, rather than general, acceptance.).

[139] See Ian R. Freckelton, *DNA Profiling: Forensic Science Under the Microscope*, 14 Crim. L. J. 23, 30 (1990); *but see* James T. Kearney, *Genetic Identification: Not Always Watertight Evidence*, 48 L. Soc'y J. 48, 49 (1990) (stating that while the *Frye* test "is known in Australia . . . it could not be said to be the law in Australia."). One court, the South Australia Supreme Court, did explicitly adopt *Frye*, in *R. v. Jarrett*, (S.C.S.A. 1994) (holding that DNA tests were admissible because they were "recognized by the scientific community as reliable and have been developed to the stage whereby the results obtained by them may be accepted with confidence.)."

[140] David E. Bernstein, *Junk Science in the United States and the Commonwealth*, 21 Yale J. Int'l L. 123, 161 (1996).

[141] EVIDENCE ACT 1995, reprinted in CIVIL LAW DIVISION, ATTORNEY GENERAL'S DEPARTMENT, COMMONWEALTH EVIDENCE LAW 5–194 (1995).

[142] David E. Bernstein, *Junk Science in the United States and the Commonwealth*, 21 Yale J. Int'l L. 123, 161–62 (1996) (arguing that courts may be able to interpret the vague language of the rule to mandate a *Daubert* type of test, much as the American courts did with Rule 702).

For the moment, the Australian rules for the admissibility of scientific evidence remain in a state of confusion.[143] The general principles that can be discerned from the various strands of case law are, on the whole, both liberal and open to divergent interpretations.[144]

In New Zealand, there is very little case law or legal commentary regarding scientific evidence, and no overarching general test is in place.[145] As with Australia, however, there has been some support for a field of expertise test[146] as well as a general acceptance test and a liberal relevancy/helpfulness standard.[147]

The New Zealand Law Commission, pursuant to its responsibility to recommend amendments to New Zealand's evidence code, issued a paper in 1991 dealing with expert and opinion evidence.[148] The Law Commission examined in considerable detail the question of whether New Zealand should adopt the *Frye* standard, noting the amount of criticism that test had received in the United States.[149] Ultimately, the Law Commission found the *Frye* test unsatisfactory, and instead recommended that a test be developed that would be responsive to reliability concerns, yet flexible enough to cover a wide range of new and evolving situations.[150] No action has been taken to implement these recommendations.

In 1995, the High Court of New Zealand decided the case of *R. v. Calder*,[151] in which it examined the contradictory New Zealand precedents supporting the

143 Id. at 162. *See* Ian R. Freckelton, *Science and the Legal Culture*, 2 Expert Evid. 107, 110–11 (1993) (noting "in Australia there has been a confusion of focuses on both reliability of a technique and assessment of its repute within the relevant scientific community").

144 In an attempt to distill some guiding doctrinal principles from the cases, some commentators have suggested the following as the approach most often used in Australian courts. Scientific evidence (1) must derive from a "field of expertise"; (2) the witness must be an expert in that field; (3) the opinion must be relevant to the fact at issue; (4) the opinion must not be a matter of "common knowledge"; (5) the opinion must not be in respect of an "ultimate issue"; (6) the expert must disclose the facts upon which the opinion is based; (7) the facts upon which the opinion is based must be capable of proof by admissible evidence; (8) evidence must be admitted to prove the assumed facts upon which the opinion is based; and (9) if adduced against a criminal defendant, the evidence must be more probative than prejudicial. *See* Sophia I. Gatowski, et al., *The Diffusion of Scientific Evidence: A Comparative Analysis of Admissibility Standards in Australia, Canada, England, and the United States, and Their Impact on Social and Behavioural Sciences*, 4 Expert Evid. 86, 89 (1996).

145 David E. Bernstein, *Junk Science in the United States and the Commonwealth*, 21 Yale J. Int'l L. 123, 163 (1996).

146 See, e.g., *R. v. B.*, 1 N.Z.L.R. 362, 367–68 (C.A. 1987) (holding that "As a precondition of admissibility the subject matter to which the expert opinion relates must be a sufficiently recognised branch of science at the time the evidence is given," and a field of expertise will be recognized when "research establishes the accuracy of knowledge in that field."); *R. v. Accused*, 1 N.Z.L.R. 714 (C.A. 1989) (applying the field of expertise test to exclude evidence of "child sexual abuse accommodation syndrome.").

147 See generally, Ian Freckelton, *Judicial Attitudes Toward Scientific Evidence: The Antipodean Experience*, 30 U.C. Davis L. Rev. 1137 (1997).

148 N.Z. LAW COMMISSION PRELIMINARY PAPER No. 18, EVIDENCE LAW: EXPERT EVIDENCE AND OPINION EVIDENCE (1991). 149 Id. at 18. 150 Id.

151 David E. Bernstein, *Junk Science in the United States and the Commonwealth*, 21 Yale J. Int'l L. 123, 165–66 (1996).

relevancy, general acceptance, and field of expertise tests, and also looked to the rules for admissibility of scientific evidence in England, Australia, Canada, and the United States. The Court concluded:

> Before expert evidence, such as that in issue in this case, can be put before the jury by a suitable qualified person it must be shown to be both relevant and helpful. To be relevant the evidence must logically tend to show that a fact in issue is more or less likely. To be helpful the evidence must pass a threshold which can conveniently be called the minimum threshold of reliability. This means the proponent of the evidence must show that it has a sufficient claim to reliability to be admitted.[152]

Thus, New Zealand has a flexible, reliability-based test similar to that of Canada. The Court recognized that it was setting forth a rather vague and general standard, but deemed that acceptable, given the need for flexibility as judges perform their gatekeeper function in dissimilar cases.

Countries in the civil law European legal tradition, such as The Netherlands, Belgium, Germany, France, and Switzerland, tend to set very low thresholds for the admissibility of expert evidence.[153] They prefer to regulate how the expert evidence, which is admitted, is assessed.[154] Furthermore, "[c]ontinental

[152] Id. at 166.

[153] Civil trials in civil law countries are profoundly different from their common law counterparts. First, a civil proceeding in a civil law country is actually a series of isolated meetings of and written communications between counsel and the judge in which evidence is introduced, testimony is given and procedural motions and rulings are made. In contrast, in the United States, although the series of meetings and written communications between counsel share some similarities to the discovery process, the presentation of evidence is usually concentrated in one proceeding. Second, in civil law countries, evidence is received and a summary record is prepared by someone other than the judge who will decide the case, whereas, in common law countries, the same individual performs both tasks. The handling of testifying witnesses is also a major contrast between civil and common law systems. In civil law systems, testifying witnesses are questioned by the hearing judge rather than by counsel for the parties. Furthermore, witnesses are not subject to intense cross-examination as in common law systems. The hearing judge makes notes of a witness' testimony and dictates a summary to the clerk. After the witness and lawyers agree about the accuracy of the summary, the summary will enter the record that goes to the deciding panel of judges. Since the deciding judges are never afforded an opportunity to observe a testifying witness' demeanor, sincerity, or recollection, and since interested witness' parties, relatives, and interested third persons are disqualified from testifying as witnesses, there is little need to discredit witnesses. The criminal trial in civil law jurisdictions is also different in character from the common law trial. In civil law jurisdictions, the main function of the criminal trial is to present the evidence to the judge and jury and to allow the prosecutor and the defendant's counsel to argue their cases, whereas in common law countries a major function of the trial is to make a record for appellate review. Although witnesses, including the defendant, can be questioned, as in civil proceedings, the questions are put to the witness by the judge and there is no developed system of cross-examination comparable to common law jurisdictions. Furthermore, like civil proceedings, there are virtually no exclusionary rules of evidence. Although there are juries in civil law criminal proceedings, these juries are usually composed of a mixture of professional judges and lay jurors. Kenneth Williams, *Do We Really Need the Federal Rules of Evidence?* 74 N. Dak. L. Rev. 1, 14–15 (1998).

[154] See HANS NIJBOER, *EXPERT EVIDENCE*, in HANDBOOK OF PSYCHOLOGY IN LEGAL CONTEXTS 557–59 (Ray Bull & David Carson eds., 1995).

European countries consider the expert, when testifying, not as a special kind of witness but rather as just a witness . . . In such countries court-appointed experts are much more common than in common law systems such as England and Wales."[155] The civil law systems feature some variation in approach. For example, France uses a system of wide open admissibility, which invests total discretion to the trial judge, while Germany uses a system involving a limited number of exclusionary rules with discretion for the most part remaining with the trial judge.[156] France literally has no rules regarding the form or scope of admissible evidence, leaving the matter entirely within the discretion of the judge.[157]

When controverted issues of fact arise, particularly facts of a technical nature, the French practice will almost in variably leave the resolution of them to an expert appointed by the court.[158] The court-appointed expert will conduct an investigation outside the court room, under no formal rules of evidence or relevance, at sessions to which the parties are convoked with full freedom to present their views and those of their experts or other representatives, orally or in writing. The result of the expertise is a report that in principle the judge need not accept, but in the absence of other evidence, almost invariably does, provided that the judge is satisfied that the expert has done what he or she was commissioned to do and that no material procedural irregularities have been committed in the course of the expertise.[159]

The German system is somewhere between the French system, which routinely admits all evidence, and the American system, with its detailed exclusionary rules. As with France, there is a general rule of admissibility in Germany, which states, "In order to search out the truth the court shall on its own motion extend the taking of evidence to all facts and means of proof that are important for the decision."[160] The Germans prefer their system of limited exclusionary rules as opposed to the more detailed alternative used by the United States. They believe that the potential damaging effects of inflammatory evidence which are, for the most part, excluded in the United States can be overcome by having professional judges serve on juries with lay persons and by requiring that the triers of fact state reasons for their results.[161]

[155] Id. at 561. See generally M. Neil Browne, et al., *The Epistemological Role of Expert Witnesses and Toxic Torts,* 36 Am. Bus. L. J. 1 (1998) (comparing the rules governing expert witnesses in several European nations).

[156] Kenneth Williams, *Do We Really Need the Federal Rules of Evidence?* 74 N. Dak. L. Rev. 1, 13 (1998).

[157] See RENE DAVID, FRENCH LAW 146 (1972); GERALD KOCK and RICHARD FRASE, THE FRENCH CODE OF CRIMINAL PROCEDURE 199 (rev. ed. 1988); Richard S. Frase, *Comparative Criminal Justice as a Guide to American Law Reform: How Do the French Do it, How Can We Find Out and Why Should We Care?,* 78 Calif. L. Rev. 539, 677 (1990).

[158] Richard W. Hulbert, *Comment on French Civil Procedure,* 45 Am. J. Comp. L. 747, 749 (1997).

[159] Id.

[160] See JOHN H. LANGBEIN, COMPARATIVE CRIMINAL PROCEDURE: GERMANY 69 (1977).

[161] Kenneth Williams, *Do We Really Need the Federal Rules of Evidence?* 74 N. Dak. L. Rev. 1, 19–20 (1998).

7

The Admissibility of Forensic Entomology Evidence

The key question in any case in which a party intends to offer forensic entomology evidence is whether that evidence will be admissible at trial. Questions as to the weight which will be given to the evidence and the most effective method for presenting it to the judge and jury, while also important, are secondary to this threshold issue, because unless the evidence is admitted, there can be no further use of it at trial.

No one owns an infallible crystal ball that will predict with absolute certainty what any given judge will do when presented with any given piece of evidence in any given procedural and factual litigation context. This is especially true in the realm of scientific evidence. We have glimpsed the decades-long struggles in the courts at various levels even to articulate a coherent doctrinal framework for analyzing the admissibility of scientific evidence. The actual application of the law to the facts in any particular case, including one involving forensic entomology evidence, is all the more fraught with opportunities for the idiosyncracies of the judge and the specific facts and equities unique to that case to tip the balance for or against admissibility. Still, the actions of many judges over many years have supplied us with considerable information as to how judges actually decide whether to admit scientific evidence.

It is apparent from a systematic reading of the cases that, irrespective of whether courts purport to use the *Frye* general acceptance test, the *Daubert* relevancy plus reliability standard, or some other doctrinal framework, in actuality they evaluate the admissibility of scientific evidence on the basis of several practical factors.[1] Although it is still important to know which standard a particular court has adopted, so that one may present the evidence in such a way as to tailor it to the elements of that standard and use the appropriate key words and phrases in argument to the court, it is more crucial on a pragmatic level to be attuned to the factors that most often are pivotal in the minds of judges.

We will look at each of these factors in the context of the types of scientific (or other expert) evidence at issue in the key cases. Within nations other than the United States as well, the factors should also be important, although in general these countries tend to be more liberal than the United States in admitting scientific evidence. This exercise will prepare us then to evaluate the prospects for admissibility of forensic entomology evidence in particular, coupled

[1] Developments in the Law, *Confronting the New Challenges of Scientific Evidence*, 108 Harv. L. Rev. 1481, 1498–1509 (1995).

with reference to the cases in which such evidence was actually before the courts.

Degree of Subjectivity

The greater the extent to which a particular theory, method, technique, or procedure depends on subjective "judgment calls" by a human being, the more reluctant judges are to rule the evidence derived therefrom admissible in court, all else being equal. Judges have often expressed concern in their judicial opinions about heavily subjective evidence, especially in the area of "scientific" evidence, because they fear that such evidence will inappropriately enjoy an elevated status in the minds of the jurors. The fear is that evidence which is in essence mostly the subjective opinion of one person will effectively be camouflaged by the trappings of scientific objectivity and be accorded more weight by the jury than it deserves.

Lawyers and the experts they hire are well aware of this potential for enhanced impact by evidence that at least outwardly appears to be scientific. They tailor their in-court presentations accordingly. Liberal use of charts, graphs, tables of numbers, diagrams, probability figures, sophisticated-looking equipment, and other similar accouterments of science can influence a lay jury to consider evidence to be more objective and reliable than it actually is. Even if the profusion of figures and complex exhibits is unlikely to be comprehensible to the jurors, the philosophy seems to be one of seeking to impress rather than to express; overwhelm the jurors with numbers and they will simply defer to the experts to tell them what it all means. Judges often attempt to counterbalance this propensity by leaning against admissibility in those cases where they believe pseudo-scientific window-dressing is being used to disguise a subjective guess.

From *Frye* itself on, the archetypal example of this principle has been polygraph evidence, in all its various forms. Many courts have excluded polygraph evidence, usually on the basis that it really amounts to an individual's subjective opinion that someone else was or was not truthful when answering certain questions, yet has the potential to mislead the jury because of the myriad charts, numbers, and impressive-looking pieces of apparatus that come along with and dress up that opinion.[2] The Supreme Court has noted that the "aura of infallibility" surrounding polygraph evidence "clothed . . . in scientific expertise" can mislead jurors into giving it more weight than is warranted.[3]

[2] See, e.g., *United States v. Alexander*, 526 F.2d 161, 167–70 (8th Cir. 1975); *United States v. Black*, 831 F. Supp. 120, 123 (E.D.N.Y. 1993); *United States v. Lech*, 895 F. Supp. 582, 585–6 (S.D.N.Y. 1995). However, a few courts have found polygraph evidence admissible. See, e.g., *United States v. Ridling*, 350 F. Supp. 90 (E.D. Mich. 1972); *United States v. Hart*, 344 F. Supp. 522 (E.D.N.Y. 1971). Recently, some courts have cited significant advancements in polygraph instrumentation and technique and have revisited and overturned earlier decisions that had ruled polygraph evidence per se inadmissible for all purposes. See *United States v. Posado*, 57 F.3d 428, 434 (5th Cir. 1995).

[3] *United States v. Scheffer*, 523 U.S. 303, 313–14 (1998).

Other types of at least quasi-scientific evidence have also suffered from judicial wariness as to their inherently subjective nature. Testimony from mental health professionals as to rape trauma syndrome, offered to establish that the victim in a rape trial behaved in a way consistent with someone who had in fact been raped, has been ruled inadmissible on this basis. Courts have stated that such evidence would create "an aura of special reliability and trustworthiness" and give "a stamp of scientific legitimacy to the truth of the complaining witness's factual testimony."[4] Judges have viewed rape trauma syndrome evidence as amounting to one person's subjective belief as to another person's veracity, albeit cloaked in expert scientific terms and credentials, and have tended to reject it for that reason.[5]

Another example of this type of judicial approach is the case of hypnotically refreshed testimony. One court excluded such evidence because the "scientific aura" surrounding the hypnosis process might be seen as placing a scientific imprimatur on the testimony and thus carried the risk of generating undue confidence in the credibility of the witness who had been hypnotized.[6]

Similarly, voiceprint evidence has been excluded by some courts for the same reasons.[7] The procedure is at bottom one person's view that two voice samples were spoken by the same individual, but this subjective opinion is shrouded in scientific garb. An impressive-looking mechanical device produces numbers, measurements, and graphs that, like polygraph charts, then are used by a person to form the basis of an opinion. One court declared that "the apparent objectivity of the machine may suggest a degree of certainty inconsistent with the subjective aspects of the enterprise," and indicated that courts should be careful when dealing with "scientific techniques in which highly subjective judgments are based upon the data received from sophisticated mechanical devices."[8]

In sum, judicial caution in this area stems from what might be termed the "Wizard of Oz" effect. A dazzling and complex array of mechanical devices, literally including smoke and mirrors, can cause a very ordinary human being to appear to be larger than life and powerful out of all proportion to reality. The potential for abuse inherent in such a distortion and magnification of one person's power is obvious. People who are not aware of the artifice may be awed by the phenomenal display of gadgetry and pyrotechnics, and may wrongly attribute enormous knowledge, wisdom, and capabilities to the "wonderful Wizard." Judges, sensitive to their duty as gatekeepers of the evidence, have been understandably reluctant to "ignore that man behind the screen."

[4] *State v. Saldana,* 324 N.W.2d 227, 231 (Minn. 1982), quoting *People v. Izzo,* 282 N.W.2d 10, 11 (Mich. Ct. App. 1979) (internal quotation marks omitted).

[5] *People v. Bledsoe,* 681 P.2d 291, 301 (Cal. 1984).

[6] *State v. Peoples,* 319 S.E.2d 177, 184–85 (N.C. 1984).

[7] See, e.g., *United States v. Addison,* 498 F.2d 741,744 (D.C. Cir. 1974); *People v. Kelly,* 549 P.2d 1240, 1243, 1245 (Cal. 1976).

[8] *Reed v. Maryland,* 391 A.2d 364, 370 (Md. 1978).

Opportunity to Decide for Oneself

Judges tend to be more amenable to admitting "scientific" evidence if the judge and the jury have the opportunity to examine the "raw materials" that form the basis for the expert's opinion on their own, with some hope of grasping the principles sufficiently to be able to decide for themselves whether the expert correctly interpreted the evidence. A chance to double-check the expert's work and conclusions is an appealing antidote to the "Wizard of Oz" effect previously mentioned. It affords the judge and jury the chance to look past the smoke and mirrors and see for themselves whether there is anything of value behind the wondrous display of wizardry. Sometimes, there might be real substance behind it all, but there also might be nothing but more facade behind the facade. In either event, judges are understandably more willing to admit unfamiliar evidence when the triers of fact can do some checking on their own and need not accept the findings and opinions of the expert witness uncritically, as if they were a matter of blind faith.

This factor has two aspects upon which it depends. First, there must be an in-court display of the underlying facts or objects for the triers of fact to observe, examine, and study. This will usually involve some form of demonstrative evidence in which the key evidentiary materials are organized and presented in a manner easily accessible by all of the jurors. For visually accessible evidence, photographic enlargements are typically employed to aid the jury in seeing the crucial features of the evidence.

Second, upon receiving a reasonable amount of in-court instruction on the principles involved, the triers of fact must be perceived by the judge as having a reasonable prospect for understanding them and applying them to the underlying facts or objects. If the subject matter can be explained clearly and logically, and does not appear to be so abstruse as to be beyond the ability of average jurors to comprehend, judges will probably deem this second prong to be satisfied.

The following examples will illustrate how this opportunity to decide for oneself can be an important factor in alleviating the hesitancy of many judges to admit novel scientific evidence. Even types of evidence with a less than sterling scientific pedigree have been accepted with a boost from the "seeing is believing" principle.

Bite mark evidence is one instance where the opportunity to look at the samples and make up one's own mind has been a key aspect of the decision to allow the evidence. In *People v. Marx*,[9] the court favorably compared dentition evidence to polygraph and other evidence, stating

> What is significantly different about the evidence in this case is this: the trier of fact... was shown models, photographs, X-rays and dozens of slides of the victim's wounds and

[9] 126 Cal. Rptr. 350 (Cal. Ct. App. 1975).

defendant's teeth . . . Thus the basic data on which the experts based their conclusions were verifiable by the court.[10]

The questionable scientific underpinnings of dentition evidence aside, it is debatable whether lay jurors are indeed capable of drawing reasonable conclusions from looking at photographic enlargements of teeth and bite marks, even after in-court "schooling." Nonetheless, the court thought it was significant, and thus it is significant for our purposes.[11]

An extreme and unfortunate example of this judicial tendency to be swayed by the opportunity for in-court examination of the evidence was *North Carolina v. Bullard*,[12] a murder case in which the defendant was convicted based in part on the testimony of a physical anthropologist that a bare footprint found at the scene of the crime matched that of the defendant. The court allowed testimony that purported to match the defendant's footprint to the footprint found at the crime scene, based exclusively on a visual comparison of their sizes and shapes. Despite the witness's admitted lack of formal training in footprint identification and the evident absence of anyone else in the United States who attempted to use or rely upon this type of analysis, the court was swayed by her detailed in-court methodology and demonstrations. The court pointed out that the witness "used photographs, models, slides and overlays that were before the court and verifiable by the jury."[13] Distinguishing this evidence from inadmissible polygraph results, the court stressed that "the polygraph, unlike the method employed in bite mark and barefoot print analyses, does not employ visual comparisons that are comprehensible to a jury,"[14] and "unlike the methods of hypnosis and polygraph testing . . ., the Court here is dealing with a scientific method which can be considered reliable based on the testimony of the expert while displaying to the jury visual aids used in making observable visual comparisons."[15]

Although the court's reasoning was questionable, the important point for our purposes remains the court's reliance on an opportunity for the triers of fact to see the evidence and decide for themselves. This "seeing is believing" tendency has manifested itself in a wide variety of other controversial evidentiary contexts as well, including voiceprint analysis, where the "hearing and seeing is believing" principle has sometimes carried the day. Courts that allow voiceprint evidence have made reference to the advantages of the opportunity for the jurors to hear the voices in question, see the spectrographs, and decide for themselves whether they agree with the expert's conclusion as to a match or non-match. For example, one court stated that expert testimony on voiceprints

[10] Id., at 356. Note, however, that polygraph evidence also can be presented in such a manner that the charts are available for in-court examination by the jury, just as with bite mark evidence.

[11] Another case following the same line of reasoning for bite mark evidence was *North Carolina v. Temple*, 273 S.E.2d 273, 279–80 (N.C. 1981).

[12] 322 S.E.2d 370 (N.C. 1984).

[13] Id., at 382.

[14] Id., at 383.

[15] Id., at 384.

"was not likely to mislead the jury," because "the jury had the opportunity in this case to hear both the voice exemplars and the recorded conversations, and to see the spectrograms."[16]

Attorneys and experts who are aware of the "seeing is believing" judicial tendency can use it to enhance their chances of gaining admission of certain types of evidence at trial. Even the oft-maligned polygraph evidence can be presented in such a way as to take advantage of this concept. Some judges would be more likely to allow polygraph evidence if the proponents availed themselves of enlarged photographs of the charts in question, side-by-side comparisons of "no deception indicated" and "deception indicated" samples, and perhaps in-court demonstration of how the equipment is used. Moreover, even if these in-court demonstrations are not necessary to secure admissibility of the evidence, it is still a good idea to use them, simply because they are apt to be more persuasive and memorable for the jury.

"Ultimate Issue" Reluctance

Another thread running through the cases is the reluctance of many courts to admit scientific evidence if doing so would directly determine the ultimate issue of guilt or innocence in the case. Of course, in order to be relevant, all evidence must in some way relate to at least one aspect of the ultimate issue at trial, but judges are especially hesitant to allow novel scientific evidence if it would seem to be outcome determinative. Conversely, they are more likely to admit such evidence in those instances in which it is offered for purposes of establishing an intermediate issue.

Judges are usually very protective of the province of the jury and the judge at trial. Expert testimony that appears to decide the case for the triers of fact tends to set off alarm bells in the minds of many judges. In part this judicial reluctance may be a remnant of an old rule, abolished by Federal Rule of Evidence 704(a), that opinion testimony was per se inadmissible to establish the ultimate issue in a case.[17] Federal courts, bound by the Federal Rules of Evidence, still may be influenced by the former rule in some implicit sense, while State courts, which are not bound by the federal rules, often explicitly state their wariness of ultimate issue opinion evidence.

For example, some federal courts that have excluded polygraph evidence have mentioned that such evidence would be determinative on the ultimate issue of guilt or innocence, even after enactment of the Federal Rules of

[16] *United States v. Smith*, 869 F.2d 348, 354 (7th Cir. 1989). See also *United States v. Williams*, 583 F.2d 1194, 1199 (2nd Cir. 1978), cert. denied, 439 U.S. 1117 (1979) in which the court declared that "it is not expected that the jury will participate only as passive listeners," because "the objective components supplied by spectrography are subject to direct evaluation by the jury."

[17] Note, however, that Fed. R. Evid. 704(b) still prohibits opinion testimony on the ultimate issue of whether the defendant in a criminal case had the requisite state of mind to commit the crime.

Evidence.[18] They fear that such evidence, which purports to reveal whether the defendant was telling the truth in denying guilt, is apt to be outcome determinative at trial, because once that question is answered, not much else of consequence remains to be concluded. These courts have distinguished polygraphs from other scientific evidence that tends rather to establish more intermediate propositions and thus does not so directly threaten to decide the case for the jury.[19]

Similarly, courts have hesitated to allow testimony regarding rape trauma syndrome on the basis of its direct impact on the ultimate issue of whether a rape did in fact occur. In *People v. Bledsoe,*[20] the court held rape trauma syndrome evidence inadmissible for purposes of establishing the ultimate issue, but stated that in other contexts this evidence "may play a particularly useful role by disabusing the jury of some widely held misconceptions about rape and rape victims, so that it may evaluate the evidence free of the constraints of popular myths."[21] The court went on to explain that

> [O]ur conclusion in this regard is not intended to suggest that rape trauma syndrome is not generally recognized or used in the general scientific community from which it arose, but only that it is not relied on in that community for the purpose for which the prosecution sought to use it in this case, namely, to prove that a rape in fact occurred.[22]

This language shows how the court was using something more than just the *Frye* test itself to determine admissibility. Other, practical factors made the difference.

When such evidence is offered only in rebuttal, courts may be more inclined to allow it. Some courts have admitted rape trauma syndrome evidence in cases where the defense attacks the complainant's credibility by proving that the complainant delayed in making a report of the incident, made inconsistent statements about it, came back to the scene of the incident to collect possessions, or continued to maintain some relationship with the defendant even after the alleged rape. In such cases, judges have allowed the prosecution to use rape trauma syndrome evidence in rebuttal of this type of evidence, to show that such behaviors are in fact consonant with the typical behavior of a rape victim.[23]

To the extent novel scientific evidence can be presented in such a way as to at least appear to be focused on something less than the ultimate issue in a case, courts will tend to be more apt to allow it. Attorneys and their experts can learn from this tendency and tailor their presentations accordingly. Some judges have actually stated explicitly that, while excluding evidence offered to

[18] The Supreme Court appeared to give some weight to this concern in *United States v. Scheffer,* 523 U.S. 303, 313–14 (1998).

[19] See, e.g., *United States v. Alexander,* 526 F.2d 161, 168–70 (8th Cir. 1975).

[20] *People v. Bledsoe,* 681 P.2d 291 (Cal. 1984). [21] Id., at 301. [22] Id.

[23] See. e.g., *State v. Middleton,* 657 P.2d 1215, 1219–20 (Or. 1983).

prove an ultimate issue, they would have admitted that same evidence to prove intermediate facts from which the jury might draw its own conclusions.[24]

To the extent that "ultimate issue reluctance" derives from judges' insistence on protecting the province of the jury as triers of fact, it also helps to explain some cases in which polygraph evidence and other "witness credibility" evidence has been rejected.[25] Polygraph cases are the classic example of evidence that has been rejected due to fears of "invading the province of the jury."[26] Similarly, courts have hesitated to allow expert testimony as to the fallibility of eyewitness identifications. These courts consider issues of witness credibility to be properly within the province of the jury, and will not permit evidence such as this to invade that province.[27]

With regard to eyewitness unreliability evidence, the Eighth Circuit's approach is representative of the stance most courts have taken. Although the Court also cited its inability to assess the scientific validity of the evidence, the court based its ruling primarily on the principle that the "evaluation of eyewitness testimony is for the jury alone."[28] The Court was concerned that the expert was proposing to testify about "the inherently untrustworthy manner with which . . . [the eyewitness] . . . identified . . . [the defendant] . . . in court," rather than about general problems inherent in eyewitness testimony. The Court held that this was an improper intrusion into the domain of the jury.

Amount of Corroborating Evidence

A key factor that can tip the scales for or against admissibility of scientific evidence, particularly of a novel sort, is whether there is other information available in the case that would corroborate the scientific evidence. Judges are quite naturally more comfortable with allowing unusual evidence if it does not stand alone and lonely in support of the proposition for which it is offered. The presence of a reasonable amount of corroborating evidence increases the likelihood that novel scientific evidence will be admitted,

[24] See, e.g., *United States v. DiDomenico*, 985 F.2d 1159, 1164–65 (2nd Cir. 1993), in which the Court upheld on appeal the trial judge's exclusion of "dependent personality disorder" evidence which had been offered to establish the ultimate fact that the defendant lacked the requisite mental state to commit the crime. The Court stated that this evidence would be admissible to establish intermediate facts that would have allowed the jury to reach its own conclusions as to the defendant's state of mind.

[25] See, e.g., *United States v. Watson*, 587 F.2d 365, 368–69 (7th Cir. 1978).

[26] See, e.g., *United States v. Alexander*, 526 F.2d 161, 169 (8th Cir. 1975). In that case, the court excluded polygraph evidence and asserted that, if polygraph evidence were admitted, the role of the jury would be "much more circumscribed." See also *United States v. Scheffer*, 523 U.S. 303, 311–12 (1998).

[27] See, e.g., *United States v. Hall*, 165 F.3d 1095, 1100–07 (7th Cir. 1999); *United States v. Hudson*, 884 F.2d 1016, 1023–24 (7th Cir. 1989), cert. denied, 496 U.S. 939 (1990); *United States v. Brown*, 540 F.2d 1048, 1054 (10th Cir. 1976), cert. denied, 429 U.S. 1100 (1977); *United States v. Brown*, 501 F.2d 146, 150–51 (9th Cir. 1974), rev'd on other grounds, 422 U.S. 225 (1975); *United States v. Amaral*, 488 F.2d 1148, 1153 (9th Cir. 1973).

[28] *United States v. Kime*, 99 F.3d 870, 883–84 (8th Cir. 1996).

because it enhances the judge's confidence that the correct result will be reached in the case.

Judges are risk-averse on matters that could lead to erroneous results and reversals on appeal, for a variety of reasons both noble and self-serving, and corroborating evidence tends to allay judicial fears of being wrong. Where the weight of the evidence in a specific case is heavily on the same side as the proposition which the novel scientific evidence is offered to support, any problem with the novel evidence is likely to be ruled "harmless error" on appeal, because it did not cause the outcome of the case to be different from what it would have been had the evidence in question not been admitted. If there is any type of error which judges find tolerable in their courtrooms, it is most definitely this benign variety, "harmless error."

For example, even questionable evidence such as the Nalline narcotics test has been allowed in cases where it was consistent with the weight of the other evidence. In *People v. Williams*[29] other evidence strongly corroborated the defendants' use of illegal drugs, including their admission of prior drug use and the presence of both old and fresh needle marks on the inside of their left forearms. In this context, the court viewed the Nalline evidence as a useful and relatively innocuous supplement to the other evidence of guilt in obtaining a conviction.[30] Voiceprint evidence has also been allowed when other evidence against the defendant "was already ample to sustain his conviction, even without the use of voiceprints."[31]

On the other hand, the principle works in reverse as well. Some dubious evidence has been excluded under this theory, due to the lack of sufficient supporting evidence. In one case, the Seventh Circuit Court of Appeals upheld the exclusion of expert defense testimony as to the unreliability of cross-racial eyewitness identification evidence, pointing out other evidence in the case that failed to corroborate such testimony.[32]

Exclusion of hypnotically refreshed recollection evidence (of the alleged victim) was similarly upheld where other evidence strongly tended to show that the witness's testimony was "vague, changeable, self-contradictory, or prone to unexplained lapses of memory."[33] The court also considered evidence of the defendant's truthfulness and lack of any history of aggressive or violent behavior and concluded that the hypnotically refreshed evidence would have jeopardized the court's chances of reaching the correct outcome in the case, i.e., acquittal. As a result, the evidence was not allowed.[34] In contrast, and usefully

[29] 331 P.2d 251 (Cal. App. Dep't Super. Ct. 1958). The Nalline test purports to reveal whether an individual has recently used certain drugs, but the scientific validity of the test is suspect.

[30] Id., at 252.

[31] *Worley v. State*, 263 So. 2d 613, 614 (Fla. Dist. Ct. App. 1972).

[32] *United States v. Watson*, 587 F.2d 365, 369 (7th Cir. 1978); see also *People v. Marx*, 126 Cal. Rptr. 350, 356–57 (Cal. Ct. App. 1975).

[33] *People v. Shirley*, 723 P.2d 1354, 1355 (Cal.), cert. denied, 459 U.S. 860 (1982).

[34] Id., at 1358–59.

illustrative of the power of the corroboration effect, was the case of *Wisconsin v. Armstrong*.[35] There, the court allowed the hypnotically refreshed testimony of an eyewitness who had supposedly seen the defendant fleeing from the crime scene in a murder/sexual assault case. The court rested its decision on the presence of other evidence that corroborated the hypnotically refreshed testimony, including traces of blood under the defendant's fingernails and toenails and on his watch band, as well as "other physical evidence."[36]

Cases such as these show how the same type of evidence can be accepted or rejected on the basis of presence or absence of other corroborating evidence. Judges want to reach the right result and to protect themselves from reversal on appeal, and so they use the corroboration principle as a guide to their evidentiary decisions when in doubt. Awareness of this tendency may be of limited utility to attorneys and experts, because litigants generally try to offer all available relevant evidence in furtherance of their cause in any event, at least to the extent it is not redundant or cumulative. However, the degree of corroboration available may at least highlight the prospects for admissibility of novel scientific evidence, and thus enable the parties better to assess the probability of prevailing on the merits at trial. This knowledge can, in turn, be useful in driving decisions on whether to settle out of court.

Perceived Unfairness to One Side

One last practical concern that can influence judicial decisions on whether to admit novel scientific evidence centers around the reluctance of judges to cripple one party's chances for winning at trial. Where scientific evidence is essentially one party's only way to win, or one party's only available means of responding to the other party's evidence, courts tend to allow that evidence in the interests of fairness to both sides.

The Supreme Court's decision in *Daubert* can in fact be viewed as an example of this principle. The Court did not want to place toxic tort plaintiffs in a position in which their claims would have failed to survive motions for summary judgment due to exclusion of unavoidably novel scientific evidence as to causation.[37] In the toxic tort context, plaintiffs are in a particularly vulnerable litigation posture. They must usually rely on innovative, even case-specific forms of statistical and epidemiological evidence to establish causation of diseases with extremely lengthy latency periods and multiple possible causes.[38] In these mass exposure cases, due to the inevitable evidentiary disadvantages plaintiffs face, courts tend to be more lenient in allowing novel scientific evidence in support of plaintiffs' claims, at least insofar as this will enable the

[35] 329 N.W.2d 386 (Wis.), cert. denied, 461 U.S. 946 (1983). [36] Id., at 390.

[37] *Daubert v. Merrell Dow Pharmaceuticals, Inc.*, 509 U.S. 579 (1993).

[38] See, e.g., David Rosenberg, *The Causal Connection in Mass Exposure Cases: A "Public Law" Vision of the Tort System*, 97 Harv. L. Rev. 851, 855–59 (1984); Jean M. Eggen, *Toxic Torts, Causation, and Scientific Evidence After Daubert*, 55 U. Pitt. L. Rev. 889, 892–93 (1994).

plaintiffs to survive summary judgment and place their evidence before the jury.

Even scientific evidence that is usually disfavored, such as testimony as to the unreliability of eyewitness identification, can benefit from the fairness principle. For example, in *United States v. Downing*[39] the Third Circuit Court of Appeals held that the lower court abused its discretion in excluding this type of evidence. The key factor was that the defendant "was convicted solely on the basis of eyewitness testimony," and thus the exclusion of this evidence removed the defense's only chance for acquittal.[40]

Similarly, in *Coppolino v. State*[41] the court allowed dubious toxicology evidence in a case in which an anesthesiologist was on trial for the murder of his wife. The toxicology test in question was devised specifically for the case at hand, and purported to detect a chemical previously considered undetectable by the medical and toxicological communities. The circumstantial evidence against the defendant was overwhelming. Nevertheless, because the prosecution would have lacked any other admissible direct evidence to convict without the toxicology test, the court ruled in its favor. In its opinion, the court specifically stated that the prosecution's case relied "almost solely" on this evidence, and thus allowed it.[42] This might seem to run counter to the previously mentioned criterion under which courts tend to exclude evidence that lacks corroboration, but in fact it is consistent with it. In this case, the court was evidently convinced that the defendant was guilty, based on all the "corroborating" circumstantial evidence, but nonetheless feared that gaps in the prosecution's admissible evidence would have led to the wrong result but for admission of the toxicology test.

In each instance, the core principle underlying all of the factors that influence admissibility decisions is that judges want to reach the correct outcome of the case at trial. If novel scientific evidence can be presented to judges in such a way that they see it as a method of furthering the ultimate goal of obtaining the right result in court, judges will tend to allow it.

Admissibility of other Scientific and Forensic Evidence

In the course of analyzing the various tests and factors courts use in determining the admissibility of scientific evidence, we have seen numerous examples of judicial opinions addressing various types of scientific and forensic evidence.

[39] 753 F.2d 1224 (3rd Cir. 1985). [40] Id., at 1226.

[41] 223 So. 2d 68, 69–71 (Fla. Dist. Ct. App. 1968), cert. denied, 399 U.S. 927 (1970).

[42] Id., at 70. The circumstantial evidence was, among other things, that: the defendant was involved in an extramarital affair; had made certain allegedly incriminating statements in the other woman's presence; was an anesthesiologist who had used succinylcholine chloride and had previously mentioned that it was undetectable in the body after death; had bought some of this drug about two months before his wife's death; his wife and her father had been a source of financial support to him but stopped helping him before his wife's death; and he remarried only six weeks after his wife's death.

Before we directly examine the admissibility of forensic entomology evidence, it should be useful to summarize the relevant case law regarding other somewhat comparable varieties of evidence. This exercise will assist us in comparing forensic entomology to other evidence and gauging the prospects for its admissibility in light of the factors that courts will likely deem important in reaching a decision.

We will first consider briefly the panoply of what might be termed "standard" forensic evidence and the treatment it has received from the courts. Then, we will examine two specific varieties of "scientific" evidence – polygraphs and DNA – in light of their arguable positions at approximately the opposite ends of the admissibility spectrum. By analyzing the factors that have led most courts to reject polygraphs and accept DNA evidence, we will be better prepared to consider the issues relevant to the admissibility of forensic entomology evidence.

"STANDARD" FORENSIC EVIDENCE

It is perhaps surprising that courts have, virtually as a routine matter of course, long admitted various types of "forensic" evidence, even in criminal cases, often with little or no reference to the *Frye* or *Daubert* tests for admissibility of scientific evidence. In many if not all instances, there seems to be no reason not to subject this type of "forensic" evidence to the tests for scientific evidence, other than judicial comfort with the tried and true, "old reliable" (or unreliable) techniques.

Some classic examples include "criminalistic" evidence relating to fingerprints, questioned documents, comparisons of firearms/bullets, comparison of imprints and pattern evidence, identification of explosive residues, examination of blood and other bodily fluids, and identification of fabric samples, hair, glass, soil and other substances. In addition to these familiar examples, forensic evidence has been admitted in the fields of: medicine (usually relating to the manner and cause of death and surrounding circumstances in cases of "questioned" death; odontology (concerned with identification of persons based on their teeth); anthropology (generally identification of persons based on skeletal remains); toxicology (dealing with the determination of toxic chemicals in human tissues and organs and their role in causing death); radiology (relating to the identification of injuries and objects such as bullets and bombs); psychology (dealing with evaluation of the psychiatric condition of a person and his or her actions in relation to a crime); climatology (using information about weather conditions to determine factors relating to bodily changes after death); and engineering (involving the determination and reconstruction of material failures in relation to accidents, disasters, and fires).[43]

[43] Henry C. Lee, *Forensic Science and the Law*, 25 Conn. L. Rev. 1117, 1118 (1993).

These and other types of forensic evidence have been admitted in court to establish one or more of the five major components of the analysis and interpretation of physical evidence (recognition, identification, comparison, individualization, and reconstruction). We will briefly describe each of these in turn.

Recognition is the capacity to separate important and relevant facts from background and irrelevant materials in a case. This involves selection of evidence from among the various investigative leads and theories of the case, using pattern recognition, observation of physical evidence, and information analysis.

Identification is a classification scheme in which items are assigned to categories, particularly with regard to items of physical evidence. This often involves scientific tests, as in the cases of drugs, arson accelerants, blood stains, and semen stains. Identification can also be accomplished through visual comparison of the structure and appearance of known and unknown samples, as is done with fingerprints, footprints, projectiles, weapons, wounds, or bite marks.

Classification involves comparison of the class characteristics of the questioned evidence with those of known standards. This can be done using chromatography, spectroscopic methods, X-rays, microscopic examination, and biological methods, using a host of techniques.

Individualization is the demonstration that a given sample is unique, or that a questioned piece of evidence and a similar known sample have a common origin. Typical examples are fingerprints and DNA evidence.

Reconstruction is the replication of an event using evidence from a crime scene, laboratory analysis, and other independent sources of information. This often involves pattern analysis, statistical analysis, and the use of inductive and deductive logic to piece together a sequence of events.

Some types of forensic evidence have been used for so many years and in so many cases that their admissibility is virtually taken for granted. An aggressive judge is certainly free in any given trial to revisit this issue in light of whatever rubric he or she has available for weighing the admissibility of scientific evidence, and in some cases, such analysis might well result in a reversal of long-established judicial acceptance. This is so because the scientific and technical underpinnings of some long-accepted forensic evidence are no better than those at issue in some examples of "novel scientific evidence" that have not been given a free ride by the courts. For this reason, the proponents of any technique must be prepared to demonstrate how it satisfies the applicable test and meets the criteria we have discussed previously.

It is interesting to note that, in the wake of *Daubert*, some courts have used the new test set forth therein to exclude some types of forensic evidence that had previously been considered admissible.[44] These cases tend to turn on the lack of a sufficient empirical basis for the evidence in question. Some examples

[44] Recall that *Kumho Tire* held that the *Daubert* test is properly applied to non-scientific expert witness testimony. 526 U.S. 137 (1999), 119 S.Ct. 1167, 143 L.Ed.2d 238.

include voice spectrography identification,[45] forensic anthropologist testimony on mistaken identity,[46] and expert testimony regarding police discipline theory.[47]

Some legal commentators have stated that a strict application of the *Daubert* test would pose a major challenge for almost all forensic evidence routinely used in criminal cases, the problem generally being a questionable scientific foundation.[48] Other than DNA identification techniques, which originated in a setting different from forensics and were subjected to rigorous scientific scrutiny during their development, most forensic methods have little or no supporting history of scientific analysis.[49] This situation developed because techniques such as hair identification, bite mark analysis, voice spectrography, handwriting analysis, and fingerprinting have been accepted by courts despite virtually no demonstration of their scientific underpinnings.[50] Thus, it is possible that some additional examples of the types of forensic evidence traditionally given a free pass by the courts may face admissibility challenges under *Daubert,* whether the evidence would traditionally be classified as "technical" or "scientific."[51]

[45] *Virgin Islands v. Sanes,* 57 F.3d 338, 341 (3rd Cir. 1995).

[46] *United States v. Dorsey,* 45 F.3d 809, 812 (4th Cir. 1995).

[47] *Berry v. City of Detroit,* 25 F.3d 1342, 1348–53 (6th Cir. 1994).

[48] Erica Beecher-Monas, *Blinded by Science: How Judges Avoid the Science in Scientific Evidence,* 71 Temple L. Rev. 55, 56 (1998); see also John L. Thornton, *Courts of Law v. Courts of Science: A Forensic Scientist's Reaction to* Daubert, 1 Shepard's Expert & Sci. Evidence Q. 475, 482 (1994), which stated that a "number of tests – e.g., handwriting comparison, hair identification – have marked empirical validity but precious little scientific foundation because they are so subjective and, furthermore, do not comport with the classical definition of a science."

[49] See Randolph N. Jonakait, *Forensic Science: The Need for Regulation,* 4 Harv. J.L. & Tech. 109, 133–34 (1991), which stated that most forensic science is not endorsed by the general scientific community, with procedures that have undergone little or no controlled testing and error rates that are either undetermined or unacceptably high. See also Michael J. Saks and Jonathan J. Koehler, *What DNA "Fingerprinting" Can Teach the Law About the Rest of Forensic Science,* 13 Cardozo L. Rev. 361, 362 (1991), which noted that "more is known about the strengths and weaknesses of DNA fingerprinting evidence than about most of the other, older, and more widely used forms of forensic science evidence."

[50] Margaret A. Berger, *Procedural Paradigms for Applying the* Daubert *Test,* 78 Minn. L. Rev. 1345, 1354–55 (1994). Many of these techniques are based on the assumption that things such as fingerprints, dentition, voice patterns, hair characteristics, and handwriting style are uniquely individual, and, as is also traditionally supposed with snowflakes, no two are alike. However, in none of these areas has this assumption been tested under controlled conditions in scientifically rigorous studies. See also Michelle McClure, *Comment, Odontology: Bite Marks as Evidence in Criminal Trials,* 11 Santa Clara Computer & High Tech. L.J. 269, 270 (1995); David L. Faigman, *The Evidentiary Status of the Social Sciences Under Daubert: Is it "Scientific," "Technical," or "Other" Knowledge?,* 1 Psychol. Pub. Pol'y & L. 960 (1995); D. Michael Risinger, et al., *Exorcism of Ignorance as a Proxy for Rational Knowledge: The Lessons of Handwriting Identification "Expertise",* 137 U. Pa. L. Rev. 731, 744–51 (1989) (reporting results of tests administered by the Forensic Science Foundation over several years, which showed at best an accuracy rate of 57%).

[51] See, e.g., *United States v. Starzecpyzel,* 880 F. Supp. 1027, 1038 (S.D.N.Y. 1995), in which hand-

POLYGRAPHS

As the subject matter at issue in *Frye*, truth-determining devices have perhaps the lengthiest and most controversial history in our legal system among the types of scientific evidence that have been offered at court. During the many years that have elapsed since *Frye*, the technology of "lie detector" tests has evolved considerably, as has the sophistication and precision of the techniques for analyzing the results. The prospects for admissibility of such evidence have undergone a concomitant evolution in the courts as judges have noted the advancements in the equipment and methodology. Whereas not long ago polygraph evidence was almost universally rejected by the courts, it is now gaining some degree of acceptance in certain jurisdictions.

For example, in *United States v. Galbreth*,[52] a federal District Court in New Mexico took a hard look at the modern polygraph instrument, the theory underlying current techniques, and the challenges made to the polygraph's reliability. The court contrasted the primitive device rejected in *Frye* and its progeny with what it viewed as the much more impressive and reliable features of the modern apparatus:

> [T]he modern polygraph machine is a sophisticated instrument capable of continuously and simultaneously measuring and recording various autonomic responses. It measures respiration at two points on the body... The polygraph machine also measures skin conductance or galvanic skin response. Electrodes attached to the fingertip or palm of the hand indicate changes in the sweat gland activity in those areas. In addition, the polygraph measures increases in blood pressure and changes in heart rate... obtained by placing a standard blood pressure cuff on the subject's upper arm. Finally, the polygraph may also measure... blood supply changes in the skin [by means of a plethysmograph] which occur as blood vessels in the skin of the finger constrict due to stimulation. Even opponents of the polygraph technique readily concede that a quality polygraph machine can accurately measure and record these responses.

The court also pointed out that the numerical method of evaluating polygraph results (first introduced in 1960), which sometimes utilizes the aid of a computer, helps to ensure a "rigorous, semi-objective evaluation of the

writing analysis has failed to meet the *Daubert* standard, yet has been admitted as "technical" evidence. See also *United States v. Velasquez*, 64 F.3d 844, 848–50 (3rd Cir. 1995), in which the court found handwriting analysis admissible so long as the same standard was applied to both prosecution and defense experts, and *United States v. Quinn*, 18 F.3d 1461, 1464–65 (9th Cir. 1994), in which the court allowed expert photogrammetry evidence that purported to estimate the height of a person in a photograph, by simply considering it technical expertise rather than scientific. In the wake of *Kumho Tire*, 526 U.S. 137 (1999), 119 S.Ct. 1167,143 L.Ed.2d 238, this dichotomy will likely disappear. For example, in *United States v. Paul*, 175 F.3d 906, 909–13 (11th Cir. 1999), the Eleventh Circuit upheld the prosecution's use of "nonscientific" handwriting analysis using *Daubert* factors. Similarly, in *United States v. Molina*, 172 F.3d 1048, 1056–58 (8th Cir. 1999), the Eighth Circuit used *Daubert* analysis to uphold introduction of expert law enforcement testimony as to the typical modus operandi of drug dealers.

[52] 908 F. Supp. 877, 883–90 (D.N.M. 1995).

physiological information contained in the charts, thereby safeguarding against examiner bias." After weighing all of the arguments and counter arguments, the court concluded that the modern polygraph passed the *Daubert* test and that the results were admissible.[53]

Although many courts continue to reject polygraph evidence, the *Galbreth* court is by no means alone in favorably reevaluating this evidence in the light of modern advancements. See *United States v. Posado*[54] in which the court stated "There can be no doubt that tremendous advances have been made in polygraph instrumentation and technique in the years since *Frye*," and *United States v. Crumby*,[55] "The science of polygraphy has made significant progress over the past decade." Other courts have, post-*Daubert*, assumed that polygraph evidence is reliable[56] and satisfies the requirements of Federal Rule of Evidence 702.[57] In light of such cases and the probable further advancements in polygraph instrumentation and methodology, it is not inconceivable that there may eventually be a widespread reversal of the previously disfavored status of polygraph evidence.

It is important to note, however, that satisfaction of Rule 702 alone does not guarantee admissibility. Several courts have ruled that Federal Rule of Evidence 403, which allows exclusion of otherwise relevant evidence, requires rejection of polygraph evidence despite its passing the Rule 702 standard. Rule 403 reads as follows:

Exclusion of Relevant Evidence on Grounds of Prejudice, Confusion, or Waste of Time. Although relevant, evidence may be excluded if its probative value is substantially outweighed by the danger of unfair prejudice, confusion of the issues, or misleading the jury, or by considerations of undue delay, waste of time, or needless presentation of cumulative evidence.

[53] Id., at 895–96. See N.M. Rule Evid. 11–707. New Mexico accepts polygraphy evidence to a greater and more unconditional extent than any other State at present. See *United States v. Scheffer*, 523 U.S. 303, 311–12 (1998).

[54] 57 F.3d 428, 434 (5th Cir. 1995). The *Posado* court opened the door within the Fifth Circuit for courts to exercise their own discretion as to the admission or exclusion of polygraph evidence. In *United States v. Cordoba*, 104 F.3d 225, 228 (9th Cir. 1997), the Ninth Circuit did the same. However, after rehearing at the trial court level, the Ninth Circuit upheld exclusion of the unstipulated polygraph results under both Rule 702 and 403. *United States v. Cordoba*, 194 F.3d 1053 (9th Cir. 1999).

[55] 895 F. Supp. 1354, 1361 (D. Ariz. 1995). See also William J. Yankee, *The Current Status of Research in Forensic Psychophysiology and Its Application in the Detection of Deception*, 40 J. Forensic Sci. 63 (1995) ("The period between 1986 and the present has been one of unparalleled advances in the psychophysiological detection of deception testing procedures and processes.")

[56] *United States v. Dominguez*, 902 F. Supp. 737, 739 (S.D. Tex. 1995) (but excluding the evidence in the instant case under Rule 403 because the government was not invited to be present at the defendant's polygraph examination). This decision includes a list of ten factors that the court stated should be considered in weighing the admissibility of polygraph evidence. Id., at 740–41.

[57] *United States v. Lech*, 895 F. Supp. 582, 585 (S.D.N.Y. 1995) (but ruling the results inadmissible in the case at hand under Rule 403 because of improper test questions).

Daubert itself warned that "A judge assessing a proffer of expert scientific testimony under Rule 702 should also be mindful of other applicable rules."[58] See *United States v. Kwong*,[59] in which the court assumed admissibility under Rule 702 but nonetheless excluded polygraph evidence under Rule 403.

Rule 403 aside, there is still considerable judicial resistance to polygraph evidence. In *United States v. Sanchez*,[60] the Fourth Circuit recently reaffirmed its per se ban on such evidence. Most State courts also continue to prohibit polygraph evidence in all cases.[61]

The Supreme Court had a recent opportunity to evaluate the current status of polygraph evidence in *United States v. Scheffer*.[62] In *Scheffer*, the Court upheld the constitutionality of Military Rule of Evidence 707(a), which renders polygraph results per se inadmissible in courts-martial proceedings.[63]

The Court held that the rule serves at least three legitimate governmental interests in the criminal trial process: (1) ensuring that only reliable evidence is introduced at trial; (2) preserving the jury's role in determining the credibility of witnesses; and (3) avoiding litigation that is collateral to the primary purpose of the trial.[64]

On the first matter, the Court noted that "the exclusion of unreliable evidence is a principal objective of many evidentiary rules," and stated that "there is simply no consensus that polygraph evidence is reliable . . . the scientific community remains extremely polarized about the reliability of polygraph techniques."[65] The Court cited conflicting scientific studies[66] as well as the

[58] 509 U.S. at 595.

[59] 69 F.3d 663, 668 (2nd Cir. 1995). See also *United States v. Sherlin*, 67 F.3d 1208, 1216 (6th Cir. 1995).

[60] 118 F.3d 192, 197 (4th Cir. 1997). See also *United States v. Messina*, 131 F.3d 36, 42 (2nd Cir. 1997) (stating that the Second Circuit has "not decided whether polygraphy has reached a sufficient state of reliability to be admissible.").

[61] See, e.g., *State v. Porter*, 241 Conn. 57, 92–95, 698 A.2d 739, 758–59 (1995); *People v. Gard*, 158 Ill. 2d 191, 202–04, 632 N.E.2d 1026, 1032, 198 Ill. Dec. 415 (1994); *In re Odell*, 672 A.2d 457, 459 (RI 1996) (per curiam); *Perkins v. State*, 902 S.W. 2d 88, 94–95 (Ct. App. Tex. 1995).

[62] 523 U.S. 303 (1998).

[63] Military Rule of Evidence 707(a) reads: "Notwithstanding any other provision of law, the results of a polygraph examination, the opinion of a polygraph examiner, or any reference to an offer to take, failure to take, or taking of a polygraph examination, shall not be admitted into evidence."

[64] Id. at 309.

[65] Id.

[66] Id. at 309–10. The Court cited, in favor of the technique, S. ABRAMS, THE COMPLETE POLYGRAPH HANDBOOK 190–191 (1968) (reporting the overall accuracy rate from laboratory studies involving the common "control question technique" polygraph to be "in the range of 87 percent"). For the contrary proposition, i.e., that the accuracy rate of the same technique is "little better than could be obtained by the toss of a coin," the Court cited W. G. Iacono and D. T. Lykken, *The Scientific Status of Research on Polygraph Techniques: The case Against Polygraph Tests*, in D. L. FAIGMAN, D. H. KAYE, M. J. SAKS, AND J. SANDERS, MODERN SCIENTIFIC EVIDENCE, § 14–5.3, p. 629 (1997). Id. Further, the Court noted issues as to the efficacy of countermeasures which would remain even if the debate about the reliability of the basic polygraph technology were resolved. Id. at 310.

reflection of this lack of scientific consensus in judicial decisions as to the admissibility of polygraph evidence.[67]

As to the issue of preservation of the jury's core function, the Court stated:

> By its very nature, polygraph evidence may diminish the jury's role in making credibility determinations... Unlike other expert witnesses who testify about factual matters outside the jurors' knowledge, such as the analysis of fingerprints, ballistics, or DNA found at a crime scene, a polygraph expert can supply the jury only with another opinion, in addition to its own, about whether the witness was telling the truth. Jurisdictions, in promulgating rules of evidence, may legitimately be concerned about the risk that juries will give excessive weight to the opinions of a polygrapher, clothed as they are in scientific expertise and at times offering, as in respondent's case, a conclusion about the ultimate issue in the trial. Such jurisdictions may legitimately determine that the aura of infallibility attending polygraph evidence can lead jurors to abandon their duty to assess credibility and guilt. Those jurisdictions may also take into account the fact that a judge cannot determine, when ruling on a motion to admit polygraph evidence, whether a particular polygraph expert is likely to influence the jury unduly. For these reasons, the President is within his constitutional prerogative to promulgate a per se rule that simply excludes all such evidence.[68]

Finally, the Court opined that per se exclusion of polygraph evidence averts the problem of devoting judicial resources to resolution of collateral matters:

> Such collateral litigation prolongs criminal trials and threatens to distract the jury from its central function of determining guilt or innocence. Allowing proffers of polygraph evidence would inevitably entail assessments of such issues as whether the test and control questions were appropriate, whether a particular polygraph examiner was qualified and had properly interpreted the physiological responses, and whether other factors such as countermeasures employed by the examinee had distorted the exam results. Such assessments would be required in each and every case. It thus offends no constitutional principle for the President to conclude that a per se rule excluding all polygraph evidence is appropriate. Because litigation over the admissibility of polygraph evidence is by its very nature collateral, a per se rule prohibiting its admission is not an arbitrary or disproportionate means of avoiding it.[69]

DNA

In contrast to the many judicial struggles with polygraphs, one of the greatest success stories in the area of modern scientific evidence is the example of

[67] Id. at 310–12 (citing a representative sampling of federal and State decisions). In the aftermath of *Scheffer*, cases such as *King v. Trippett*, 192 F.3d 517, 522 (6th Cir. 1999), have continued to exclude polygraph evidence in a variety of procedural contexts, including as impeachment evidence.

[68] Id. at 313–15 (citations omitted).

[69] Id. at 315 (citations omitted). Of course, similar issues must be litigated as a necessary predicate to all forms of expert scientific evidence other than those that are expressly allowed by statute or are a proper subject for judicial notice. Perhaps the Court intended to say that polygraph evidence requires more extensive evidentiary maneuvering than other types of expert evidence, because of its lack of general acceptance and its complexity.

deoxyribonucleic acid (DNA) as a tool for identification and individualization. In fairly short order, DNA evidence has been accepted by the overwhelming majority of courts that have considered the issue, both in the United States and in other nations.[70] As stated in *Commonwealth v. Crews*,[71] "By 1990 more than 2,000 U.S. court cases in 49 states and the District of Columbia had used DNA tests." Even more remarkable, the theory and techniques underlying DNA evidence are becoming so well-established that some courts are now taking judicial notice of their reliability, and some state legislatures have passed statutes recognizing this evidence as reliable and admissible, thereby eliminating the need to prove the issue at trial.[72]

The great volume of scientific corroboration of the underlying principles is central to the widespread acceptance of DNA evidence. Modern science has firmly established the principles that people have DNA, that it is found in all cells with a nucleus, that an individual's DNA is unique (except in the case of identical twins), and that proper testing procedures can identify a person on the basis of his or her DNA. There can be no serious question today as to the scientific legitimacy of the essential principles involved.

Still, forensic use of DNA evidence in criminal cases has been disallowed under certain circumstances.[73] These circumstances are case-specific and generally involve errors in methodology (both collection/maintenance of samples and testing protocols)[74] and questionable probability estimates from experts as to the likelihood that someone other than the defendant could have the same DNA profile. Naturally, mistakes and sloppiness in the application of any technique in a particular case can be expected to lead to exclusion of evidence, so it is hardly surprising that the improperly collected or handled DNA evidence has been excluded. These cases are noteworthy principally for the proposition that even the most well-rooted scientific evidence can be defeated in any particular case if the methodology in that specific instance was suspect. In other words, failures of execution trump successes of theory. Aside from these cases, it is in the area of statistical claims that DNA evidence has been particularly singled out for criticism.

[70] Jennifer Callahan, Survey: *The Admissibility of DNA Evidence in the United States and England*, 19 Suffolk Transnat'l L. Rev. 537 (1996).

[71] 640 A.2d 395, 400 n.3 (Pa. 1994).

[72] See, e.g., *United States v. Martinez*, 3 F.3d 1191, 1197 (8th Cir. 1993); *Taylor v. State*, 889 P.2d 319, 338–39 (Okla. Crim. App. 1995); HOWARD COLEMAN AND ERIC SWENSON, DNA IN THE COURTROOM: A TRIAL WATCHER'S GUIDE 113–20 (1994).

[73] See, e.g., *Armstead v. State*, 673 A.2d 221, 227 n.6 (Md. App. 1996) (listing examples of situations in which DNA evidence has been excluded). Jennifer Callahan, Survey: *The Admissibility of DNA Evidence in the United States and England*, 19 Suffolk Transnat'l L. Rev. 537 (1996).

[74] See, e.g., *Taylor v. State*, 889 P.2d 319, 323 n.6 (Okla. Crim. App. 1995); See also *Ex Parte Perry v. State*, 586 So. 2d 242 (Ala. 1991). It is important to note that there are multiple techniques for DNA testing, so one court's rejection of such evidence must be considered in light of the specific methodology at issue in that case.

Current DNA testing does not examine a person's entire DNA sequence, nor anything close to it, but only a few alleles out of the approximately three million available. The composition of this sample is then compared to information from a database of DNA from selected portions of a given population, and a probability estimate is generated therefrom.[75] The risk of unfairly overwhelming the jury with the supposed phenomenally large odds against the DNA coming from anyone other than the defendant has led some courts to hesitate in admitting such evidence.[76] Where the proffered "odds" are so long that the numbers exceed the current population of the world, some judges are understandably leery of allowing such mind-boggling evidence to reach the jury. Thus, for DNA evidence, as with polygraph evidence, Rule 403 has been used to exclude evidence that passed the Rule 702 test, on the basis of unfair prejudicial impact and risk of misleading the jury or confusing the issues. As we shall see in the next chapter, forensic entomology evidence can be prone to the same type of problems regarding statistical evidence, and we shall examine the relevant case law in greater detail then.

Forensic Entomology Case Law

ADMISSIBILITY IN GENERAL

A thorough review of the reported case law, including all federal and State courts, indicates that forensic entomology evidence has fared quite well in the courtroom. Indeed, the courts have generally assumed without much controversy that forensic entomology evidence as a category satisfies both the *Frye* and *Daubert* tests for expert testimony under Rule 702.[77] Nevertheless, the case law does yield several important points, and we will now address these in detail, and in light of all the other case law we have already examined.

[75] See Michael J. Short, Comment, *Forensic DNA Analysis: An Examination of Common Objections Raised to the Admission of DNA Fingerprinting as Illustrated by* State v. Pierce, 19 U. Dayton L. Rev. 133, 143 (1993).

[76] See, e.g., *Commonwealth v. Lanigan*, 596 N.E.2d 311 (Mass. 1992); compare with the same case on re-hearing in which more conservative statistical estimates were used, 641 N.E.2d 1342 (Mass. 1994). See also *State v. Johnson*, 922 P.2d 294, 300 (Ariz. 1996); *State v. Bloom*, 516 N.W.2d 159, 160 (Minn. 1994); *State v. Vandebogart*, 652 A.2d 671, 675 (N.H. 1995); *State v. Cauthron*, 846 P.2d 502, 517 (Wash. 1993); and *State v. Anderson*, 881 P.2d 29, 47 (N.M. 1994).

[77] Rule 702, as amended effective December 1, 2000, now reads:

> **Testimony by Experts.** If scientific, technical, or other specialized knowledge will assist the trier of fact to understand the evidence or to determine a fact in issue, a witness qualified as an expert by knowledge, skill, experience, training, or education, may testify thereto in the form of an opinion or otherwise, if (1) the testimony is based upon sufficient facts or data, (2) the testimony is the product of reliable principles and methods, and (3) the witness has applied the principles and methods reliably to the facts of the case.
>
> This amounts to a codification of *Daubert* and its progeny, and thus should not significantly change the legal analysis.

Unlike controversial forms of analysis such as polygraphs, voiceprints, and hypnosis, forensic entomology rests upon a foundation of scientific principles that courts have found to be quite adequate for purposes of Rule 702 expert opinion testimony. The body of peer-reviewed scientific literature governing the underlying concepts is both large and long-established. There has been a great deal of controlled-conditions research on the array of insect species that oviposit on the tissue of dead organisms. The size, morphological characteristics, and rates of development of the larvae of these various species have been studied under a variety of conditions and are generally well-accepted within the entomological community. Indeed, the scientific facts and principles described in detail elsewhere in this book are firmly rooted in solid empiricism.

As a result, the courts that have considered forensic entomology evidence have not found much cause for concern in any aspect of the underlying science. Predicated upon a detailed, reasoned, methodical explication of the science by a well-qualified expert witness, judges have routinely held that Rule 702 is satisfied. For example, in both *People v. Reynolds*[78] and *State v. Thibodeaux*,[79] State appellate courts mentioned the fact that expert testimony in the field of forensic entomology was used in their respective cases, without any commentary or citation indicating any controversy relative to such evidence.

The only reported instance in which anything resembling forensic entomology evidence was excluded under a Rule 702 analysis, *State v. Miller*,[80] rested on very unusual facts. In *Miller*, Dr. Rodriguez, a forensic anthropologist called as an expert witness by the prosecution, estimated that the victim's decomposed, maggot-ridden corpse had been in an icehouse for 18 to 20 days, an assertion vigorously disputed by the defense. This witness arrived at his estimate based on his examination of insect larvae removed from the body, and on photographs of the body's degree of decay. The trial court allowed this testimony over defense objections.

On appeal, the Supreme Court of South Dakota noted that, although Dr. Rodriguez had earned a doctorate in anthropology eight months before testifying, he had no degree in entomology or zoology.[81] This alone was not

[78] 257 Ill. App. 3d 792, 798, 629 N.E. 2d 559 (1994). In this case, forensic entomology evidence was offered by the defense, and was admitted. "Defendant's case consisted solely of the expert testimony of Dr. Bernard Greenberg, who is a consultant in the field of forensic entomology and has been studying maggots for 40 years. Based on his experience and a photograph of a maggot taken from the victim's body, Dr. Greenberg opined that the time of death was 1 p.m. on June 2, 1989."

[79] Supreme Court of Louisiana, case number 98–KA-1673, decided September 8, 1999. Here, it was the prosecution that offered forensic entomology evidence, which was admitted. "Dr. Lamar Leek, professor of entomology at Louisiana State University, testified as an expert in the field of forensic entomology. He examined the insect samples taken from [the] body. He testified that flies will lay eggs on a carcass within a couple of hours, but will not lay eggs after dark. Therefore, he determined that the eggs were laid before nightfall on July 19, 1996, and calculated the age of the fly larvae (maggots) to be between 24–48 hours old at discovery."

[80] 429 N.W. 2d 26 (S.D. 1988). [81] Id., at 35.

dispositive, as the Court acknowledged that "the field of forensic anthropology is recognized as a scientific discipline whose practitioners have been accepted as experts to testify regarding identification of human remains, and time, place, or manner of death."[82] However, Dr. Rodriguez derived his estimates by extrapolating from information he had personally obtained in a "unique study involving exposure of human corpses in various Louisiana locations. No similar experiments have been carried out by other scientists, and the accuracy and applicability of his methods are thus unverified."[83]

The Court applied the *Frye* "general acceptance" test, as controlling the issue of admissibility of evidence involving scientific principles, and found Dr. Rodriguez's evidence lacking. The Court concluded, "The forensic anthropologist's testimony that his methodology is completely untested by other scientists would necessarily mandate our ruling that his estimate of the time of . . . death is inadmissible."[84]

The *Miller* case can easily be distinguished from all the decisions in which forensic entomology evidence was allowed. As a threshold matter, the evidence was presented as forensic anthropology, not forensic entomology, by a person with education and training in anthropology, not entomology. Even as forensic anthropology evidence, however, the key obstacle to admissibility was the witness's reliance on untested, unverified personal observations rather than on controlled experiments and peer-reviewed research. It is unlikely that "scientific" evidence of any type would be ruled admissible under similar circumstances, and thus the *Miller* decision in no way challenges forensic entomology evidence per se.

In fact, the Supreme Court of Kansas in *Taylor v. State*[85] went so far as to declare, with approval, "The district court specifically found the identification of maggot growth on a body as an investigative tool in ascertaining the interval between death is not a recent or novel development." In that case, forensic entomology evidence had been used at trial, and one of the issues on appeal was whether the alleged subsequent discovery of additional forensic entomology evidence was sufficient to require a new trial. The appellate court noted that the forensic entomology information necessary to perform the maggot test in question had been known and was available in 1982 when the appellant was tried for murder, and the expert witness's testimony provided substantial evidence to support the trial court's finding that the test could have been performed in 1982. As a result, there was no requirement for a new trial.[86]

[82] Id., at 36. (Citations omitted.) [83] Id., at 35–36.

[84] Id., at 37–38. The Court contrasted this untested, unverified technique with that of electrophoresis, a "widely used form of testing of documented reliability," and found the difference "striking." Id., at 38. [85] 251 Kan. 272, 288, 834 P.2d 1325 (1992).

[86] Id., at 288. This opinion contains a rather detailed description of the methods of forensic entomology, as well as a quotation from the expert witness in the trial court to the effect that "the use of insects as forensic indicators to determine the time of death has been used since the mid-1200's in China, but it is only within the past decade or so that there has been a suffi-

A similar situation was present in *Seebeck v. State*,[87] where the Supreme Court of Connecticut reviewed a petition for a new trial based on allegedly newly discovered forensic entomology evidence. The Court implicitly accepted the basic validity of forensic entomology evidence without discussion, and concluded that the new evidence did not indicate any material change from the forensic entomology evidence presented at trial. The Court further held that a key expert witness had not materially changed his opinion or his methodology subsequent to trial.[88]

QUALIFICATION AS AN EXPERT WITNESS

The *Miller*[89] case which we have already discussed is a good example of the way in which a trial judge should consider whether a given witness is qualified to render expert opinion testimony in court. In an appropriate case, judicial scrutiny of the expert's training, experience, and education might reveal that he or she does not possess any greater level of knowledge than that possessed by the general public on the specific subject matter in question, and on that basis declines to recognize that witness as an expert in that particular subject. For the most part, however, courts are willing to use the wide discretion afforded them under the rules of evidence to grant expert recognition to forensic entomology witnesses who possess a reasonable amount of education, training, and/or experience in entomology.

To illustrate, let us consider the case of *State v. Hart*.[90] There, an employee of the Cincinnati Zoo, Ms. Stein, was allowed to testify as an expert witness as to the life cycle of the greenbottle fly, the type of fly found on the victim's body. Ms. Stein testified as to the time required for these flies to reach maturity after oviposition, and this evidence was used to ascertain the approximate date of death. The Court of Appeals of Ohio noted that Rule 702 allows a witness to qualify as an expert on the basis of knowledge, skill, experience, training, or education, and that Ms. Stein had worked at the Cincinnati Zoo for five years. Her duties there included the raising of greenbottle flies as food for predatory insects, and she had completed "a number of college-level courses in entomology and biology."[91]

cient data base for the broad spectrum of species." Id., at 287. The same case subsequently came before the federal district court on a petition for habeas corpus, in *Taylor v. Hannigan*, 1998 U.S. Dist. LEXIS 6904 (D. Kan. 1998), where the prisoner Taylor argued unsuccessfully that "newly discovered" entomological evidence warranted a new trial. The court found that the evidence was not sufficiently new or outcome determinative to justify a grant of habeas corpus. The court noted that the "new" evidence actually had been available at time of trial, that it was contradicted by other expert testimony at trial, and that it was not such convincing proof of Taylor's innocence that the court believed it would have resulted in a different verdict at trial. Id. at *71–77.

87 246 Conn. 514, 714 A.2d 1161 (1998).

88 Id., at 537–38. Additionally, the Court noted that other evidence in the case corroborated the forensic entomology evidence as to the date of death, and thus, even if one expert's opinion had changed, the result of the trial was not likely to be different. Id., at 538.

89 429 N.W. 2d 26 (S.D. 1988).

90 94 Ohio App. 3d 665, 641 N.E.2d 755 (1994).

91 Id., at 678.

The Court found no evidence that the trial judge abused his discretion in qualifying Ms. Stein as an expert witness on the life cycle of the greenbottle fly based upon her technical knowledge and practical experience, and noted that an expert's credentials, or lack thereof, go to the weight, and not the admissibility, of his or her testimony.[92] Other courts have similarly disposed of challenges to a trial court's recognition of an expert witness in forensic entomology.[93]

FACTUAL PREDICATE FOR THE EXPERT'S OPINION

Related to the issue of the expert's qualifications is the requirement that there must be a factual basis for the expert's opinion. In *State v. John*,[94] the Supreme Court of Connecticut considered a challenge to the trial court's decision to allow a forensic entomologist to testify and render an expert opinion based upon photographs of larvae on the victim's body. The appellants claimed that because the expert did not have actual specimens but rather relied upon photographs of the victim's body, and because the expert's opinion was based on allegedly inaccurate weather data, there was an insufficient and inaccurate factual basis for his opinions.

The Court recognized that there must be "some facts" shown as a foundation for an expert's opinion, but that there is no rule that establishes the precise facts that must be shown before an expert opinion may be received in evidence.[95] The Court then noted that the expert in this case, a forensic entomologist named Wayne Lord, testified at trial that a meaningful opinion could be derived from the photographs of larvae on the victim's body. The expert also testified that he and other experts in the field of entomology commonly rely on photographs such as the ones he used to identify the size and characteristics of the larvae and their corresponding stage of development. The Court concluded that "The trial court was entitled to rely upon this showing in ruling on the admissibility of Lord's testimony, and therefore, it did not abuse its wide discretion." It also noted that the fact that there was conflicting evidence regarding the weather on the days in question would not affect the admissibility of the expert's opinion, but would only be a factor in influencing the weight the triers of fact would afford his opinion.[96]

The *John* Court also alluded to the fact that Lord's testimony satisfied the *Frye* test, in that he showed that reliance on autopsy photographs to identify

[92] Id. See also *Kennedy v. Collagen Corp.*, 161 F.3d 1226, 1230–31 (9th Cir. 1998).

[93] See, e.g., *Pasco v. State*, 563 N.E.2d 587 (Sup. Ct. Ind. 1990), in which the Supreme Court of Indiana ruled that the trial court did not err in finding a person qualified as an expert in forensic entomology. The Court cited the facts that the witness was employed with Purdue University as an entomologist and a consultant in forensic entomology, was completing a master of science degree in forensic entomology at Purdue, had lectured and published articles in the area of forensic entomology, was a member of various entomological societies, and had conducted approximately 90 forensic entomology examinations. Id., at 594.

[94] 210 Conn. 652, 557 A.2d 93, cert. denied, 493 U.S. 824 (1989).

[95] Id., at 677.

[96] Id., at 678.

the developmental stages of larvae "is generally accepted in the field of forensic entomology."[97] The Court, as an aside, added its own observation as to the legal status of forensic entomology evidence in general:

> Neither defendant claims that the expert testimony of an entomologist to establish the time of death is inadmissible per se. We note, however, that other courts have allowed such evidence. See, e.g., *People v. Clark*, 6 Cal. App. 3d 658, 664, 86 Cal. Rptr. 106 (1970); *Knoppa v. State*, 505 S.W.2d 802, 803 (Tex. Crim. App. 1974).[98]

CHAIN OF CUSTODY FOR EVIDENCE

An important variant of the requirement for a factual basis in support of an expert's opinion was at issue in *Pasco v. State*,[99] specifically, chain of custody. The appellant argued the lack of proper chain of custody regarding the evidence relied on by the forensic entomologist. The evidence showed that a forensic pathologist had conducted an autopsy on the victim's body, during which he discovered examples of flies and of fly larvae. He took samples of the larvae and placed them in tubes and then in a box, which he transferred to a police officer who transported the package to Purdue University. During this travel, the package containing the samples was never out of the officer's custody, and upon his arrival at Purdue, he delivered them to the home of a senior research associate in medical veterinary entomology. She testified that she opened the package and found four preserved samples and four live samples, which she placed in a locked drawer in the laboratory until the forensic entomologist returned. Both she and the police officer signed a chain-of-custody form documenting the handling of the samples.[100]

The appellant challenged the chain of custody because the police officer testified that he received a box containing four tubes, while the senior research associate testified that she received a box containing eight tubes. However, the Supreme Court of Indiana stated that, in order to show a proper chain of custody, "the State is required only to present evidence which strongly suggests the exact whereabouts of the evidence at all times," and need not exclude all possibility of tampering, but only must provide "reasonable assurance that the evidence remained in an undisturbed condition."[101]

The Court noted that the chain-of-custody form was signed by both individuals, and that it corroborated the research associate's testimony. The Court then concluded, "We cannot say that this discrepancy amounts to a breakdown in the chain of custody. There is sufficient evidence in the record showing a proper chain of custody by the State."[102] Thus, the Court ruled that there was a sufficient factual predicate underlying the expert testimony of the forensic entomologist who based his opinion upon the samples in question, despite a

[97] Id. [98] Id. [99] 563 N.E.2d 587 (Sup. Ct. Ind. 1990). [100] Id., at 594. [101] Id.
[102] Id., at 595.

rather sizable discrepancy in the testimony of the last two individuals who handled the evidence samples.

RISK OF UNFAIR PREJUDICE OR INFLAMING THE JURY

Paradoxically, one potential obstacle to admissibility of forensic entomology evidence stems from the very same highly effective methods that are used to present it to the judge and the jury. Recall that a key factor that judges use to help them weigh admissibility issues is the extent to which the evidence can be presented to the triers of fact, in court, in such a way as to enable them to decide for themselves whether the expert witness reached the correct conclusions. Judges tend to be more accepting of evidence that can be demonstrated, often by photographic enlargements, to the jury, for a "seeing is believing" independent review of the expert's opinion. In the case of forensic entomology evidence, however, there is a tension between this tendency and judicial concerns about unfair prejudicial effect.

To put it bluntly, the problem centers around the potential for jurors to be particularly outraged, disgusted, and inflamed by photographs of dead people being devoured by maggots. Although such photographs may well be necessary to demonstrate for the jurors the scientific bases for the expert's conclusions, they also virtually inevitably carry a risk of overwhelming the sensibilities of ordinary lay persons, even those who are not by nature unusually squeamish. It is not uncommon for people to have somewhat of an aversion to insects, particularly unattractive ones such as maggots. It is also not uncommon for people to be very uncomfortable with the sight of blood and open wounds. Thus, when the phobia with regard to insects is combined with exposure to an enlarged color image of a partially decomposed dead person's body, there is a genuine prospect that some jurors will be so revolted that they will be disproportionately influenced by the sight and cast their votes inappropriately. Emotions may get the better of rationality under such circumstances.

As we have seen, Federal Rule of Evidence 403, and its State analogues, allow judges to exclude this type of relevant but potentially inflammatory evidence if they determine that its evidentiary value "is substantially outweighed by the danger of unfair prejudice, confusion of the issues, or misleading the jury." Other major common law nations also have a rule similar to this, which, in light of the generally more liberal admissibility standards in those nations, probably presents the greatest challenge to forensic entomology evidence.

Some American courts have considered this issue with regard to forensic entomology evidence, and have concluded that the evidence should be allowed notwithstanding its admittedly gruesome aspects. Of course, this is a case by case determination that will hinge on the specific facts and methods used in each instance, but we will examine the manner in which some courts have handled the issue to enable us better to assess the prospects for admissibility in any given case.

For example, in the early case of *Knoppa v. State*[103] a medical examiner testified that from his examination of the body of the deceased he was able to estimate the time of death "from the state of decomposition and from the presence and size and age of the maggots."[104] His testimony was accompanied by nine black-and-white photographs of the corpse, two of which showed the right side and face of the deceased and the maggots that were found thereon. The court rather summarily upheld the admission of the photographs into evidence, relying on a prior holding that found "nothing particularly gruesome" about the pictures and stating that such photographs are "admissible always to show the surrounding circumstances in a trial for murder" and as necessary evidence in corroboration of an out-of-court confession.[105]

The *Knoppa* decision contains no discussion of any admissibility issues on the basis of expert testimony regarding the forensic entomology evidence, essentially taking that matter for granted. This may have been attributable to the fact that the entomology evidence was adduced by a medical examiner together with long-accepted observations as to rate of decomposition of a corpse, and thus seemed to fit comfortably within old familiar forms of evidence.

The more recent case of *State v. Klafta*[106] makes an important point related to this issue. In *Klafta*, an entomologist testified as to the time and circumstances of the death of a 16-month-old child named Heather, and used photographs and exhibits in support. The principal issue at trial was the defendant's intent in abandoning the child for an extended period, and therefore evidence of the child's condition both before and after death was relevant. The prosecution established these facts with evidence from multiple sources, including the entomologist's testimony and exhibits. On appeal, the convicted raised an issue of unfair prejudice, particularly with regard to the forensic entomology evidence.

On appeal, the majority of the Supreme Court of Hawaii found that there was no abuse of discretion by the trial court in allowing the "photographs, exhibits and testimony with respect to the maggot infestation" of the dead infant, Heather.[107] The court pointed out the obvious, that probative evidence always prejudices the party against whom it is offered, for the very reason that it tends to prove the case against that person. The court then stated:

> The jury, in determining the issue of appellant's responsibility . . . was entitled to know Heather's condition by the persons who found her, by the doctors who examined her, and by an expert on entomology to explain the time-range and how the infestation developed. It is true that the evidence of maggot infestation is revolting to a person of ordinary sensibilities, but the testimony of even one witness as to Heather's condition,

[103] 505 S.W. 2d 802 (Tex. Crim. App. 1974). [104] Id., at 803.
[105] Id., at 804, quoting *Bunn v. State*, 154 Tex. Crim. 279, 226 S.W. 2d 646.
[106] 73 Haw. 109, 831 P.2d 512 (Sup. Ct. Haw. 1992). [107] Id., at 113.

when found, is just as revolting. It is possible to conceive of a case where so much cumulative evidence is admitted that its total prejudicial effect demonstrates an abuse of discretion by the trial judge, but this is not such a case . . . Rather it is a case where the prosecution properly painted a complete picture of Heather when found.[108]

In *North Carolina v. Trull*,[109] the North Carolina Supreme Court held that the trial court did not err by admitting eight photographs that showed a murder victim's body "in an advanced state of decomposition with maggot infestation" where the photographs assisted in illustrating the testimony of the forensic entomologist and two other witnesses. The defendant had challenged admissibility of the photographs on the basis that, because "he did not place maggots on the victim and the decomposition was not the result of any of his alleged actions, the photographs were irrelevant, inflammatory, and unduly prejudicial," but the court found no merit in this argument.[110] The court further held that it was within the trial court's discretion to permit the jurors to reexamine the photographs in the courtroom during their deliberations.

Only one of the eight photographs was actually used by the forensic entomologist during his testimony; that photograph helped him explain his determination of the date and time of the victim's death. A second photograph was used during the testimony of a State Bureau of Investigation agent to illustrate evidence-gathering techniques, including placing bags over the victim's hands and using a stain to enhance blood found on the victim's body. The remaining six photographs were autopsy photographs which depicted different views of the victim's neck wounds, injured artery, and genitalia; they were used to illustrate the testimony of a forensic pathologist. The court concluded that all of the photographs were relevant and had probative value in that they assisted in illustrating the testimony of the forensic entomologist, the investigative agent, and the forensic pathologist.[111]

As to whether any unfairly prejudicial effect of the photographs outweighed their probative value, the court emphasized the great discretion enjoyed by trial judges in weighing such evidentiary matters. The court stated, "Generally, gory or gruesome photographs are admissible so long as they are used for illustrative purposes and are not introduced solely to arouse the jurors' passions."[112] The court had no difficulty deciding that there was no violation of that rule in this case.[113]

Other courts have wrestled with this issue, and have concurred, based on the facts in each individual case.[114] However, this result should not be taken for

[108] Id., at 115–16. [109] 349 N.C. 428, 509 S.E.2d 178 (Sup. Ct. N. Carol., 1998).

[110] 349 N.C. 444, 509 S.E.2d 189. [111] 349 N.C. 444, 509 S.E.2d 190.

[112] 349 N.C. 444, 509 S.E.2d 189. [113] 349 N.C. 445, 509 S.E.2d 190.

[114] See, e.g., *Ohio v. Hart*, 94 Ohio App. 3d 665, 641 N.E.2d 755 (1994), affirmed 1996 Ohio App. LEXIS 2641. That case involved a 93-year-old woman who was discovered in her home, horribly murdered. Although it overturned the defendant's conviction on other grounds, the appellate court ruled that the trial court properly admitted a number of photographs of the

granted. It is worthwhile to consider the dissenting opinion in *Klafta* for purposes of noting potential trouble spots.

One judge of the Supreme Court of Hawaii dissented, arguing that the manner in which the forensic entomology evidence was presented was unfairly prejudicial under the Hawaiian version of Federal Rule of Evidence 403. This dissenting judge pointed out that "The graphic evidence of maggot infestation consisted of not just one color photograph, but several, depicting Heather's naked body eaten by the maggots [including the vaginal area]."[115] This judge was also offended by the prosecution's continuation of this "horrific" manner of evidence presentation by a slide show demonstrating large-screen color magnifications of the life cycle of a fly, including the mouth parts, as well as the introduction of two vials of maggots and two slides containing maggots obtained from the dead baby's body. In this judge's opinion, the entomology evidence under the circumstances was unfairly prejudicial and cumulative.[116]

Some further guidance on this issue may be gleaned from the case of *Commonwealth v. Auker*.[117] In *Auker*, the Supreme Court of Pennsylvania considered the testimony and photographic evidence[118] from an entomological expert as to the approximate time of death. Among other assertions, the appellant argued that the trial court abused its discretion in admitting into evidence photographs of the victim covered with insects, under a Rule 403 theory. The Court held that there was no abuse of discretion in allowing the photographs into evidence. The Court stated that, when a trial judge is confronted with

victim's body which depicted her decomposed and fly-ravaged body. The court noted that the mere fact that a photograph is gruesome does not render it inadmissible per se, and that the photographs that were admitted were relevant and not cumulative. Among other things, the photographs were admissible because they were used to illustrate the testimony of an expert witness as to circumstances of the victim's death based on forensic entomology principles. Id., at 678–79. On its second trip to the State appellate court, the issue of the photographs was dealt with in the same manner. The court noted that the trial court limited the number of photographs that the prosecution could introduce into evidence. Although those photographs that were admitted "were, for the most part, graphic," the court found that they were relevant in providing the jury with a clearer understanding of the circumstances surrounding the victim's death, and were demonstrative of testimony provided by several witnesses including an entomology technician at the Cincinnati Zoo, who offered insight into the type and life cycle of the flies found in the victim's home relevant to the date of death. The court stated that, while trial courts should "take great care to limit this type of gruesome evidence," the trial court had not abused its discretion in allowing the photographs in this case to be admitted into evidence. 1996 Ohio App. LEXIS 2641, at *11–12.

115 73 Haw. 126. 116 Id., at 126–27.

117 545 Pa. 521, 681 A.2d 1305 (Sup. Ct. Penn. 1996).

118 The trial court admitted 13 color photographs of stained, knifed clothing and one small color photograph of insects in the body bag without the body. Additionally, two black-and-white photographs of insects on the corpse were admitted. Of these, the first was of a totally jeans-clad lower body from below the knees down to the sneakers, and the other was of the body from the position of the sneakers so that the decomposition of the upper body was not clearly visible. Id., at 545.

"gruesome or potentially inflammatory photographs," the test in Pennsylvania is whether they are of such essential evidentiary value that their need "clearly outweighs the likelihood of inflaming the minds and passions of the jurors." If such inflammatory photographs are merely cumulative of other evidence, they will be inadmissible.[119]

In support of the holding in favor of admissibility, the *Auker* Court mentioned several factors: (1) the photographs of the insects on the body were black and white; (2) these photographs were presented to assist the jury in understanding the expert testimony concerning the presence of various insects and the use of entomology in determining the date of the victim's death; (3) the date of death was a critical and disputed fact in the case; and (4) the trial court took adequate precautions to ensure that the photographs did not inflame the passions of the jury.[120] On this last point, the trial judge warned the jury of the nature of the photographs prior to presentation of the photographs to the jury. The judge also limited the period of time for viewing them, and did not allow the jury to take any of the photographs with them into the deliberation room. Under the totality of the circumstances, the appellate Court held:

> The pictures are unpleasant; however, their disturbing nature does not control the question of admissibility. By viewing the deterioration of the victim's body, the jurors were better able to understand and evaluate the testimony concerning the victim's date of death, which was a critical issue in the case. Appellant has failed to demonstrate that the trial court abused its discretion in determining that the need for the black-and-white photographs outweighed the likelihood of inflaming the minds and passions of the jurors, particularly in light of the court's cautionary remarks and its limitations on viewing the pictures.[121]

For cases litigated in England, Canada, Australia, New Zealand, and other common law countries, this issue is apt to be the main obstacle to admission of forensic entomology evidence. As demonstrated in the previous chapter, these nations have not developed full-scale methods for assessing the admissibility of scientific evidence on the order of *Frye* or *Daubert*, but each has some variant of the Rule 403 probative value versus prejudicial impact test. The trial judge in these countries has even more gatekeeping discretion than those in the United States, and that discretion is most likely to be exercised against forensic entomology evidence where the judge believes the Rule 403 calculus is unfavorable to the evidence.

Civil law nations such as France and Germany, similarly, invest enormous discretion in the trial judge. In the absence of detailed evidentiary rules, judges in these countries as well will be apt to admit forensic entomology evidence routinely unless they perceive that considerable unfairness and undue prejudice will thereby be generated.

[119] Id., at 544–45. (Citations omitted.) [120] Id., at 546. [121] Id.

For these reasons, litigators in nations other than the United States should pay particular attention to the points raised in this section. There is a very high probability that forensic entomology evidence will be ruled admissible, unless serious Rule 403-type issues exist and are vigorously argued before the court.[122]

STATISTICAL ISSUES

One other issue frequently encountered during the presentation of forensic entomology and other scientific evidence is a potential source of difficulty in court. The use of statistical evidence and probability estimates can be problematical for forensic entomology evidence, as we saw earlier in the case of DNA evidence. In each instance, the fear is that the jurors either will be overwhelmed with seemingly insurmountable odds in favor of one side, or will be unable to detect any flaws that might exist in the mathematical and probabilistic calculations and the assumptions that were made in obtaining the numbers. In part, the reluctance of some judges to admit probabilistic evidence may stem from the presumption that many lay persons are essentially incapable of handling even simple mathematics, let alone discerning fallacies in more complex mathematical evidence.[123] We will examine one example of a controversy involving mathematical evidence within the context of forensic entomology.

In *People v. Clark*,[124] an entomologist (Dr. Rees) examined the larvae of the black blow fly found on the victim's body. At trial, he testified that these larvae pass through three "instars" or larval stages and that the larvae were at a certain specific stage of development when removed from the corpse. The entomologist also testified that the minimum growth period for the larvae between inception and maturity (at 99 degrees Fahrenheit) is four days, while the maximum period (at 58 degrees Fahrenheit) is 15 days. By applying a series of arithmetic averages or norms, he concluded that the eggs had been oviposited approximately three and a half days before the body was discovered.[125]

[122] Of course, this presupposes that there are no major flaws in the methodology actually used by the entomologist, and that the witness has the appropriate credentials to testify as an expert.

[123] This presumption as to the mathematical deficiencies of many members of the general public is not entirely devoid of a basis in fact. See generally, JOHN ALLEN PAULOS, INNUMERACY : MATHEMATICAL ILLITERACY AND ITS CONSEQUENCES; A. K. DEWDNEY, 200% OF NOTHING: AN EYE-OPENING TOUR THROUGH THE TWISTS AND TURNS OF MATH ABUSE AND INNUMERACY.

[124] 6 Cal. App. 658 (1970).

[125] Id., at 662. Similar methods were used, without any challenge from the defendant on appeal, in *Taylor v. State*, 251 Kan. 272, 288, 834 P.2d 1325 (1992). There, the expert entomologist witness calculated average daily temperatures and an average temperature for the entire period of time in question based on the National Weather Service report for the City of Salina, Kansas, and then made two calculations based on experimental data reported in a study of rearing greenbottle fly larvae. One calculation was based on constant temperature rearing and the other on changing temperatures. He made a series of calculations from the developmental graphs in the reported study, on the basis of gross weight of the maggot. He then found temperature curves that bracketed the average temperature during the

The defense raised no challenge to the forensic entomology evidence per se, nor to the witness's qualifications to testify as an expert in entomology. However, the defense argued that the witness's use of averages in determining the age of the larvae amounted to a substitution of mathematical probability for relevant evidence as the foundation for his opinion. This would have been contrary to the principles established in *People v. Collins*.[126]

The *Clark* court noted, on appeal, that a trial judge has wide discretion to admit or reject opinion evidence, and found no abuse of that discretion in this case. The court mentioned that the witness, "through accepted scientific methods," was able to ascertain the exact stage of development of the larvae, and that although he did not use actual climatological conditions to determine when the eggs were oviposited, he explained that, in his opinion, these conditions were not an important factor in this case. The court concluded:

> In short, when Dr. Rees' testimony is considered as a whole, it is apparent that his opinion is reasonably accurate from a scientific standpoint and that his use of norms and averages went to the weight of the opinion, not to its admissibility. And, since the jury was accurately instructed in this respect, the defendant cannot be heard to complain in this appeal. The case of *People v. Collins*... is clearly distinguishable, and no further comment is necessary.[127]

The court also found that there was other evidence that made it "reasonably certain" that the victim had been killed on the night in question, and thus is was "hardly likely that the jury was unduly influenced by Dr. Rees' opinion."[128] This is another example of the principle, outlined earlier, which holds that judges are more likely to allow scientific evidence, particularly of a rather novel variety, if it is corroborated by other evidence in the case.[129] One may question the

Footnote 125 (*cont.*)
period in question and concluded that, under constant temperature, the average development probably would have been a maximum of six days from the immature to the third instar. He used this information to estimate the date of death. A second witness, a forensic anthropologist, also testified at trial and disagreed with the entomologist's determination of the number of days from death to discovery of the body, based on temperature differences between the microclimate scene where the body was found and that of the city as a whole.

126 68 Cal. 2d 319, 66 Cal. Rptr. 497, 438 P.2d 33 (1968). This is a widely-cited case in which the Supreme Court of California dealt with novel issues pertaining to statistical probability evidence. The prosecution had used such evidence in an attempt to show, at trial, the extremely low probability that anyone other than the defendant had been the perpetrator of a certain crime, based on various outward physical characteristics of the defendant, his vehicle, and other factors.

127 *People v. Clark*, 6 Cal. App. 658 (1970), at 664.

128 Id., at 665.

129 See also *John v. Warden*, Case Number CV 90 848 S, Superior Court of Connecticut, 1994. In this unreported decision, the court considered a dispute among the three entomologists who testified as expert witnesses at trial. In evaluating the appellant's ineffective assistance of counsel claim, the court found that the defense attorney had thoroughly explored this area of dispute when he questioned these witnesses, and, in any event, there was an "abundance of other available evidence bearing on the time of death," which corroborated the opinion of two of the three expert entomologist witnesses. Id.

wisdom of relying on the witness's own testimony to justify his own rather questionable methodology, as was done in this case to dispose of a challenge to the evidence based on errors in the climatology data, but, for our present purposes, the most important aspect of this case is the issue of mathematical evidence. Judicial deference to the expert witness, as in *Clark*, is likely to result in the admissibility of reasonably standard mathematical evidence such as the average of the minimum length or weight of larvae, and simple extrapolations or interpolations from plotted curves in most cases. There should rarely if ever be the need for more sophisticated or advanced mathematics in support of forensic entomology evidence.

To address the other possible problem area in this category, it is safe to say that forensic entomology evidence, unlike DNA evidence, is not particularly prone to distortion through extreme probabilistic claims. However, in any given case it is possible to couch some of the conclusions of the expert entomologist witness in terms of probabilities, and in the event that takes place, it is important to keep in mind the potential legal difficulties inherent in such evidence.

MISCELLANEOUS CASES IN WHICH FORENSIC ENTOMOLOGY EVIDENCE WAS ADMITTED

In addition to the cases cited in the preceding sections of this chapter, there are several other reported decisions in which the admission of forensic entomology evidence at trial is mentioned. In these cases, the admissibility of this evidence was not a matter of significant controversy. Still, it is worthwhile to be aware of these cases, because they provide further indications that forensic entomology is a widely used, broadly accepted evidentiary concept.

In *Smith v. Secretary of New Mexico Department of Corrections,*[130] an entomologist testified that a maggot collected from a murder victim's body was at a stage of development consistent with having hatched approximately two and a half or three to eight days earlier.[131]

In *United States ex rel. Coleman v. Ryan,*[132] an expert forensic entomologist provided evidence as to the time and circumstances of death of a victim of murder and aggravated kidnapping. The co-author of this book studied the development of the fly larvae found on the victim's body and concluded that her body had been deposited in a certain building on either May 29 or May 30, approximately three weeks before the discovery of the body.[133]

In *Hilbish v. Alaska,*[134] a forensic entomologist estimated from the generations of flies on the murder victim's body that the victim had been dead approximately 10 weeks before his body was discovered and seized.[135]

130 50 F.3d 801 (10th Cir 1995). 131 Id. at 806, n. 6.

132 U.S. District Court for Northern District of Illinois, Eastern Division, 1998 U.S. Dist. LEXIS 8456; see *People v. Coleman,* 168 Ill. 2d 509; 660 N.E.2d 919 (Sup. Ct. Ill. 1995).

133 Id. at *5. 134 891 P.2d 841 (Ct. App. Alaska, 1995). 135 Id. at 847.

In *Smolka v. Florida*,[136] the appellate court found that the circumstantial evidence used to establish guilt of first degree murder was insufficient to sustain the conviction, and reversed. A medical-veterinarian entomologist, based on his examination of maggot larvae removed from the victim's body, estimated her time of death to be sometime after sunset, at 8:30 p.m., on July 10, and no later than 10 a.m. on July 11. He could not, however, rule out the possibility of her being killed before 8:30 p.m., nor could he testify to how long her corpse was lying in the grass where it was eventually found. He was only able to determine approximately when flies had laid the eggs which developed into the maggots. The court noted:

> He stated that the eggs could have been deposited within minutes of Betty Anne's death, but that the flies would not deposit their eggs during nighttime hours. The eggs are only laid during daylight hours but could have been deposited on the body in a car, a house, or almost any other closed quarters.[137]

The court did not discuss the merits of forensic entomology in general, or in any way suggest that this evidence should have been excluded at trial. Rather, the court concluded that there was not enough certainty in the State's entirely circumstantial case to support the conviction.[138]

In *Milburn v. Ohio*,[139] another murder case, the date of death was substantiated by the report of an entomologist who rendered an opinion, based on the growth of fly larvae found on the corpse, that the death had occurred on or before a certain date.[140]

In *Wilson v. Kentucky*,[141] the issue was ineffective assistance of the accused by his defense counsel at trial. In rejecting this claim, the appellate court noted, as to the defense attorney's performance:

> He effectively challenged the credibility of the entomologist who had estimated the time of death. He got the witness to admit that he had been told the date of the victim's disappearance prior to his examination of the corpse and resulting opinion.[142]

Clearly, the forensic entomology evidence was admitted at trial despite the efforts of the defense attorney on cross-examination. The *Wilson* court was simply alluding to the fact that the witness's opinion was arguably tainted by prior information as to the date the victim may have been killed. This would not affect the admissibility of evidence, but could influence the weight to be given the expert's testimony by the judge or jury.

Finally, in *Miller v. Leapley*,[143] the defendant in a murder case presented an alibi defense based upon the testimony of his expert witness, a forensic entomologist. The appellate court found that there was contradictory evidence as to the time of death that was sufficient to support the jury's verdict, despite the

[136] 662 So. 2d 1255 (Ct. App. Fla., 5th Dist., 1995). [137] Id. at 1265. [138] Id. at 1267.
[139] 130 Idaho 649, 946 P.2d 71 (Ct. App. Idaho, 1997). [140] 130 Idaho 653, 946 P.2d 75.
[141] 836 S.W.2d 872 (Sup. Ct. Ky., 1992). [142] Id. [143] 34 F.3d 582, 584 (8th Cir. 1994).

scientific incompetence of some of the prosecution's evidence. There was no suggestion that the forensic entomology evidence was suspect, but the conflicting evidence was deemed adequate to establish both opportunity and motive for the defendant to commit the crimes.

The following chapter will build upon the foundation we have set down, and examine some practical aspects of the introduction of forensic entomology evidence in litigation. We will discuss some strategic and tactical considerations with a view toward first optimizing the prospects for getting such matters successfully admitted into evidence and then moving beyond that threshold and using forensic entomology evidence to achieve the maximum possible advantage at trial.

8

The Introduction and Optimal Use of Forensic Entomology Evidence at trial

The myriad legal issues discussed in the previous chapters as to the admissibility of scientific evidence can and do arise at various stages in the litigation process. Generally, challenges to the scientific validity of an expert's proposed testimony should be raised prior to trial, in a preliminary hearing under Federal Rule of Evidence 104 or the State equivalent.[1] Failure to raise such challenges pre-trial can be outcome-determinative on appeal.[2] It is therefore definitely the best practice for litigants who wish to oppose forensic entomology evidence to make their challenges a matter of record during a Rule 104 pre-trial hearing.[3] However, these evidentiary battles can also be fought at other stages, including a motion in limine[4] during trial, or during a motion for summary judgment.[5]

We will now consider several factors that are important in influencing a trial judge to admit forensic entomology evidence in court. This discussion is predicated on the material in chapters 6 and 7, particularly with regard to the considerations that have been shown to be significant in the case law. Because of the

[1] See Erica Beecher-Monas, *Blinded by Science: How Judges Avoid the Science in Scientific Evidence*, 71 Temple L. Rev. 55, 76 (1998).

[2] See *Hose v. Chicago N.W. Trans. Co.*, 70 F.3d 968, 973–74 (8th Cir. 1995) (holding that because the defendant failed to make a pre-trial challenge to proffered polsomnogram evidence relating to sleep disorders, the District Court did not abuse its discretion in admitting the expert testimony and rejecting a last-minute challenge at trial.); see also *Waitek v. Dalkon Shield Claimants Trust*, 934 F. Supp. 1068, 1079–80 (N.D. Iowa 1996).

[3] On occasion, defendants who had failed to mount challenges to scientific evidence at trial have attempted to raise such a challenge for the first time on appeal. Under these circumstances, appellate courts generally examine the trial courts' determinations for "plain error" and conclude that the defendants are not entitled to any relief. See, e.g., *United States v. Sherwood*, 98 F.3d 402, 408 (9th Cir. 1996); *Hose v. Chicago N.W. Trans. Co.*, 70 F.3d 968, 973 n.3 (8th Cir. 1995).

[4] A motion in limine is a procedural device used to determine certain matters outside the presence of the jury, as provided under Federal Rule of Evidence 104. See *Isely v. Capuchin Province*, 877 F. Supp. 1055, 1058 n.4 (E.D. Mich. 1995) (stating that although motions in limine generally should be filed well in advance of trial, they are "permissible at any point").

[5] *Daubert v. Merrell Dow Pharmaceuticals, Inc.*, 727 F. Supp. 570 (S.D. Cal. 1989). The issue in *Daubert* itself arose on the defendants' motion for summary judgment at trial. Defendants presented epidemiological evidence that Bendectin does not cause birth defects, and the plaintiffs countered with expert reanalysis of the defendants' evidence. It was plaintiffs'reanalysis evidence that the trial court found inadmissible, and the judge granted the defendants' motion for summary judgment, because at that point there was no genuine issue of material fact that needed to be litigated and defendants were entitled to judgment as a matter of law.

relative paucity of judicial opinions directly on point for forensic entomology evidence, we will, by necessity, draw analogies from some of the precedents involving other types of scientific evidence.

The guidelines that follow are meant to be pragmatic, workable principles that will directly enhance the practitioner's prospects for success in litigation. Naturally, there is no set of rules that can guarantee a positive outcome in every instance, given the vagaries of the judicial system, the unique facts that exist in each case, and the unpredictable interplay of personalities, prejudices, and politics that has a role in all human endeavors. But one can derive some useful lessons from the prior experiences of others, and legal precedents can point the way toward the path of least resistance in future cases. We therefore seek to learn from others' mistakes in the hopes of avoiding such pitfalls ourselves.

Choosing the Expert

As we have seen, the legal standard governing who can qualify as an expert witness at trial is fairly liberal, both in the United States and in other major common law and civil law nations. Simply put, the person need only possess more knowledge on a particular specific subject than the general populace, whether by virtue of education, training, or experience.[6] Some of the legal precedents we have discussed illustrate that courts have even qualified people other than entomologists to testify as expert witnesses regarding forensic entomology evidence, so long as these "experts" knew something about the principles involved.[7] Therefore, in any given case it should be a relatively easy matter to find a witness who can satisfy the threshold test and qualify to testify as an expert.

Unless absolutely necessary, however, no one should be satisfied with a witness who merely reaches the bare minimum legal threshold for an expert. The Rule 702 standard is only the floor, not the ceiling. There is a great deal of

[6] See Federal Rule of Evidence 702; *Rushing v. Kansas City Southern Railway Co.*, 185 F.3d 496, 507 (5th Cir. 1999). The proposed amendment to Rule 702 should not change the situation, inasmuch as it essentially codifies the holding of *Daubert* and its progeny.

[7] See *State v. Miller*, 429 N.W. 2d 26 (S.D. 1988); *State v Hart*, 94 Ohio App. 3d 665, 641 N.E.2d 755 (1994). See generally *Pasco v. State*, 563 N.E.2d 587 (Sup. Ct. Ind. 1990), for an example of the types of factors to which a court may look in determining whether to qualify a witness as an expert in forensic entomology. But see *United States v. Paul*, 175 F.3d 906, 911–12 (11th Cir. 1999) (upholding the trial court's refusal to accept an expert based on lack of qualifications, including an acceptable amount of knowledge, skill, training, and experience in the relevant subject matter); *Munoz, et al. v. Orr, et al*, 200 F.3d 291, 300–02 (5th Cir. 2000) (upholding refusal to allow expert testimony and reports on statistical analysis of disparate impact employment discrimination due to lack of reliability in the methodology used and lack of objectivity on the part of the expert); *Seatrax, Inc. v. Sonbeck Int'l, Inc.*, 200 F.3d 358, 373 (5th Cir. 2000) (upholding the trial judge's exclusion of expert testimony on complex matters of profits, losses, and trademark infringement, due to the witness's "lack of formal training or education in accounting, and his failure to conduct an independent analysis" of the relevant sales figures).

room between floor and ceiling in the courtroom, and it is in this capacious area that many trials are won and lost. There is a major practical difference between legally qualifying to testify as an expert and being able to express the science and the methodology in a clear, understandable way, that is to say, to teach the subject effectively and efficiently.

Even experienced teachers are sometimes unaccustomed to the courtroom milieu in which they have a "classroom" of lay persons with little or no foundational knowledge, and only one "class session" in which to teach the material, with no chance to assign homework, conduct quizzes, allow study time, or hold review sessions. The courtroom, with its formal procedures and many procedural and practical strictures, is not an ideal environment in which to teach. There is a real talent necessary to take a group of unprepared strangers, and, in one sitting, while working within all the constraints of the legal rules, present new and complex material to them in such a way that they understand and remember it.

Given all these rather formidable challenges, the most effective expert witness, strictly from a teaching ability standpoint, would most likely be one who is an educator by profession. University professors, for example, generally have the benefit of extensive teaching experience at an advanced level. As a result of their multitude of experiences as educators, it is fair to say that, as a group, they tend to have certain important teaching skills. For example, they know how to organize difficult materials in logical, sequential, manageable sub-units so as to maximize learning potential. They are adept at the use of visual aids in all their variety. They understand the need to explain jargon and define technical or scientific terms at the outset, and to ensure that their audience is able to comprehend this specialized language. They are aware of the effectiveness of an approach that begins with first principles and gradually moves from the simple to the complex, building upon the foundation of material previously discussed. They know the importance of summarizing and reviewing material periodically during a lecture or presentation. And they are experienced in handling questions and responding extemporaneously to unexpected points raised by their students.

All of these qualities will enhance a professor's effectiveness as a witness on the stand at trial. The same skills and techniques that are second nature for a good teacher also enable a witness to convey complex material to a lay jury in a single session, with the greatest probability that the jurors will understand and retain the essence of the subject matter. Of course, professors do not hold a monopoly on these skills and techniques, and non-teachers can sometimes prove to be very proficient in explaining complicated matters to a jury. But as a general rule, those who teach for a living will tend to be better teachers in the courtroom as well, and it is vitally important that an expert witness be able to teach the jury what they need to know to understand the scientific evidence in the case at hand. Trial attorneys will want the best possible teacher as a witness for this reason. They know that a witness who is confusing, disorganized, and cannot be understood by the jury is not going to improve their prospects for winning in court. Thus, litigators prize an effective teaching witness as one of their most valuable assets; such experts are a rare and precious commodity.

In addition, not all experts are created equal in terms of how much weight their opinions will garner in the minds and hearts of a jury. It is one thing to be able to teach difficult subject matter to the jury, but it can be quite another thing altogether to be perceived by the jury as someone worthy of belief and trust. These intangibles can be vitally important, because no matter how effectively the jurors are taught the underlying scientific principles in the case, to some extent they will ultimately be asked to rely on the expert's opinion on a matter beyond the ken of a lay person. At what quite properly might be termed the "moment of truth" in the deliberation room, it will be for each juror to decide, on the basis of a combination of ineffable and subjective factors perhaps not even consciously known to the jurors themselves, whether they will adopt the expert's view on the key issue. What are some of the variables that can make an expert in forensic entomology more apt to gain the confidence of the jury?

Certainly, a person with one or more advanced degrees in entomology will, other things being equal, be more influential with the jury than an individual with no formal credentials when testifying about entomological matters.[8] A university professor, particularly one with tenure and one who has earned certain professorial honors/status, will usually enjoy an advantage in this regard, in addition to his or her other desirable qualities as an effective teacher. Similarly, an expert who actually works or has recently worked extensively in the field will tend to be a more persuasive witness than someone with knowledge limited to "book learning." In this regard, obviously, it is desirable to have a witness with as large a number of years of actual experience in the area as possible. Likewise, an expert who has published extensively on the subject of forensic entomology, and/or has delivered papers at scholarly gatherings will enjoy more credibility than one who has not.

Professional honors and membership or, ideally, leadership status in professional organizations and professional boards is another positive factor that differentiates among experts. In many fields of expertise, board certification is an important criterion that can be a tie-breaker in the minds of the jurors as they evaluate the claims of competing experts. The medical profession, for example, has long used board certification as a seal of approval for its internal purposes, and this can be a useful addition to an expert witness's credentials in court as well. For forensic entomologists, the American Board of Forensic

[8] We have shown that courts may qualify a person as an expert in a subject despite the fact that their training and experience may have been in a tangentially related field. However, the more closely the expert's training and experience fits the subject matter about which he or she testifies, the more likely it is that jurors will perceive the expert as an authority worthy of deference in that area.

Entomology[9] is a fairly recent development that now offers board certification within that specialty. It is possible that certification by this board eventually may become a desirable part of the credentials for expert witnesses in this field.

Another consideration is the expert's track record as a witness in other cases. Certainly one of the most reliable indicia of a person's effectiveness as an in-court witness is his or her prior performance in that same role. This is a subjective factor, of course, which to a great extent is a matter of opinion about which various observers can differ, but it is worthwhile to ask attorneys who handled similar cases their views on which experts were particularly effective witnesses in the courtroom. Some particularly glaring problems such as an expert's prosecution for perjury or failure to gain judicial recognition as an expert in prior cases might be gleaned from a computer search of judicial decisions, or anecdotally by inquiring of experienced litigators in the field. In any event, an expert is likely to gain in effectiveness upon repeated experience as a witness "on the hot seat" in actual trials. He or she will be more at ease in the courtroom environment, and will be more familiar with the processes of direct and cross-examination.

Finally, a less objective but still very significant factor is whether the witness will present a likeable and personable image for the jurors so that they identify with the witness and to some extent bond with him or her. Jurors naturally dislike witnesses who patronize them and who ostentatiously talk down to them as if speaking to little children. An expert who exudes a condescending air of superiority will generate resentment and resistance among the jurors. Conversely, an effective expert will strive to be clear, logical, objective, organized, and interesting, and will take pains to show the jury that there is a normal human being behind all those credentials who is present in court to help the jury, not to preach to them or show off.[10]

These two criteria – ability to teach complex material and demonstrable professional and personal credentials – will be crucial in determining whom to call as an expert witness. It would be optimal, of course, if the proponent of forensic entomology evidence, or the opponent of it, could obtain one expert witness who possesses all of these qualities. This would have the obvious advantage of economy, because the party would only need to hire one person rather than two. Further, it can be easier for a jury to learn from and identify with one witness. There is no need to assess the credentials and the credibility of multiple experts, and there is more time to become comfortable with the one witness, with more opportunity to form a positive impression of him or her. Therefore, one witness can be better than two. But this is not always the case.

It is very important for the litigator to evaluate each potential expert witness on the basis of both primary criteria. Will this person be capable of teaching our

[9] The official website for the American Board of Forensic Entomology can be accessed at http://web.missouri.edu/cafnr/entomology/

[10] THOMAS A. MAUET, TRIAL TECHNIQUES, 314–15 (5th ed., 2000).

jury the essential elements of forensic entomology in court? Will the jury view this person as someone to trust, someone whose achievements and status make them worthy of deference, someone to whom the jurors will give the benefit of the doubt in case it comes down to a battle of the experts?[11]

In many instances, no one individual will be the optimal witness on both counts. Ability to teach is not always linked to the most illustrious professional credentials. At times, the most accomplished experts are so familiar with their subject matter that they take too much for granted, and they fail to grasp the fact that the jurors are totally unaccustomed to the subject the expert has lived with and been immersed in for so many years. Such experts may reflexively speak in "expertese," a jargon-filled foreign language so riddled with undefined terms of art that no lay person could hope to comprehend what is being said. The jury may be impressed with such a witness, but they probably will not know what it is they are supposed to be impressed about. Also, some subject-matter experts are not teachers by profession, and do not have the skills or experience necessary to explain what they do to others who lack their familiarity with the basics. They can be extraordinarily adept in the science and its application, yet be incapable of conveying the gist of their work to a lay person.

Conversely, some very effective teachers lack the professional and experiential credentials that can affect the weight their testimony is given by the jury and thereby tip the scales of justice. Not all entomology teachers, even at the university level, have real-world experience in forensic entomology practice. Of those who do, some have much more extensive experience than others, and over a wider range of situations. And, as always where human beings are involved, some teachers have attained and amassed more honors, degrees, professional recognition, publicity, and publications than others. Among publications, too, there is a hierarchy of prestige. A person who has authored one or more books on the subject of entomology or forensic entomology will enjoy an edge over one who has not. Likewise, in the area of journal articles, it is preferable that the witness has authored a number of published pieces in well-established, scholarly, peer-reviewed journals such as the Annals of the Entomological Society of America. Authorship of articles in forensic journals might be deemed by some as less noteworthy, and publications in non-scholarly, non-specialized, popular magazines even less so, at least in the opinion of other experts in the field.

As a practical matter, how does one locate potential experts for use in any given case who might possess all or most of the aforementioned desirable

[11] It is not uncommon for experts to testify on both sides of a particular issue. In such an event, it should be apparent that the jurors will have to choose which expert to believe, and which expert to disbelieve. The factors we have listed generally should be part of that decision making process, but it is exceedingly difficult to know precisely what factors are most significant in swaying the "battle of the experts" in any given case.

characteristics of an efficacious witness at trial? One of the best sources of information is other litigators who have handled forensic entomology evidence in court; they will have first-hand knowledge of the advantages and disadvantages of one or more experts in the field. Also, it is worthwhile to contact nearby universities, because jurors may be more receptive to local experts than to imported "hired guns."[12] One may also obtain lists of experts in various disciplines from organizations such as the Association of Trial Lawyers of America,[13] the Technical Advisory Service for Attorneys,[14] and the Defense Research Institute.[15] Specifically on point, the American Board of Forensic Entomology[16] maintains a partial list of persons available to testify as forensic entomologist expert witnesses.

In some nations, including England and Wales, the forensic scientists who assist the police during the investigation stage of each criminal case are drawn from an independent agency, such as the Forensic Science Service.[17] Until recently, the Forensic Science Service was a branch of government, but even now, as an independent agency, it provides experts to work with the police during preliminary investigations.[18] Such experts develop a significant degree of skill and familiarity with specific procedures as they repeatedly deal with certain types of cases. Naturally, the experts who actually worked on any given case would tend to be very knowledgeable and credible witnesses at trial, at least as fact witnesses. For potential teaching witnesses, a list of other experts in the applicable area of concentration could also be obtained from the Forensic Science Service. Such agencies can serve as a useful and time-saving clearinghouse for information regarding experts.

We should note that some wily litigators will kindly offer to stipulate that a well-credentialed witness qualifies to testify as an expert without the need of testimony as to his or her background and accomplishments. It may come as a shock to some to learn that such lawyers are not generally making this offer from a spirit of magnanimity, friendliness, and helpfulness. Smart trial attorneys know the power of credentials, such as those listed previously, to impress a jury and influence the outcome of a case. Therefore, if they can prevent the jury from hearing a lengthy, detailed, eye-watering recitation of these stellar credentials in court, they can significantly reduce the probability that the jury will decide that an opposition witness is the "expert's expert" in their case. It is simply good trial tactics to attempt to stipulate away the need to present the opposition expert's qualifications and credentials.

[12] THOMAS A. MAUET, TRIAL TECHNIQUES, 316–17 (5th ed., 2000).

[13] http://www.atlanet.com/ [14] http://tasanet.com/ [15] http://www.driewb.org/

[16] http://web.missouri.edu/cafnr/entomology/

[17] Peter Alldridge, *Scientific Expertise and Comparative Criminal Procedure,* 3 Int'l J. Evid. & Proof 141, 148–49 (1999).

[18] See David E. Bernstein, *Junk Science in the United States and the Commonwealth,* 21 Yale J. Int'l L. 123, 170–73 (1996).

If the jurors do not hear the factors that earned a witness his or her expert status, they will have no basis other than whatever they can glean in-court from the witness's testimony and demeanor to attribute any special authority to that witness. They certainly would have no reason to consider that witness more of an authority on the subject matter in question than the witness proffered by the cunning counsel for the other side. If the expert for the other side is able to expound in court on his or her glittering credentials, that expert would very likely enjoy, in the minds of the jurors, a significant advantage over the relatively anonymous expert for the opposition.

In light of this, the proponents of a stellar expert witness should never consent to a stipulation as to that witness's qualifications. Again, the legal minimum requirements are a floor, not a ceiling, and litigators must make full use of all the available space between floor and ceiling. All the credentials that make an expert an expert should be explained to the jury, in great detail, to prove to them that this witness is the one to whom they should give their trust.

Prestige is, of course, a highly subjective factor and will depend on the eye and mind of the beholder as much as on the actual merits of any given witness. The criteria we have mentioned are intended to be helpful guidelines, but the decision on which expert to employ will always involve a degree of educated guesswork and judgment calls, with all the maddening imprecision inherent therein.

If one is not so fortunate as to locate one person who is both an efficacious teacher and an embodiment of the trappings of prestige in the field, and if the available financial resources are sufficient to make it affordable, it is advisable to use two experts. It is not uncommon for trials to feature two or even more experts per side on an important matter such as forensic entomology evidence. In such cases, one witness, sometimes known as the "teaching witness," first testifies about the underlying scientific and technical principles. Usually, this witness will not have actually collected or examined the evidence in the case at hand, but is present only to set down a foundation. This teaching witness in effect teaches the jurors what they need to know to understand the opinion testimony that follows. Next, the witness or witnesses who gathered and analyzed the evidence in the case before the court will testify. It is here that expert opinion under Rule 702 will be offered as to a key issue in the case, so this is where it is especially important that the witnesses possess those prestige factors that will tend to make the jury attribute to their testimony the greatest possible weight. Particularly if there are experts on both sides of the issue, it is crucial to obtain an illustrious, well-credentialed person to offer this opinion testimony.

Where two or more experts are used by one party, it is essential that they be aware of one another's proposed testimony and in agreement on all key points and conclusions. It would be devastating at trial if the opposition were able to turn the other side's witnesses against each other on cross-examination. Thus, thorough pre-trial coordination and preparation of all witnesses are key.

Beware, however, of such close cooperation between your witnesses that you leave them open to allegations that they were told what to say or conspired in any way to "get their stories straight." An effective litigator will ensure that his or her witnesses know that what is ultimately expected of them is the truth, and that all preparation is geared toward determining the most effective way of discovering and presenting the truth to the court.

Establishing the Foundation

As we saw in the previous chapter, forensic entomology evidence has been admitted in a sizable number of criminal trials, in various jurisdictions.[19] There is no reported case in which forensic entomology evidence has been excluded as a matter of law. Therefore, it behooves the litigator/proponent of forensic entomology evidence to have a summary of this case law available to cite as precedent in the event of an opposition motion to exclude such evidence at trial.

Naturally, a primary factor in the successful introduction of forensic entomology evidence in these cases was the presentation of a sufficient foundation, in each trial, to persuade the judge that the evidence passed muster under whatever test was used in that jurisdiction at that time. This foundational work must never be taken for granted. Only evidence that is so firmly rooted in our legal tradition that it is the proper subject for judicial notice, or has been codified by the State legislature, is exempt from this requirement. It may seem redundant and needlessly burdensome to go through the laborious establishment of the legal predicate in case after case, but this is what the law requires. Moreover, even if the judge were to allow the evidence with less than a complete foundation, a wise litigator would nonetheless want to spell it all out for the jury, lest they fail to understand and appreciate the importance of the evidence when they enter into their deliberations. To reiterate this key point, there is a huge difference between mere admissibility of evidence and the weight it will have in influencing the jury's decision.

In every case, therefore, a qualified expert will have to testify as to the scientific underpinnings of forensic entomology. He or she will set forth the facts regarding those insect species that feed on corpses, including their life cycle stages and the factors that affect their growth and development. This "teaching witness" will generally use diagrams and/or photographs to show the jury what each relevant insect species looks like at each stage of development, as well as charts or tables that illustrate the effects of temperature or other factors on the rate of development of the larvae of these species. Because it is so important that the jurors comprehend and remember this new and unfamiliar (and pos-

[19] See, e.g., *People v. Clark*, 6 Cal. App. 658 (1970); *Commonwealth v. Auker*, 545 Pa. 521, 681 A.2d 1305 (Sup. Ct. Penn. 1996); *State v. Klafta*, 73 Haw. 109, 831 P.2d 512 (Sup. Ct. Haw. 1992); *Pasco v. State*, 563 N.E.2d 587 (Sup. Ct. Ind. 1990); *State v. John*, 210 Conn. 652, 557 A.2d 93, cert. denied, 493 U.S. 824 (1989); *State v Hart*, 94 Ohio App. 3d 665, 641 N.E.2d 755 (1994).

sibly repugnant) information, the teaching witness must be meticulous, organized, and thorough, using all the available techniques of an effective teacher. This should not be a cursory or hasty once-over. The jury has only one chance to learn all of the key aspects of forensic entomology they will need to comprehend and appropriately evaluate the evidence they receive at trial, and the teaching witness has to make the most of this solitary opportunity.

Note that, due to the unbroken record of successes enjoyed by forensic entomology evidence in court to date, there has been no need to attempt to introduce it under the non-scientific portions of Rule 702. Nevertheless, it is worth keeping in mind that some courts have found other types of evidence inadmissible as scientific evidence, but have still allowed it as technical or other specialized knowledge.[20] If a litigator should encounter a judge in a particular case who for some reason is skeptical regarding the scientific pedigree of forensic entomology, it is possible that the litigator might be able, in the alternative, to persuade the judge to allow the entomological evidence in as another acceptable, if non-scientific, form of forensic evidence, akin to fingerprint analysis. In such a case, the litigator should cite the precedents set forth herein to support the proposition that courts have recognized a different, and apparently more relaxed, standard for admitting non-scientific expert opinion evidence. However, after the *Kumho Tire*[21] decision, this distinction should be far less important, especially in federal courts.

As we have seen, it may not make much difference in practice whether any given court uses the *Daubert* or the *Frye* test, or some other method of evaluating the admissibility of scientific evidence.[22] It is important to know which test is going to be applied, however, so that the testimony can be couched in the appropriate terms and emphasis can be placed on the relevant aspects of the evidence. But whatever formalistic test is used, essentially the same factors will be part of the equation leading to admissibility or inadmissibility.

For example, even in *Daubert* jurisdictions it is a good practice to introduce some evidence of the "general acceptance" by the scientific community of the particular type of forensic entomology methods used in the case at hand as one relevant factor in support of its validity and reliability. The teaching witness will generally discuss some examples of peer-reviewed publications in which the fundamental theory and principles of the methodology in question were rigorously analyzed and found to be scientifically acceptable. This testimony should include a discussion of experiments conducted under controlled conditions, to illustrate the scientific validity of forensic entomology as a tool for estimating time and place of death. And certainly if the method used in a given

[20] See, e.g., *Iacobelli Constr. V. County of Monroe*, 32 F.3d 19 (2d Cir. 1994); *United States v. Webb*, 115 F.3d 711, 716 (9th Cir. 1997); and *United States v. Jones*, 107 F.3d 1147, 1157 (6th Cir. 1997).

[21] 526 U.S. 137, 119 S.Ct. 1167,143 L.Ed.2d 238 (1999).

[22] See Developments in the Law, *Confronting the New Challenges of Scientific Evidence*, 108 Harv. L. Rev. 1481, 1498–509 (1995).

case was in any way unusual or innovative, there is the need thoroughly to establish its scientific pedigree.

Following this, either the same or another qualified expert witness will testify as to the actual forensic entomology evidence in the case at hand. It is a sound practice to discuss the chain of custody and all the precautions that went into safeguarding the integrity of the samples, lest any doubt creep into the jurors' minds on this point. As made famous in the criminal trial of O.J. Simpson, even the best scientific evidence, including DNA identification, can be nullified by the specter of tainted samples or sloppy handling procedures. It is best to eliminate such concerns preemptively and forestall any attempts by the opposition to discredit the evidence. A rigorous, complete chain of custody paper trail and careful, no-nonsense procedures in sampling, transporting, storing, and analyzing the evidence will be a valuable insurance policy in favor of success at trial.

Building on this foundation of procedural and methodological soundness, the expert witness will then explain, in detail, how the evidence was analyzed. The culmination of this phase of the testimony will be the expert's actual opinion on the issue for which they have been brought to court, which will usually be the approximate date and time of death of some person. It is crucial for its ultimate impact on the jury that the offering of this opinion be both a legal and a dramatic climax in the case. This requires a sequential, building-block approach to the testimony in which piece after piece of the evidentiary puzzle is revealed until the stage is fully set for the "big question" that will conclude the witness's direct examination.

It is during this portion of the testimony that the expert should use whatever evidence is available to show the jury how he or she arrived at the ultimate opinion. Sometimes this will include vials of preserved larvae, which can be passed among the jurors for them to examine. As a general practice, it is wise at least to offer such an opportunity, however unlikely it may be that the jurors will truly understand the ramifications of what they are seeing, because then no one can argue that the proponent of the evidence had something to hide and as a consequence refused to show the jury the samples in question.

The presentation of actual evidentiary samples should be supplemented with enough photographs of the corpse and the insect samples to enable the expert witness to show the jury in specific detail the essential points upon which his or her opinion is based. As discussed in the previous chapter, judges are more likely to admit evidence when the jury is allowed to see it for themselves and in some sense make up their own minds, i.e., "seeing is believing."[23]

Ideally, these photographs and exhibits will be high-quality color enlargements, big enough to be seen clearly from the jury box by even nearsighted

[23] See, e.g., *People v. Marx*, 126 Cal. Rptr. 350, 356 (Cal. Ct. App. 1975); *State v. Temple*, 273 S.E.2d 273, 279–80 (N.C. 1981); *State v. Bullard*, 322 S.E.2d 370 (N.C. 1984); *United States v. Smith*, 869 F.2d 348, 354 (7th Cir. 1989).

jurors, and positioned such that no juror's view will be impeded by glare, poor lighting, or physical obstructions. Also ideally, there will be enough photographs, from enough different angles, to depict as unmistakably as possible every important factor that entered into the formation of the expert's opinion. As a practical matter, however, it will often be imprudent to give full vent to these concepts, due to Rule 403 risks. This is a countervailing factor in tension with the aforementioned ideal features of demonstrative evidence that must be considered in concert with the "more is better" approach. We will discuss the Rule 403 issues shortly.

A potential obstacle to admissibility of forensic entomology evidence derives from one of its primary strengths, that is, its power to determine the time of death of a particular person. In many homicide cases, the time of death is a critically significant factor, and forensic entomology evidence can either support or destroy an alibi claim by the defendant as well as indicate the circumstances surrounding a homicide committed gradually, as through prolonged neglect. As such, forensic entomology evidence often goes quite directly to the "ultimate issue" in a homicide case. As noted previously, some courts are hesitant to allow novel scientific evidence when it purports to determine squarely the ultimate issue at trial.[24] However, this tendency can be overcome through careful preparation.

For example, the effective litigator will corroborate the expert's opinion, wherever feasible, with other evidence in the case. We have seen that courts are more comfortable admitting novel scientific evidence where it does not stand alone, but rather is bolstered by other evidence as well.[25] However, in cases where forensic entomology evidence is essentially all the ammunition that one side has available in its arsenal of evidence, it may be possible to persuade the trial judge to allow it as a matter of fairness to both sides. That principle also has been a factor influencing some courts in their determinations as to whether to allow novel scientific evidence, as discussed previously.[26]

We have noted also that courts are more likely to allow novel scientific evidence where it can be shown that there is not an excessive degree of subjectivity involved in the techniques and methodology in question.[27] Forensic

[24] See, e.g., *United States v. Alexander*, 526 F.2d 161, 168–70 (8th Cir. 1975); *People v. Bledsoe*, 681 P.2d 291, 301 (Cal. 1984); *United States v. DiDomenico*, 985 F.2d 1159, 1164–65 (2nd Cir. 1993); *United States v. Watson*, 587 F.2d 365, 368–69 (7th Cir. 1978).

[25] See, e.g., *People v. Williams*, 331 P.2d 251 (Cal. App. Dep't Super. Ct. 1958); *Worley v. State*, 263 So. 2d 613, 614 (Fla. Dist. Ct. App. 1972); *State v. Armstrong*, 329 N.W.2d 386 (Wis.), cert. denied, 461 U.S. 946 (1983); *United States v. Watson*, 587 F.2d 365, 369 (7th Cir. 1978).

[26] See, e.g., *United States v. Downing*, 753 F.2d 1224 (3rd Cir. 1985); *Coppolino v. State*, 223 So. 2d 68, 69–71 (Fla. Dist. Ct. App. 1968), cert. denied, 399 U.S. 927 (1970).

[27] See, e.g., *United States v. Alexander*, 526 F.2d 161, 167–70 (8th Cir. 1975); *United States v. Black*, 831 F. Supp. 120, 123 (E.D.N.Y. 1993); United States v. Lech, 895 F. Supp. 582, 585–86 (S.D.N.Y. 1995); *People v. Bledsoe*, 681 P.2d 291, 301 (Cal. 1984); *United States v. Addison*, 498 F.2d 741,744 (D.C. Cir. 1974); *People v. Kelly*, 549 P.2d 1240, 1243, 1245 (Cal. 1976); *Reed v. State*, 391 A.2d 364, 370 (Md. 1978).

entomologists should be able to base their testimony on objective, observable, quantifiable, measurable criteria as to the identifying features that are the hallmarks of any given larval stage of the relevant insect species, and show these features to the court with exhibits and full accompanying explanations. Likewise, forensic entomologists can point out the rigorous laboratory studies concerning the factors influencing the rate of maturation of these same species, with emphasis on the scientifically controlled conditions that bolster the validity of the research. Such reliance on objective facts should suffice to persuade judges that forensic entomology does not rest excessively on subjective judgment calls.

Ironically, one of the factors that illustrates the objectivity of forensic entomology evidence also is another potential trouble spot. Forensic entomologists sometimes rely on probabilities and statistics in formulating their opinions, especially with regard to the effect of temperature variation on the maturation rate of the larval stages of certain species of insects. As discussed earlier, for some courts, overly heavy reliance on mathematical or statistical evidence can lead to exclusion of evidence.[28] This should not be an insurmountable obstacle for forensic entomology evidence, inasmuch as it draws on various other foundations for its support and, unlike DNA evidence, does not purport to estimate the probability that a specific defendant was the guilty party in any given case.[29] It is worth keeping this in mind, however, when preparing for trial. Statistics and probabilities should be counterbalanced with all the other criteria that also are important in aiding the expert to the formation of his or her opinion.

Danger of Inflaming the Jury

As we saw in the previous chapter, Rule 403 or some variant thereof is an important consideration wherever graphic evidence is used at trial, both in the United States and in other common law and civil law nations.[30] Regardless of the applicable analytical framework or lack thereof, each jurisdiction allows for exclusion of otherwise admissible evidence based on a probative value/prejudicial impact assessment.

Forensic entomology evidence is especially prone to Rule 403[31] type

[28] See *People v. Collins,* 68 Cal. 2d 319, 66 Cal. Rptr. 497, 438 P.2d 33 (1968).

[29] See, e.g., *People v. Clark,* 6 Cal. App. 658 (1970); *Taylor v. State,* 251 Kan. 272, 288, 834 P.2d 1325 (1992).

[30] See generally, Sophia I. Gatowski, et al., *The Diffusion of Scientific Evidence: A Comparative Analysis of Admissibility Standards in Australia, Canada, England, and the United States, and Their Impact on Social and Behavioural Sciences,* 4 Expert Evid. 86 (1996).

[31] Federal Rule of Evidence 403, which has analogues in every State and in other nations, reads:

> **Exclusion of Relevant Evidence on Grounds of Prejudice, Confusion, or Waste of Time.** Although relevant, evidence may be excluded if its probative value is substantially outweighed by the danger of unfair prejudice, confusion of the issues, or misleading the jury, or by considerations of undue delay, waste of time, or needless presentation of cumulative evidence.

challenges, due in part to the squeamishness many lay persons have concerning insects. For whatever reason, a revulsion bordering on a phobia towards insects seems to be a rather common phenomenon. It is probably true that this revulsion is stronger than usual when, as with forensic entomology evidence, the insects in question are maggots and other larvae rather than the more attractive and popular adult butterflies or beetles.

The problem is exacerbated by the unfortunate but unavoidable fact that forensic entomology evidence juxtaposes unattractive larval insects with grisly images of dead and decaying human bodies. Surely, the tendency for people to be disgusted by insects must be rivaled by the revulsion many feel at the sight of blood, open wounds, rotting flesh, or a combination of all three. But it is difficult if not impossible to establish for the jury a frame of reference for what they are seeing without at least one or two photographs taken from a sufficient distance to show that there was in fact a human body being consumed by insects. Given this, subsequent close-up photographs depicting larvae on the body will have substantial potential for affecting the jurors in a powerful, visceral way.

This is a two-edged sword. On one side, this type of evidence is perhaps uniquely capable of moving a jury to desire to ensure that justice is done and that whoever was responsible for creating such a nightmarish scene is appropriately convicted and punished. Litigators rarely have such a potent weapon in their arsenal at trial. The dramatic, emotional impact of this type of visual evidence is highly desirable from a litigator's perspective. But the other side is Rule 403, which is designed to prevent this very type of emotional impact where it rises above a certain level and threatens to override rationality and fairness.

We have discussed cases in which the gut-level power of forensic entomology evidence has been in issue vis-à-vis a Rule 403 challenge.[32] The evidence has been allowed in each case, but the reasons why are important to keep in mind.

Implicit in the case law is a principle of moderation. Color photographs may be used, but if black-and-white shots can illustrate the same features just as well, it may be advisable to use them. Obviously, color images are more apt to press the jury's "inflame" button, and so litigators and their expert witnesses should use them only where there is a viable argument to be made that black-and-white photographs would be inadequate to represent accurately what the jury needs to see.

The number of photographs used at trial can also tip the scales of the Rule 403 balancing test. Each photograph should have some identifiable and objective value to add to the body of evidence before the jurors. If two photographs

[32] See, e.g., *Knoppa v. State*, 505 S.W. 2d 802 (Tex. Crim. App. 1974); *State v. Klafta*, 73 Haw. 109, 831 P.2d 512 (Sup. Ct. Haw. 1992); *State v. Hart*, 94 Ohio App. 3d 665, 641 N.E.2d 755 (1994); *Commonwealth v. Auker*, 545 Pa. 521, 681 A.2d 1305 (Sup. Ct. Penn. 1996).

show essentially the exact same thing, it will be difficult to persuade the judge that the second photograph is not cumulative and redundant. Additionally, at some point, the total number of photographs in evidence might be deemed excessive under Rule 403, with the overall probative value substantially outweighed by the risk of inflaming the jury. Therefore, it would be prudent to exercise some self-restraint and carefully select a representative sample of photographs sufficient to illustrate all key aspects of the expert's decision-making process, but limited enough to avert the specter of piling on excessive and inflammatory evidence. Whatever photographs are voluntarily withheld in this manner can always be maintained in reserve and subsequently introduced in rebuttal, if events at trial make them necessary to counter some assertions from the opposition.

Similarly, enlargements of some or all of the photographs is probably justifiable as necessary to enable the expert witness simultaneously to show every juror the specific points to which he or she is referring throughout the testimony. Standard-size prints could not serve the same purpose, because they would have to be passed from juror to juror and the entire panel could not view the same exhibit at the same time. However, it would again be advisable to limit the number of enlargements used, because an image that is revolting in a 3-by-5 inch print would tend to be much more so when greatly enlarged into a blow-up, or when projected onto a big screen in a darkened courtroom. A rule of reason should prevail. A litigator who can articulate an objective rationale in support of each enlargement used should successfully weather a Rule 403 challenge, but there is some indeterminate point beyond which a judge will stop the show.

Finally, there is the matter of the length of time to which the jury will be exposed to the photographs. Some courts have found it significant that the exhibits were only shown to the jurors for a relatively brief period of time in open court.[33] It may be tempting to allow the jurors to take the photographs into the deliberation room with them, where they can study them at great length and let the maximum impact sink in, but that can be risky. The argument in favor of admissibility of demonstrative evidence is strongest where it can be shown that the exhibits are required to supplement and clarify an expert's testimony, point by point, as it is delivered. Once the testimony is over, there is less reason for the jurors to retain the exhibits in their possession and more risk that the exhibits will primarily serve a visceral rather than fact-finding function. As a general rule, it is best not to attempt to send the photographs into the deliberation room with the jury. It creates an issue on appeal, and is probably not worth the risk.

Even when forensic entomology photographs are limited to use in open court, it is advisable to make the duration of the jury's exposure to them a

[33] See *Commonwealth v. Auker*, 545 Pa. 521, 681 A.2d 1305 (Sup. Ct. Penn. 1996).

matter of record. Particularly where there are color slides or big color enlargements of especially gruesome images, it is a good practice to keep track of the number of minutes and seconds each image is visible by the jury, and to announce this for the record after each image is shown. This will preserve the matter for appeal in case the opposition raises the issue by alleging that some or all of the grisly photographs were displayed for such an excessive length of time that Rule 403 was violated. A precise record of the duration of jury exposure to each image in question will assist the appellate court in making its determination, and will preemptively eliminate any suggestion by the opposition that the photographs were visible for longer than they actually were.

Opposing Forensic Entomology Evidence

All of the factors examined in the previous sections of this chapter have their flip-side. A litigator and/or expert witness who wishes to oppose the admissibility of forensic entomology evidence, or to undermine the weight the jury will give to such evidence once admitted, can use these same factors to do so.

The avenue of attack most likely to prevail in any given case is probably a focus on the failure of the person or persons involved to use the proper procedures, whether in collecting, preserving, or analyzing the samples. It is highly improbable that any litigator or expert witness will succeed in persuading a judge that forensic entomology evidence as a general rule fails the applicable admissibility standard per se. As we have noted, there is no precedent in support of that position, and quite a bit in direct opposition.[34] But, just as with DNA evidence in the O.J. Simpson criminal trial, a theory's sterling scientific pedigree can be no match for inept or careless implementation of the procedures and techniques in a particular case. A failure of execution can open the door to a successful attack on even the most well-established scientific evidence.

It would be very helpful in mounting such an attack to have the assistance of an expert witness, of course. A person who is well-versed in forensic entomology theory and practice will be ideally situated both to identify flaws in the techniques and methods used, and to assess the significance of those flaws in undermining the opposition expert's ultimate opinion in the case. However, where resources are scarce or time is short, a defense attorney can still learn enough about the proper methods to probe the prosecution's expert on cross-examination and discover points of weakness. If the attorney is aware of the relevant factors (especially temperature) as described in this book, and the importance of controlled and accurately recorded conditions both before and

[34] See, e.g., *People v. Clark*, 6 Cal. App. 658 (1970); *Commonwealth v. Auker*, 545 Pa. 521, 681 A.2d 1305 (Sup. Ct. Penn. 1996); *State v. Klafta*, 73 Haw. 109, 831 P.2d 512 (Sup. Ct. Haw. 1992); *Pasco v. State*, 563 N.E.2d 587 (Sup. Ct. Ind. 1990); *State v. John*, 210 Conn. 652, 557 A.2d 93, cert. denied, 493 U.S. 824 (1989); *State v Hart*, 94 Ohio App. 3d 665, 641 N.E.2d 755 (1994).

after collection of the samples, he or she can focus on these issues and expose areas of vulnerability.

Courts are split as to whether problems of execution go to admissibility or go only to the weight of the evidence. In either event, if there was evidence of a break in the chain of custody[35] of the insect samples after collection, or if the samples were not maintained under known conditions, there is room to make these issues a matter of emphasis at trial. It may also be possible to undermine the expert's opinion by showing that the expert already had extrinsic information as to the date of the victim's death before he or she formulated an opinion on that issue. For example, if the forensic entomologist had been told the date of the victim's disappearance in advancing of evaluating the entomology evidence, his or her opinion might have been inappropriately influenced.[36] This could diminish the weight given to the expert's testimony by the trier of fact. If a litigator or expert witness can score a number of valid points against the opposition's methods, they may succeed in negating much or all of the impact of that evidence.

The other primary area ripe for attack is in the Rule 403 category. At trial, or on appeal, there will virtually always be an opportunity to mount a Rule 403 challenge to forensic entomology evidence. For the reasons previously discussed in this chapter, this type of evidence is perhaps uniquely prone to Rule 403 problems, particularly where the proponent fails to exercise the self-control we have advocated in this chapter.

The prospects for a successful Rule 403 challenge to forensic entomology evidence will be at their zenith where the proponent of such evidence piles it on with every means available. If the proponent inflicts a lengthy series of gigantic color photographic enlargements on the jury, depicting the "gory details" as only forensic entomology evidence can, he or she opens the door to the opposition.

For example, the opponent can argue persuasively that several, or possibly even all, of the photographs add nothing new of relevance to the case, and are thus cumulative and should be excluded. This argument can be strengthened by making an offer to stipulate to everything that the photographs can fairly be said to illustrate objectively. Such an offer to stipulate enhances the opponent's subsequent Rule 403 challenge by supporting the argument that the photographs are being offered in an improper attempt to inflame the jury rather than for any objective evidentiary value, and that the proponent's refusal to agree to the stipulation reinforces that point. If the proponent is not prepared to respond by articulating objective reasons why a stipulation will not suffice to allow the expert witness to testify effectively on relevant issues, this challenge may prevail. The key inquiry will be whether the exhibits are helpful in explain-

[35] See *Pasco v. State*, 563 N.E.2d 587 (Sup. Ct. Ind. 1990).

[36] See *Wilson v. Kentucky*, 836 S.W.2d 872 (Sup. Ct. Ky., 1992).

ing relevant factors to the jurors. The more photographs the proponent attempts to introduce, without a clear and objective value-added for each one, the easier it will be for the opponent to challenge the evidence in this manner.

Likewise, if the proponent insists on using nothing but color photographs, the opponent may be able to persuade the court that the exhibits' potential to inflame the jury substantially outweighs their probative value. An offer to stipulate to whatever the actual objective factual content of the photographs might be can again bolster such a claim. Is it truly necessary for every photograph, or any particularly graphic photograph, to be in "living color" for the jury to be able to see and understand what the witness is discussing? That will be the battleground for this type of attack.

Finally, the opponent of forensic entomology evidence may be able to make a viable Rule 403 challenge if the proponent wishes to send the exhibits into the deliberation room with the jury, or otherwise expose the jurors to some or all of the photographs for an extended period of time. It may be possible to sway the judge against the evidence if there is no objective reason articulated as to why the exhibits must be before the jury for such a protracted period. Particularly if the exhibits were on display or in the possession of the jury at a time when the expert witness was not making reference to what the photographs purported to depict, the opponent has a reasonable chance to prevail. Under such circumstances, the proponent's argument that the photographs are essential to aid and complement the expert's testimony is considerably weakened.

Even if any individual one of these challenges fails, it may still be possible to raise a successful challenge on appeal. Conceivably, the totality of the evidence might indicate a Rule 403 violation if the proponent of the evidence was stretching the limits of admissibility on multiple points. This is why it is unwise for the proponent of forensic entomology evidence to pull out all the stops at trial in terms of the number, color, size, and duration of exposure to the jury of gruesome photographic evidence. Particularly where such photographs are supplemented by the in-court introduction of actual samples of larvae from the corpse, at some point a judge is apt to find that enough is enough, or, more accurately, more than enough is too much. The cumulative impact of an aggressive approach to too many of these issues may appear to an appellate court to be exactly the type of passionate as opposed to fact-based evidence that Rule 403 was drafted to control.

Conclusion

A familiar cliché indicates that we would love to be "a fly on the wall" in some room where fascinating events unfolded. If only we could magically transform ourselves into a fly, we could discover truths that are otherwise hidden from us. As a lowly, oft-despised, tiny fly we would uncover the secrets we otherwise could only dream about.

In the preceding chapters, we have covered in great detail what is to many a decidedly unpleasant subject. We have examined the species of flies that visit human bodies after they die. For most people, flies themselves are repulsive and to be avoided wherever possible. When combined with decaying human corpses, flies become nothing short of anathema. But murder is itself the most heinous of crimes and should disgust civilized people. Forensic entomology takes the raw materials of flies and decay, refines them in the crucible of science with the fire of law, and produces justice and truth. That is a form of alchemy, with the power to transform lives and right wrongs.

Forensic entomology has joined, as a member in good standing, the family of widely accepted forms of scientific evidence. Under the various tests governing the admissibility of scientific evidence worldwide, forensic entomology has the scientific and legal qualifications necessary to pass. But this is not automatic. This form of evidence will be admissible at the trial level, and will be upheld on appeal, if it is properly researched and supported scientifically, and if it is presented to the court with the appropriate legal foundation and with the requisite safeguards against improper use. The primary purpose of this book is to provide both the entomologist and the trial lawyer with the information necessary to navigate successfully the hazards that threaten forensic entomology evidence.

The state of the art is still evolving, and this evolution will continue. Entomologists will learn more about the lifecycles and behavioral patterns of the various insect species that can interact with corpses. The law of scientific evidence in general, and forensic entomology in particular, will be modified, either gradually or abruptly, by legislation and by judicial decisions. The direction this will take is largely unpredictable. Perhaps, as with DNA evidence in some jurisdictions, forensic entomology evidence will one day be recognized as admissible by statute or rise to the level of an acceptable subject for judicial notice. But whatever path science and the law may take, the principles set forth in this volume will prepare scientists and litigators to anticipate and adapt to any such changes, and to use them effectively.

Ultimately, forensic entomology is the union of two very different disciplines. Only rarely do the legal and scientific realms coalesce to create a partnership. When they do, the result can be a powerful synergy.

In this book, we have seen how, in a sense, forensic entomology allows us to become the proverbial fly on the wall. This modern melding of science and law enables human beings to use the fly as a key that unlocks mysterious doors. We can place ourselves at the scene of a murder and learn important facts about the crime. We can traverse time and space to recreate events previously known only to perpetrator and victim. Vicariously, we can be the fly on the wall that witnessed the killing.

But more than this, forensic entomology enables that fly, in effect, to take the witness stand and testify as to the details it saw. The fly on the wall at the scene of the crime can enter the courtroom and tell the world the truth. The fly can set the innocent free and convict the guilty. Not bad for an insect.

Index

Bold page numbers indicate figures